Chemistry: The Key to our Sustainable Future

Minu Gupta Bhowon • Sabina Jhaumeer-Laulloo
Henri Li Kam Wah • Ponnadurai Ramasami
Editors

Chemistry: The Key to our Sustainable Future

Springer

Editors
Minu Gupta Bhowon
Department of Chemistry, Faculty of Science
University of Mauritius
Réduit, Mauritius

Sabina Jhaumeer-Laulloo
Department of Chemistry, Faculty of Science
University of Mauritius
Réduit, Mauritius

Henri Li Kam Wah
Department of Chemistry, Faculty of Science
University of Mauritius
Réduit, Mauritius

Ponnadurai Ramasami
Department of Chemistry, Faculty of Science
University of Mauritius
Réduit, Mauritius

ISBN 978-94-007-7388-2 ISBN 978-94-007-7389-9 (eBook)
DOI 10.1007/978-94-007-7389-9
Springer Dordrecht Heidelberg New York London

Library of Congress Control Number: 2013953577

Printed on acid-free paper

Springer is part of Springer Science+Business Media (www.springer.com)

Preface

The second International Conference on Pure and Applied Chemistry (ICPAC 2012) was held from 2 to 6 July 2012 at Hilton Mauritius Resort and Spa, Wolmar, Flic en Flac, in Mauritius. The theme of the conference was “Chemistry: The Key for our Future”. ICPAC 2012 was attended by 150 participants from 25 countries. The conference featured 80 oral and 80 poster presentations. The keynote address was given by Prof. Robert Huber, the 1988 Chemistry Nobel Prize winner.

The participants of ICPAC 2012 were invited to submit full papers. This book is a collection of the papers selected during a subsequent peer review.

The book consists of 25 chapters covering a wide range of topics from fundamental to applied chemistry.

We would like to thank all those who submitted full manuscripts for consideration and the reviewers for their timely help in assessing these manuscripts for publication.

We would also like to pay a special tribute to all the sponsors of ICPAC 2012.

We hope that this collection of papers will serve as a useful resource for researchers.

Department of Chemistry
University of Mauritius, Réduit, Mauritius
June 2013

M. Gupta Bhowon
S. Jhaumeer-Laulloo
H. Li Kam Wah
P. Ramasami

Contents

Contributors

Ahlam J. Abdulghani Department of Chemistry, College of Science, University of Baghdad, Baghdad, Iraq

Muhammad B. Abubakar National Centre for Petroleum Research and Development (Energy Commission of Nigeria), Abubakar Tafawa Balewa University, Bauchi, Nigeria

Adeleke Adeniyi Department of Chemistry, Lagos State University, Lagos, Nigeria

Mansoor Ahmad Research Institute of Pharmaceutical Sciences, Department of Pharmacognosy, University of Karachi, Karachi, Pakistan

Nessar Ahmed School of Healthcare Science, Manchester Metropolitan University, Manchester, UK

Zainab Z. Ahmed Department of Chemistry, College of Science, University of Baghdad, Baghdad, Iraq

Ummeh W. Ahsun Aleemiah College (Girls), Phoenix, Mauritius

Khalil H.A. AL-Bayati Department of Physics, College of Science for Women, Baghdad University, Baghdad, Iraq

Safeenaz B. Alladin Department of Chemistry, Faculty of Science, University of Mauritius, Réduit, Mauritius

Nada I.I. AL-Zubaidi Department of Physics, College of Science, Diyala University, Diyala, Iraq

A. Bialecki Laboratoire de Chimie des Substances Naturelles et des Sciences des Aliments, Faculté des Sciences et Technologies, Université de La Réunion, La Réunion, France

Raheemot Balogun Department of Chemistry, Lagos State University, Lagos, Nigeria

Jamal Benjilali National Institute of Medicinal and Aromatic Plants, Taounate, PAMSN, University Sidi Mohamed Ben Abdellah, Fez, Morocco

Dalila Bousta National Institute of Medicinal and Aromatic Plants, Taounate, PAMSN, University Sidi Mohamed Ben Abdellah, Fez, Morocco

Marie Chan Sun Department of Medicine, Faculty of Science, University of Mauritius, Réduit, Mauritius

Dušan Chorvát Jr. International Laser Center, Bratislava, Slovakia

Klaudia Czaniková Department of Composite Materials, Polymer Institute, Slovak Academy of Sciences, Bratislava, Slovakia

Hanane El-Hajaji Faculty of Sciences Dhar el Mehraz, LIMOM, University Sidi Mohamed Ben Abdellah, Fez, Morocco

Latifa El Mansouri National Institute of Medicinal and Aromatic Plants, Taounate, PAMSN, University Sidi Mohamed Ben Abdellah, Fez, Morocco

Amal El Youbi-Hamsas National Institute of Medicinal and Aromatic Plants, Taounate, PAMSN, University Sidi Mohamed Ben Abdellah, Fez, Morocco

A. Emelia Department of Chemistry, Lagos State University, Lagos, Nigeria

Jevisha Erriah Department of Medicine, Faculty of Science, University of Mauritius, Réduit, Mauritius

Abdellah Farah National Institute of Medicinal and Aromatic Plants, Taounate, PAMSN, University Sidi Mohamed Ben Abdellah, Fez, Morocco

Thimme A. Gowda Haranahalli Ramaswamy Institute of Higher Education, Hassan, Karnataka, India

Said O.S. Hassane Faculty of Sciences and Technology, University of Comoros, Moroni, Comoros

Brian Horsfield Organic Geochemistry Section, GeoForschungsZentrum, Telegrafenberg, Potsdam, Germany

Aminah Ibrahim Department of Chemistry, Lagos State University, Lagos, Nigeria

William K. Isharaza Department of Biochemistry, Mbarara University of Science and Technology, Mbarara, Uganda

Noor Jahan Dow College of Pharmacy, Dow University of Health Sciences, Karachi, Pakistan

Gautam Jaiswar Department of Chemistry, Dr. B. R. Ambedkar University, Agra, India

Aliyu Jauro National Centre for Petroleum Research and Development (Energy Commission of Nigeria), Abubakar Tafawa Balewa University, Bauchi, Nigeria

Peter Kasák Department of Composite Materials, Polymer Institute, Slovak Academy of Sciences, Bratislava, Slovakia

Igor Krupa Center of Advanced Materials, Qatar University, Doha, Qatar

Mohammed Lachkar Faculty of Sciences Dhar el Mehraz, LIMOM, University Sidi Mohamed Ben Abdellah, Fez, Morocco

Hiteyeshi Lallbeeharry Sugar Industry Labour Welfare Fund, Port-Louis, Mauritius

Henri Li Kam Wah Department of Chemistry, Faculty of Science, University of Mauritius, Réduit, Mauritius

Mohamad F. Mahomoodally Department of Health Sciences, Faculty of Science, University of Mauritius, Réduit, Mauritius

Roman Marsalek Department of Chemistry, Faculty of Science, University of Ostrava, Ostrava, Czech Republic

Ishmael B. Masesane Department of Chemistry, University of Botswana, Gaborone, Botswana

Ofentse Mazimba Department of Chemistry, University of Botswana, Gaborone, Botswana

Mehjabeen Department of Pharmacology, Federal Urdu University of Arts, Science and Technology, Karachi, Pakistan

Riyadh M. Mungur Department of Physics, Faculty of Science, University of Mauritius, Réduit, Mauritius

Fawzia B. Narod Department of Science Education, Mauritius Institute of Education, Réduit, Mauritius

Jared C. Ogunde Scientific Advisory and Information Network (SAIN) and Chemistry Aid Kenya, Nairobi, Kenya

Mária Omastová Department of Composite Materials, Polymer Institute, Slovak Academy of Sciences, Bratislava, Slovakia

Aggrey Omolo Scientific Advisory and Information Network (SAIN) and Chemistry Aid Kenya, Nairobi, Kenya

Yusuf Omotayo Department of Chemistry, Lagos State University, Lagos, Nigeria

Olawale Osifeko Department of Chemistry, Lagos State University, Lagos, Nigeria

Olabisi Owoade Department of Chemistry, Lagos State University, Lagos, Nigeria

Ewa Pavlová Institute of Macromolecular Chemistry, Academy of Sciences of the Czech Republic, Prague, Czech Republic

Atar S. Pipal Department of Chemistry, Dr. B. R. Ambedkar University, Agra, India

Deerajen Ramasawmy Department of Management, Faculty of Law and Management, University of Mauritius, Réduit, Mauritius

Kanisha Ramchurn Department of Chemistry, Faculty of Science, University of Mauritius, Réduit, Mauritius

Jugjeet S. Ramkissoon Department of Health Sciences, Faculty of Science, University of Mauritius, Réduit, Mauritius

Shahram Ranjbar Department of Physical Chemistry, Razi University, Kermanshah, Iran

Antony J. Rest Chemistry Video Consortium, Educational Techniques Group Trust of the Royal Society of Chemistry (UK) and Chemistry Aid, University of Southampton, Southampton, UK

Soonil D.D.V Rughooputh Department of Physics, Faculty of Science, University of Mauritius, La Réduit, Mauritius

Jacqueline Smadja Laboratoire de Chimie des Substances Naturelles et des Sciences des Aliments, Faculté des Sciences et Technologies, Université de La Réunion, La Réunion, France

Said H. Soidrou National Institute of Medicinal and Aromatic Plants, Taounate, PAMSN, University Sidi Mohamed Ben Abdellah, Fez, Morocco

Anwar H. Subratty Department of Health Sciences, Faculty of Science, University of Mauritius, Réduit, Mauritius

Farid Taherkhani Department of Physical Chemistry, Razi University, Kermanshah, Iran

Ajay Taneja Department of Chemistry, Dr. B. R. Ambedkar University, Agra, India

Sukarma Thareja Department of Chemistry, Christ Church College, CSJM Kanpur University, Kanpur, Uttar Pradesh, India

Raymond G. Wallace Educational Techniques Group Trust of the Royal Society of Chemistry (UK), Chemistry Aid, and School of Science and Technology, Nottingham Trent University, Nottingham, UK

Milton A. Wesuta Department of Biochemistry, Mbarara University of Science and Technology, Mbarara, Uganda

Heinz Wilkes Organic Geochemistry Section, GeoForschungsZentrum, Telegrafenberg, Potsdam, Germany

Asieh Yahyazadeh Department of Chemistry, University of Guilan, Rasht, Iran

Chapter 1
Elastomeric Actuators Based on Ethylene-Vinyl Acetate and Carbon Nanotubes

Klaudia Czaniková, Mária Omastová, Igor Krupa, Peter Kasák, Ewa Pavlová, and Dušan Chorvát Jr.

Abstract The development of new types of visual-aid tablet for visually impaired people requires the development of cheap, but still very effective photoactuating materials. This requirement can be satisfied by the use of new kind of elastomers filled by nanofillers, such as carbon nanotubes. Nanocomposites based on commercial ethylene vinyl-acetate (EVA) copolymer and multiwalled carbon nanotubes (MWCNT) were prepared by casting from solution. The non-covalent surface modification of MWCNT was carried out by special, newly synthesized compatibilizer cholesteryl 1-pyrenecarboxylate (PyChol). In order to mimic Braille character, special home-built silicone punch and die moulds were used. The Braille element based on EVA/MWCNT-PyChol composite displays reversible, multiple changes of dimension in the direction of the irradiation during/upon illumination by red and blue light-emitted diode (LED). Transmission electron microscopy (TEM) showed a good dispersion of the MWCNT-PyChol within the matrix. The Braille element behaviour under illumination was analysed by atomic force microscopy

K. Czaniková (✉) • M. Omastová • P. Kasák
Department of Composite Materials, Polymer Institute, Slovak Academy of Sciences, Dúbravská cesta 9, 845 41 Bratislava, Slovakia
e-mail: upolklcz@savba.sk; Klaudia.Czanikova@savba.sk; Maria.Omastova@savba.sk; upolmaom@savba.sk; peter.kasak@savba.sk

I. Krupa
Center of Advanced Materials, QAPCO Polymer Chair, Qatar University, P.O. Box 2713, Doha, Qatar
e-mail: igor.krupa@savba.sk

E. Pavlová
Institute of Macromolecular Chemistry, Academy of Sciences of the Czech Republic, Heyrovsky Sq. 2, 162 06 Prague 6, Czech Republic
e-mail: pavlova@imc.cas.cz

D. Chorvát Jr.
International Laser Center, Ilkovičova 3, 812 19 Bratislava, Slovakia
e-mail: chorvat@ilc.sk

M. Gupta Bhowon et al. (eds.), *Chemistry: The Key to our Sustainable Future*,
DOI 10.1007/978-94-007-7389-9_1,

(AFM) and by nanoindentor. Nanoindentor, even if the purpose of its original use is different, can be effectively applied for the determination of the actuation stroke, the sample dimensional changes in the direction of irradiation.

1.1 Introduction

Revolutionary technologies are needed to improve the lives of visually impaired and blind people. Current haptic representation in refreshable displays is technically inadequate and very expensive, thus limiting their use in daily life [1]. Resolution, scalability to larger displays, and portability are deficient [2–4]. Mechanical actuation by optical excitation is a much-sought after technology [5]. The devices which utilise an effective photoactuating material are able to convey information in the form of Braille text, maps and graphics to the visually impaired to improve their mobility and quality of life [1, 6]. Part of the 7 RP Nano-Optical Mechanical Systems (NOMS) project was to prepare photo-actuators which must not display fast actuation response during illumination, but have to provide fully reversible actuation. The main disadvantage of available electronic Braille devices is that they show only a single line of text and cannot display graphical images, mathematical equations, maps, music, and so on [7]. The typical Braille cell is illustrated in Fig. 1.1. It has to be pointed out that in the proposed design not only embossed points are displayed, as it is in the cases of printed Braille characters. In this case, all six points are potentially displayable. The required Braille character is then formed by moving the individual Braille elements up.

So far various prototype electronic Braille cells have been constructed using conjugated polymers such as polypyrrole [8], ionic polymer-metal composite bending type actuators [9], or electrostrictive elastomers [10]. However, according to the best of our knowledge, commercially available devices work on the basis of

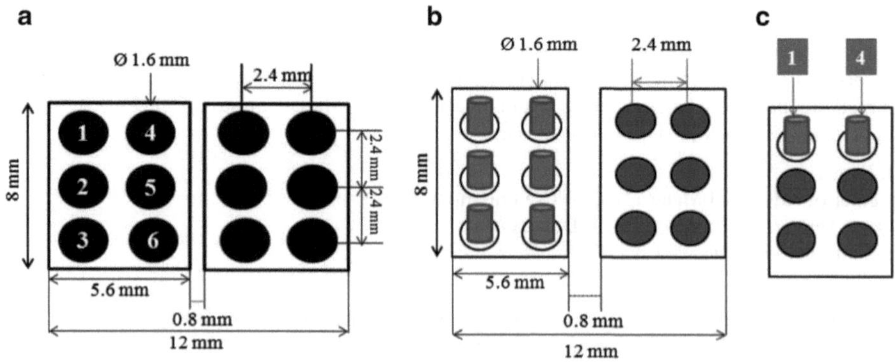

Fig. 1.1 (**a**) Two Braille characters, each one consisting of six raised dots arranged in two columns containing three elements, (**b**) cross-section of the two Braille characters and (**c**) example of Braille symbol – Braille glyph for letter C (raised elements at positions 1–4)

piezoelectric phenomenon. The following parameters have been reported for standardization of Braille devices – the pin matrix density should be up to 1 cell/mm^2, actuating speed have to be more than 50 Hz, and energy density about 10 W/cm^2 [11, 12]. New type of Braille display was presented based on dielectric elastomer. The tactile display is organized with a dual-layer array of tactile cells which generates vertical motion of the Braille pins. The elastomer actuator is compressed in the thickness direction while it is expanded in the lateral direction when a voltage is applied [13–20]. In another study the Braille tablet made by two adjacent Braille cells (consisting of six stimulating pins arranged in a 3×2 array format) were read by the visually impaired persons. The data for two types of recognition rates were obtained as Hit Recognition Rate (HRR) and Number Recognition Rate (NRR). The HRR corresponds to the rate movement of the Braille dots and NRR correctly reading of the Braille dots. The results represented at actuating frequency of the Braille pins was 15 Hz and HRR increases up to about 80 % and the NRR indicates a maximum of 41 %. The obtained results are much better than originally expected [21]. Carpi et al. [22] described the developing push–pull hydrostatically coupled dielectric elastomer actuators. Silicone elastomers membranes filled with oil created bubbles with diameter of 6 mm and were driven up to a voltage of 2.25 kV, applied across a silicone film with a thickness of 42 μm. Specific interest to miniaturize such kind of actuators was motivated by an intention to develop a novel tactile display.

Commercially available refreshable tactile display providing access to high-resolution graphics (pictures, graphs, tables, diagrams etc.) and more than two lines Braille texts are still missing. Only Braille displays using piezoelectric elements work very reliably but on the other hand these displays are still very expensive and noisy [7, 23]. Revolutionary technologies are needed to improve the lives of blind and partially sighted people in order to increase the change of obtaining more information. The nanotubes-polymeric materials are potential candidates for creating new type of actuator because of expected decreased manufacturing costs and true photo-actuation which can be used for construction of haptic display for visually impaired people. The photomechanical actuation is preferred to electromechanical transduction due to the following reasons: wireless actuation, low noise, easy scaling up and down.

The main aim of NOMS project was to develop the prototype of a high resolution, refreshable, tactile visual-aid tablet and demonstrate its capability to depict Braille text and basic graphical information. This tablet have several features, such as full text and graphical capability, fully integrated electronic circuitry, capability to connect to a PC, rapid refresh rate, portability, manufacturability and low cost. Two fundamental requirements are height raise and force output. The minimum stroke height is 0.25 mm and force/pressure about 0.2 N [24]. The promising solution is the use of actuators based on carbon nanotubes-polymeric materials that can be activated by light.

Smart materials can react to a stimulus such as light, temperature, pH, mechanical stress, etc. Dimensional changes in polymers can happen reversibly, dependently on intensity and time of illumination with light [25]. The reversible shape change can be achieved by conformation changes in the case of photochromic molecules, for example, azobenzene undergoes a trans–cis isomerization controlled by the polarization of the light [26].

Liquid-crystal elastomers inherently possess photo-actuating behaviour due to the photoisomerization conformational changes of rod-like dye molecules when small amounts of nanofillers-carbon nanotubes were incorporated into the matrix. The photo-actuation mechanism in the case of liquid-crystal elastomers filled with multiwalled carbon nanotubes (MWCNT) composites containing very low content of MWCNT was explained by the absorption of light in the UV-vis or near IR region, and the light was rapidly converted into local heat. The local heat is then efficiently transferred to the stretched polymer chains near the MWCNT. Subsequent contraction of the stretched polymer chains leads to the photo-mechanical actuation [27, 28].

Ethylene vinyl acetate (EVA) is a commercial elastomer whose properties depend on the ethylene/vinyl acetate ratio [29]. The domain structure of EVA copolymers consists of partially crystalline polyethylene blocks and flexible vinyl acetate blocks [30]. This is mainly focused on EVA containing non-covalently modified multiwalled carbon nanotubes by newly prepared surfactant, arbitrarily here called PyChol and on their photo-actuation response. The composites were prepared by casting from solution due to better dispersion of MWCNT-PyChol within the dissolved polymeric matrix against the mixing with molten polymer [31]. The cast composite foil was used to prepare Braille element using specially designed moulds. The height of the Braille element temporarily increases after illumination and this process is fully reversible. After switching off the light, the Braille element returns to its original shape and height. New developed methods, namely the method using the atomic force microscopy (AFM) and nanoindentor were used to investigate the photo-actuation behaviour of the prepared composites. As far as we know, these methods for characterisation of photoactuation were not reported in literature till now. Here it must be mentioned, that almost all the reported results which can be found in literature are based on the characterisation of photoactuating behaviour of materials in the form of strips. For this characterisation, various setups, usually home-made ones were created, utilising the measurements of the force change at the fixed length during illumination [32–36]. The reports on the testing of photoactuation of Braille characters are very rare, the newest papers in this field have been presented by Camargo et al. [37].

1.2 Methodology

1.2.1 Materials

Tetrahydrofuran (THF, POCH S.A. 99.5 %, Poland) was dried and freshly distilled from sodium/benzophenone. A commercial ethylene-vinyl acetate copolymer (EVA, Levapren 500, Lanxess, Germany) containing 50 wt% of vinyl acetate was

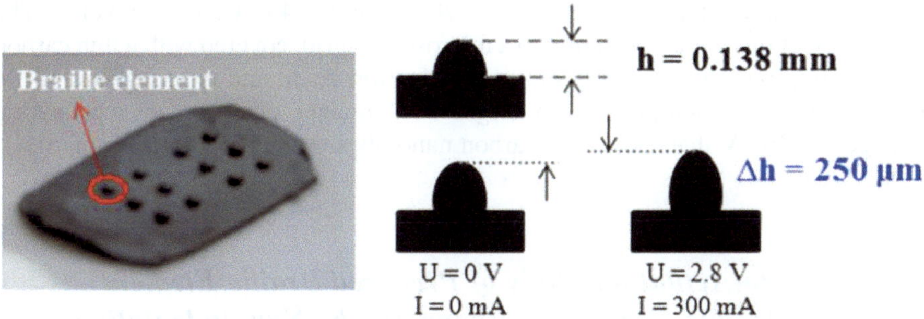

Fig. 1.2 The prepared two Braille characters and detail of the Braille element based on an EVA composite filled with non-covalent modified carbon nanotubes

used as a matrix. MWCNT (Nanostructured & Amorphous Materials, Inc.; Houston, TX 77084, USA) were used as the filler. The purity of the MWCNT was 95 %, the outside diameters were in the range of 60–100 nm, the lengths were in the range of 5–15 μm and the surface area was 64 m^2/g.

1.2.2 Preparation of Composites

The EVA/MWCNT nanocomposites were prepared by casting from solution. Non-covalent surface modification of MWCNT was done using special compatibilizer cholesteryl 1-pyrenecarboxylate (PyChol). The weight ratio MWCNT/Py-Chol was chosen as 1/5 after first testing with lower amount of modifier. The used amount of the carbon nanotubes was selected according to the published results [5]. The solution was sonicated for 2 h at amplitude of 20 % (~35 μm, ~60 W/cm^2, Hielscher 400 S) and a duty cycle of 100 %. After sonication, 10 g of EVA was added and the final solution was stirred and subsequently poured into a Teflon-coated Petri dish and allowed to dry. The sample was dried in the oven and additional drying was performed in a vacuum oven for 6 h at 70 °C. The EVA/MWCNT-PyChol composite foil was prepared by compression moulding (Fontijne SRA-100, The Netherlands) for 15 min at a pressure of 2.4 MPa and temperature of 60 °C. Special custom-made punch/die moulds were applied to the EVA/MWCNT-PyChol nanocomposite to achieve the shape of a Braille element [38].

The composite material was placed between punch/die moulds and loaded by 200 g weight in the oven at 60 °C and subsequently cooled down in ice water in order to freeze the structure. The final shape of the two Braille characters and also one Braille element are shown in Fig. 1.2.

1.2.3 Transmission Electron Microscopy (TEM)

TEM was performed with a Tecnai G2 Spirit Twin 12, FEI, and thin samples were prepared by ultramicrotome (Ultracut UCT, Leica) under cryo-conditions (the sample and knife temperatures were −70 °C and −45 °C, respectively). The ultrathin sections were transferred to a microscopic grid, covered with a thin carbon layer to improve their stability under the electron beam and observed in a TEM microscope. All micrographs are bright field images taken at an accelerating voltage of 120 kV that show dark carbon nanotubes in the light polymer matrix.

1.2.4 Photo-Actuation Study of Prepared Braille Element by Atomic Force Microscopy and by Nanoindentation

In this paper we present two newly developed methods for characterisation of the photoactuation behaviour of nanocomposites. Despite the fact that both utilised equipment are commonly used for totally different types of material characterisation, they can be also adapted for the characterisation of the photoactuating behaviour.

The first of these methods, the Atomic Force Microscopy (AFM), is a well-established tool for the study of structural and physical properties of macromolecules at the surface, as well as high-precision 3D topography. It allows characterization of the surface both in dry and wet conditions with nm resolution, depending on the size of tip. Here, we applied AFM in contact mode (Smena Solver P47H, NT-MDT, Russia) to study the deformation changes of Braille element under illumination using light-emitted diodes (LEDs). However, it must be pointed out that this method enables only the qualitative characterisation of the photoactuation process. Simply said, we can obtain only the information whether the material is photoactuating or not. This fact is caused by the restricted amplitude of the cantilever movement, as can be seen later.

Two types of LEDs were used - red LED (Philip Luxeon, $\lambda = 627$ nm) and blue LED (Philip Luxeon, $\lambda = 470$ nm) at applied current of 150 mA or 300 mA. A Si cantilever (length 100 μm and width 35 μm) with a force constant of 11 $N{\cdot}m^{-1}$ and a tip curvature of 10 nm (NT-MTD, Russia) was used. The changes in the position of the AFM tip in the vertical direction were recorded and plotted against time. The hole in alumina foil is used for focusing the light to the Braille element illumination, where the CNT are aligned in order to achieve the actuation. Scheme of a AFM setup is shown in Fig. 1.3.

Nanoindentor Hysitron TriboLab® Nanomechanical Test Instrument equipped with a Scanning Probe Microscope (SPM) and a Berkovich probe was used for the characterisation of the photoactuating behaviour of materials. The actual use of nanoindentor is the characterisation of mechanical properties of the surfaces. However, similarly as in the case of AFM, it can be adapted for the photoactuation measurements. The TI 750 Ubi nanomechanical test instrument is a dedicated scanning nanoindentor. The principal components in a nanoindentation experiment are the sensors and actuators used to apply and measure the mechanical load and indentor displacement, and the indentor tip. The latter component is conventionally made of diamond, formed into a symmetric shape. The force and displacement are recorded as the indentor tip is pressed into the test material's surface (in our case on

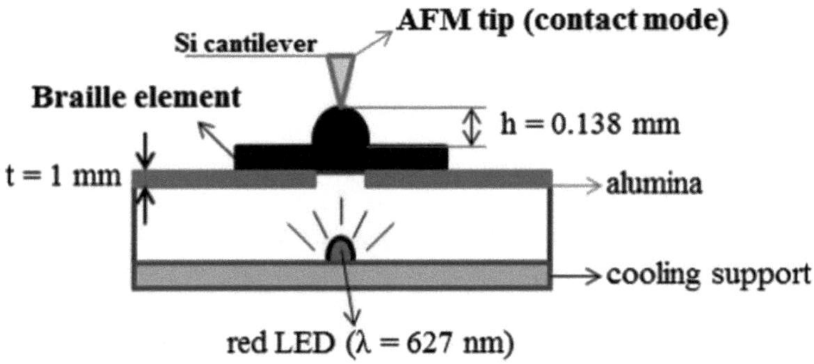

Fig. 1.3 Sketch of the setup for AFM measuring height changes for Braille element during illumination by *red LED*

the Braille element) with a prescribed loading and unloading profile. For our purpose, we only used the fact that it is possible to determine accurately the height of the Braille element on the top before and after illumination. This gives us the information about total actuation deformation of the material, and, what time is needed to reach that maximum. In this case we do not obtain the whole dependence deformation versus time, as it was possible in the case of AFM measurement.

1.3 Results and Discussion

1.3.1 A Dispersion Study of Carbon Nanotubes Within Polymeric Matrix

The proposed non-covalent surface modification of carbon nanotubes (CNT) is based on the van der Waals interaction between the nanotubes and various molecules that consist of aromatic rings through π-π stacking. The main advantage of this procedure is that CNT are not broken during treatment as well as it does not disturb delocalized π electrons and thus, it does not change the inherent electrical conductivity of CNT. Moreover, this kind of non-aggressive treatment does not lead to the breaking of nanotubes, as it usually happens during modification by strong acids [39].

Specially developed surfactant, based on pyrene molecules and long alkyl or cholesteryl groups, was used for CNT surface modification to ensure affinity and good compatibility of the CNT surface with polymeric matrix and good filler dispergation. For better dispergation of CNT we modified the carbon nanotubes non-covalently using PyChol surfactant.

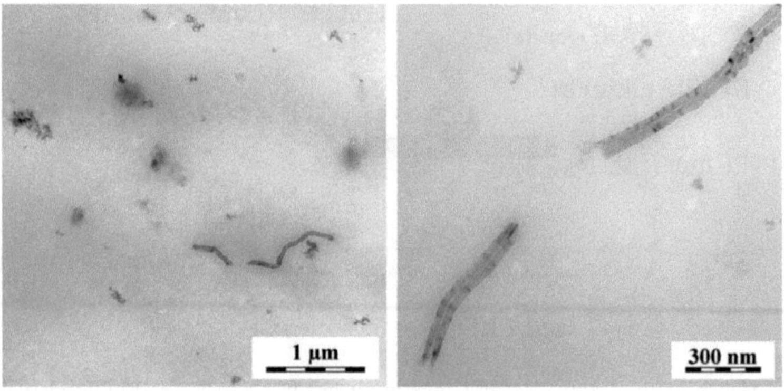

Fig. 1.4 TEM images of the Braille element based on an EVA composite containing 0.1 wt% MWCNT

Table 1.1 The power of the red LED, illumination and relaxation times (T_{illum} and T_{relax}) height changes of the Braille element based on an EVA/0.1 wt% MWCNT-PyChol composite at two different applied currents

Applied current (mA)	Power of LED (mW)	T_{illum}^{*} (s)	T_{relax}^{*} (s)	Height changes (µm)
150	3.5	35	65	2.52
300	6.6	6	30	2.54

The extent of MWCNT dispersion within the EVA polymeric matrix was characterized using TEM. Figure 1.4 depicts a good dispersion of carbon nanotubes due to cholesteryl 1-pyrenecarboxylate compatibilizer used for CNT non-covalent surface modification. Single carbon nanotubes and a minimal amount of their agglomerates were observed, but also a small amount of amorphous carbon nanotube impurities were detected within EVA matrix. The dispersion study of the nanocomposite with unmodified MWCNT was also done by TEM (not shown here), in this case worse dispersion was obtained compared to nanocomposite prepared with compatibilizer.

1.3.2 The Photo-Actuation Study of Braille Element by AFM and Nanoindentation

The photo-actuation response during/after illumination of the Braille element was investigated by AFM method. Two Braille characters were prepared using special titanium punch and die moulds, as depicted in Fig. 1.2. The Braille characters were cut to individual Braille elements for characterization of photo-actuation response.

Using red LED diode when the applied current was set to 150 mA (power 3.5 mW) the original height of Braille element (BE) was increased about 2.52 µm after 35 s of illumination. After switching off the light, the time for the BE relaxation to original shape was 65 s (Table 1.1). A faster response was obtained when the power of the red LED was increased to 6.6 mW (300 mA). In this case, the

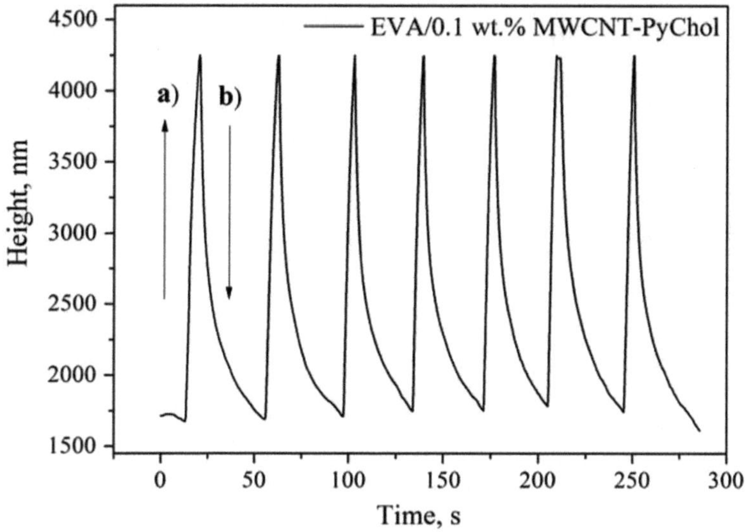

Fig. 1.5 An AFM recording of the height changes over time for a Braille element based on an EVA/0.1 wt% MWCNT-PyChol composite (**a**) upon illumination (*red LED*, at an applied current of 300 mA) and (**b**) after switching off the *red LED*

Braille element grew to 2.54 μm within 6 s and then BE returned back to its original position after 30 s. Figure 1.5 depicts the actuation during illumination with red LED at applied current of 300 mA (a) within 6 s, and after switching off the light-emitting source (b) the Braille element relaxed to the original height.

Table 1.1 summarized the results obtained during AFM study of BE illumination by red LED at 150 mA and 300 mA applied currents, illumination and relaxation times, measured height changes (μm) of the Braille element based on an EVA/0.1 wt% MWCNT-PyChol composite. The original height of the Braille element was 0.138 mm.

In the next study, the photo-actuation response of the Braille element was measured using a blue diode ($\lambda = 470$ nm) at applied current of 300 mA, see Fig. 1.6.

In the case of blue LED at 300 mA applied current the power of this LED is 9.6 mW. Figure 1.6 depicts the measured photo-actuation response of Braille element during illumination by blue LED. The results show high reproducibility of photo-actuation. Figure 1.6 represents 19 cycles of reversible actuation of the characterized BE as an example of BE behaviour during illumination. The maximum deformation of BE observable by AFM due to cantilever movement upon illumination was obtained after 6 s of illumination in both cases when the applied currents were set 150 mA or 300 mA. As can be seen in Fig. 1.6, the Braille element returned to its original position after 15 s (b), which means that the relaxation time was approximately half of that measured after illumination using red LED at the same set current of 300 mA. The main advantage of using blue LED with high

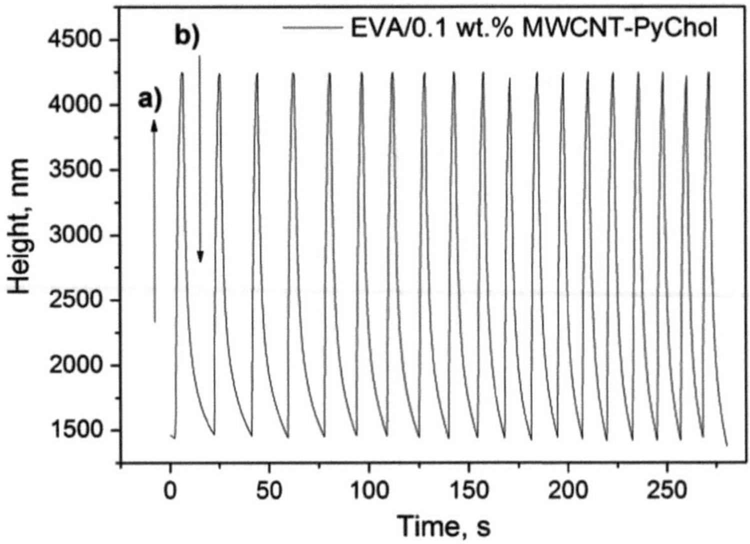

Fig. 1.6 An AFM recording of the height changes over time for a Braille element based on an EVA/0.1 wt% MWCNT-PyChol composite (**a**) upon illumination (*blue LED*, at an applied current of 300 mA) and (**b**) after switching off the *blue LED*

power was faster actuation and relaxation responses. The photo-actuation measurement by AFM was realizable only over a range of −3.0 μm to +3.0 μm, which is the maximum amplitude of the cantilever movement. AFM method was used here to observe the photo-actuation responses of new types of prepared composites and to obtain information about the rates of actuation and relaxation for prepared Braille elements. On the other hand, it was not possible to determine the maximum amplitude of the actuation and relaxation for these samples due to the limited amplitude of the cantilever movement.

Due to this limitation, the method based on the nanoindentor was introduced to determine the maximal deformation changes under illumination for the Braille element. As mentioned above, this method could not determine the dimensional profiles over time like AFM, but it could give us information about the real height changes of the Braille element before and during illumination. In this case, the Braille element was illuminated from the bottom using (as depicted in Fig. 1.3) a red or blue LED. One Braille element was illuminated and the surrounding area of Braille element was shielded with alumina foil. The photo-actuation response of the Braille element under illumination was measured at various current settings. An appropriate photo-actuation of the composite material was observed. An expansion was obtained during illumination. The results are presented in Table 1.2 for the Braille element measured at 200 and 300 mA following illumination with red or blue LEDs.

A maximum deformation change upon illumination using blue LED about 15.2 μm was obtained at applied current of 300 mA. This deformation was much higher than that following illumination of the Braille element using a red LED.

Table 1.2 The power of the red and blue diodes, height changes of the Braille element based on an EVA/0.1 wt% MWCNT-PyChol composite as measured by nanoindentation at different applied currents

Applied current (mA)	Power of red LED (mW)	Height changes (μm)	Power of blue LED (mW)	Height changes (μm)
200	4.4	6.4	5.8	9.8
300	6.6	9.7	9.6	15.2

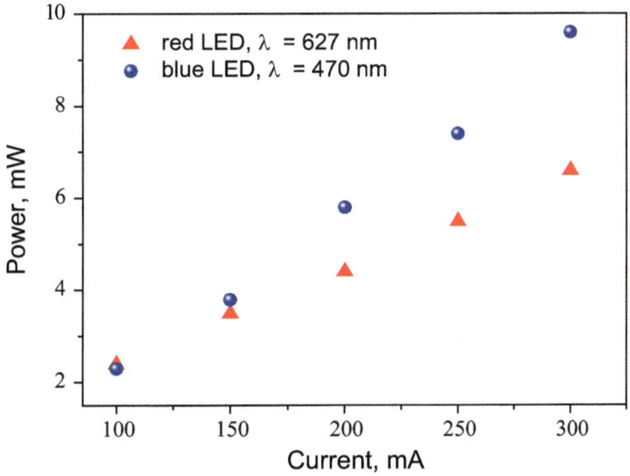

Fig. 1.7 Power outputs (mW) against different applied current (mA) for *red LED* (λ = 627 nm) and *blue LED* (λ = 470 nm)

These data were also in good agreement with the faster actuation and relaxation responses previously observed by AFM when using a blue LED instead of a red one. The power (mW) for both diodes was measured at different applied currents (mA) as depicted in Fig. 1.7.

Higher actuation obtained with blue LED diode (Table 1.2) is associated by increased scattering and by higher absorption at lower wavelength [40] compared with red LED diode (at applied current of 300 mA) by nanoindentation technique. The same effect by AFM method was obtained, as faster relaxation speed was investigated when driven by blue LED (9.6 mW) than red LED one (6.6 mW) at applied current of 300 mA. Information about the rates of actuation and relaxation times were investigated by AFM. The precise, actual height changes of Braille element during illumination were obtained by nanoindentation method. The height of the Braille element temporarily increases after illumination and this process is fully reversible as shown by AFM study. After switching off the light, the Braille element returns to its original shape and height. The photo-thermo-mechanical actuation mechanism is explained by presence of carbon nanotubes in elastomeric nanocomposite, which convert light to heat and increase the heat transfer efficiency. It is caused by inherent high conductivity of carbon nanotubes that originates from their delocalized π-bonded skeleton [41]. The accumulated heat from CNT is released to polymeric matrix and triggering the actuation. Even by AFM study it was not possible to determine the maximum amplitude of the actuation and relaxation, but it showed that prepared nanocomposite is perfectly stable, and fully reversible actuation was determined after hundreds of cycles. The new developed nanocomposites based on EVA copolymer and well dispersed carbon nanotubes are promising materials for creation of actuators which are activated by light. Therefore photoactuation study of nanocomposites is in progress in our laboratory and other, more complex methods will be used in the near future for detailed characterization of Braille element prepared from EVA and carbon nanotubes.

1.4 Conclusions

The EVA/0.1 wt% MWCNT-PyChol composites were prepared by casting from solution. The reason for the selection of this filler content was motivated by the works published very recently, where this filler content was reported as sufficient for the evocation of photoactuating effect, maintaining the properties of a neat polymeric matrix. To improve the dispersion of carbon nanotubes within the EVA polymeric matrix, newly synthesized cholesteryl 1-pyrenecarboxylate compatibilizer (PyChol) was used. The ratio CNT/PyChol equal to 1/5 was determined from the optimisation process, testing the dispersibility and the dispersion stability over long time in the low molecular solvents. Very good dispersion of modified fillers within the EVA matrix was demonstrated by TEM. The photo-actuation behaviour of Braille element prepared from nanocomposite was investigated by the newly developed methods, utilising commercial devices such as AFM and nanoindentor, which are originally employed for different types of characterisations. However, after a small adaptation, these methods were able to characterize the photoactuation response of investigated materials upon illumination by red and blue LEDs. An expansion was detected upon illumination of the bottom of the Braille element. AFM method was chosen to determine the actuation and relaxation times, both of which depended on the type and power of the LEDs used. This method is very simple and offers fast information about qualitative behaviour of materials under irradiation. As for the characterisation by nanoindentor, this one provided information about the total deformation amplitude of material.

Acknowledgements This work was supported by project NOMS, which is funded by the European Commission under contract no. 228916. The authors are grateful to Centro Nacional de Microelectrónica (CNM) – Spain for provision of special punch and die silicon moulds. This contribution/publication is also the result of the implementation of the following project: Centre for materials, layers and systems for applications and chemical processes under extreme conditions Stage II, which is supported by the Research & Development Operational Program and funded by the ERDF. This work was also partly financially supported by VEGA 2/0149/14, and VEGA 2/0119/12. IMC in Prague are grateful for the institutional support under RVO: 61389013.

References

1. Campo EM, Roig J, Roeder B, Wenn D, Mamojka B, Omastov M, Terentjev EM, Esteve J (2011) Nano opto-mechanical systems (NOMS) as a proposal for tactile displays. Proc SPIE 8107, Nano-Opto-Mechanical Systems (NOMS): 81070H. doi:10.1117/12.897529; http://dx.doi.org/10.1117/12.897529
2. Pasquero J (2006) Survey on communication through touch. TR-CIRM, Centre for Intelligent Machines, McGill University, Canada. Technical Report: TR-CIM 06.04
3. Bonet-Borrás C (2004) Braille the pleasure of reading: we blind people want to continue reading with our fingers. UPGRADE, Lect Notes Comput SC 5:77–83
4. Parkes D (1988) Nomad: an audio-tactile tool for the acquisition, use and management of spatially distributed information by partially sighted and blind persons. In: Tatham AF, Dodds AG (eds) Proceedings of the second international symposium on maps and graphics for visually handicapped people, King's College, University of London, pp 24–29
5. Camargo CC, Campanella H, Marshall JE, Torras N, Zinoviev K, Terentjev EM, Esteve J (2011) Localised actuation in composites containing carbon nanotubes and liquid crystalline elastomers. Macromol Rapid Commun 32:1953–1959
6. Mamojka B, Teplický P (2012) Aplications of optically actuated haptic elements. In: Miesenberger K, Karshmer A, Penaz P, Zeglar W (eds) ICCHP 2012, Part II, Springer, Berlin Heidelberg, LNCS 7383, pp 659–663
7. KGS Corporation, http://www.kgs-jpn.co.jp/eindex.html
8. Madden JD, Cush RA, Kanigan TS, Hunter IW (2000) Fast contracting polypyrrole actuators. Synth Met 113:85–92
9. Kato Y, Sekitani T, Takamiya M, Doi M, Asaka K, Sakurai T, Someya T (2007) Sheet-type Braille displays by integrating organic field-effect transistors and polymeric actuators. IEEE Trans Electron Devices 54:202–209
10. Lee S, Jung K, Koo J, Lee S, Choi H, Jeon J, Nam J, Choi H (2004) Braille display device using soft actuator. In: Bar-Cohen Y (ed) Proceedings of SPIE smart structures and materials: electroactive polymer actuators and devices (EAPAD), vol 5385, pp 368–379
11. Jungmann M, Schlaak HF (2002) Taktiles display mit elektrostatischen Polymer-Aktoren. Internationales wissenschaftliches Kolloquium, Tagungsband, p 47
12. Asamura N, Shinohara T, Tojo Y, Koshida N, Shinoda H (2001) Necessary spatial resolution for realistic tactile feeling display, Robotics and Automation. In: Proceedings 2001 ICRA. IEEE international conference on, vol. 2, pp 1851, 1856, 21–26 May 2001. doi:10.1109/ROBOT.2001.932878
13. Choi HR, Lee SW, Jung KM, Koo JC, Lee SI, Choi HG, Jeon JW, Nam JD (2004) Tactile display as a Braille display for the visually disabled, Intelligent Robots and Systems, 2004 (IROS 2004). In: Proceedings 2004 IEEE/RSJ international conference on, vol. 2, pp 1985, 1990, 28 Sept–2 Oct 2004. doi:10.1109/IROS.2004.1389689
14. Koo IM, Jung KM, Koo JC, Nam J, Lee YK, Choi HR (2008) Development of soft actuator based wearable tactile display. IEEE Trans Robot 4:549–559
15. Pelrine R, Kornbluh R, Pei Q, Joseph J (2000) High-speed electrically actuated elastomers with over 300% strain. Science 287:836–839
16. Chakraborti P, Karahan Toprakci HA, Yang P, Di Spigna N, Franzon P, Ghosh T (2012) A compact dielectric elastomer tubular actuator for refreshable Braille displays. Sensor Actuat A-Phys 179:151–157
17. Carpi F, Frediani G, De Rossi D (2011) Hydrostatically coupled dielectric elastomer actuators: new opportunities for haptics. Mater Res Soc Symp Proc 1312:3–12
18. Gorny LJ, Zellers BC, Lin M, Liu S, Zhang QM (2010) The development of compact electroactive polymer actuators suitable for use in full page Braille displays. In: Proceedings of SPIE – the International Society for Optical Engineering Article number 76420B
19. Pei Q, Yu Z, Niu X, Brochu P (2010) Electroactive polymers for rigid-to-rigid actuation and Braille e-books. SPIE Newsroom. doi:10.1117/2.1201002.002632

20. Carpi F, Frediani G, De Rossi D (2010) Hydrostatically coupled dielectric elastomer actuators. IEEE/ASME Trans Mech 15:308–315
21. Choi HR, Koo IM, Jung K, Roh S, Koo JC, Nam J, Lee YK (2009) A Braille display system for the visually disabled using a polymer based soft actuator. In: Carpi F, Smela E (eds) Biomedical applications of electroactive polymer actuators. Wiley, Oxford
22. Carpi F, Frediani G, Tarantino S, De Rossi D (2010) Millimetre-scale bubble-like dielectric elastomer actuators. Polym Int 59:407–414
23. HyperBraille, http://www.hyperbraille.de/?lang=en
24. Fricke J (1997) Substituting friction by weak vibration on a tactile pin array, Developments in Tactile Displays (Digest No. 1997/012). In: IEE Colloquium on, vol., pp 3/1, 3/3, 21 Jan 1997. doi:10.1049/ic:19970082
25. Courty S, Mine J, Tajbakhsh AR, Terentjev EM (2003) Nematic elastomers with aligned carbon nanotubes: new electromechanical actuators. Europhys Lett 64:654–660
26. Yu Y, Nakano M, Ikeda T (2003) Photomechanics: directed bending of a polymer film by light. Nature 425:145
27. Ajayan PM, Terrones M, Guardia A, Huc V, Grobert N, Wei BQ, Lezec H, Ramanath G, Ebbesen TW (2002) Nanotubes in a flash – ignition and reconstruction. Science 296:705
28. Loomis J, King B, Burkhead T, Xu P, Bessler N, Terentjev EM, Panchapakesan B (2012) Graphene-nanoplatelet-based photomechanical actuators. Nanotechnology 23:045501
29. George JJ, Bhowmick AK (2009) Influence of matrix polarity on the properties of ethylene vinyl acetate-carbon nanofiller nanocomposites. Nanoscale Res Lett 4:655–664
30. Martinez-Garcia A, Reche AS, Martin-Martinez JM (2004) Improved adhesion of EVAs with different vinyl acetate contents treated with sulphuric acid. J Adhes Sci Technol 18:967–982
31. Czaniková K, Krupa I, Ilčíková M, Mosnáček J, Kasák P, Mičušík M, Chorvát D Jr, Omastová M (2011) In: Proceedings of SPIE – Bellingham, USA: SPIE – International Society for Optical Engineering, 8107, art. n. 810707
32. Lu S, Panchapakesan B (2007) Photomechanical responses of carbon nanotube/polymer actuators. Nanotechnology 18:305502
33. Hu Y, Chen W, Lu L, Liu J, Chu C (2010) Electromechanical actuation with controllable motion based on a single-walled carbon nanotube and natural biopolymer composite. ACS Nano 4:3498–3502
34. Hu Y, Wang G, Tao X, Chen W (2011) Low-voltage driven sustainable weightlifting actuator based on polymer-nanotube composite. Macromol Chem Phys 212:1671–1676
35. Harvey CLM, Terentjev EM (2007) Role of polarization and alignment in photoactuation of nematic elastomers. Eur Phys J E 23:185–189
36. Ahir SV, Squires M, Tajbakhsh AR, Terentjev EM (2006) Infrared actuation in aligned polymer-nanotube composites. Phys Rev B 73:085420
37. Camargo CJ, Campanella H, Marshall JE, Torras N, Zinoviev K, Terentjev EM, Esteve J (2012) Batch fabrication of optical actuators using nanotube-elastomer composites towards refreshable Braille displays. Micromech Microeng 22:075009
38. Czaniková K, Krupa I, Ilčíková M, Kasák P, Chorvát D Jr, Valentin M, Slouf M, Mosnacek J, Mičušik M, Omastová M (2012) Photo-actuating materials based on elastomers and modified carbon nanotubes. J Nanophotonics 6:063522
39. Spitalsky Z, Krontiras ZA, Georga SN, Galiotis C (2009) Effect of oxidation treatment of multiwalled carbon nanotubes on the mechanical and electrical properties of their epoxy composites. Compos Part A-Appl S 40:778–783
40. Uğur G, Chang J, Xiang S, Lin L, Lu J (2012) A near-infrared mechano responsive polymer system. Adv Mater 24:2685–2690
41. Ahir SV, Huang YY, Terentjev EM (2008) Polymers with aligned carbon nanotubes: active composite materials. Polymer 49:3841–3854

Chapter 2
Identification of Volatile Compounds from Flowers and Aromatic Plants: How and Why?

A. Bialecki and Jacqueline Smadja

Abstract When working on volatile compounds from plants, the objectives are multiples and can be summarized in four points: (1) Research of bioactive molecules, (2) Chemotaxonomic studies, (3) Applications in perfume industry, (4) Plant-insect interactions. Each of these four points will be discussed and illustrated by one or several examples of research projects conducted in the Chemistry Laboratory of Natural Substances and Food Sciences. The first two points exclusively concern volatile compounds generated by essential oils extracted from endemic or indigenous plants of Reunion, Mauritius and Madagascar islands. The two last points are dedicated to volatiles found in the airspace (headspace) surrounding flowers. This paper will also present a selection of sampling methods for volatile compounds that range from conventional, inexpensive, solvent-free, quick sampling methods to innovative methods, as well as an overview of detection and identification methods of volatiles including GC-FID and GC-MS.

2.1 Introduction

Aromatic plants are often confused with medicinal plants because they secrete chemicals which sometimes have pharmacological effects. But rigorously, aromatic plants are considered as plants which secrete volatiles by, at least, one vegetative or reproductive organ, often leaves but also roots, stems, bark, seeds, fruits and flowers. These volatiles may act as aroma and flavour molecules due to their interactions with human receptors. The primary functions of these compounds released into the atmosphere are to defend plants against herbivores and pathogens or to provide a reproductive

A. Bialecki (✉) • J. Smadja
Laboratoire de Chimie des Substances Naturelles et des Sciences des Aliments, Faculté des Sciences et Technologies, Université de La Réunion, 15 Avenue René Cassin, CS 92003, 97 744 Saint-Denis cedex 9, La Réunion, France
e-mail: anne.bialecki@univ-reunion.fr; jacqueline.smadja@univ-reunion.fr

M. Gupta Bhowon et al. (eds.), *Chemistry: The Key to our Sustainable Future*,
DOI 10.1007/978-94-007-7389-9_2, © Springer Science+Business Media Dordrecht 2014

advantage by attracting pollinators and seed dispersers [1]. Today, a total of 1,700 volatile compounds have been described from more than 90 plant families [2]. These volatiles constitute about 1 % of plant secondary metabolites known to date and are typically classified into four major categories: terpenoids, phenylpropanoids/benzenoids, fatty acid and amino acid derivatives [3]. Knowledge of the identity and relative amounts of the volatile substances released by plants is of great importance to several fields of basic and applied research in biology, chemistry and many other disciplines. Obtaining this knowledge requires overcoming many analytical challenges posed by these complex mixtures, because they present large variations in component amounts, chemical structures and functionalities.

After a short presentation of the functions of plant volatiles, the chemical compounds classes in plant volatiles, and a discussion on Sect. 2.3, this chapter will cover several practical approaches to plant volatiles analysis: isolation techniques, separation and detection techniques, and compound identification procedures. A few examples will be presented next, to highlight some of the research topics focused on plant volatiles and developed by the Chemistry Laboratory of Natural Substances and Food Sciences (LCSNSA).

2.2 Plant Volatiles

2.2.1 Functions of Plant Volatiles

It is recognized that these compounds stored in specialized secretory structures such as glandular trichomes or resin ducts [4, 5] are not only emitted by plants in response to abiotic stress such as light and temperature changes, flooding and drought, ultraviolet radiation and oxidants, but they are also used as a sophisticated "language" by plants to have a dialogue with other organisms: microbes, animals, and even other plants [1, 3, 4, 6–10].

Some compounds may attract beneficial insects such as pollinators, whereas others are involved in different modes of defense: direct defense, indirect defense and inter-plant priming.

Direct defense involves the production of compounds that inhibit microbial growth, also kill or repel herbivores. *Indirect defenses* involve the production of compounds that minimize infestations of herbivores by attracting natural enemies preying upon or parasitizing herbivores according to the proverb "The enemy of my enemy is my friend". Finally, chemical volatile signals released from injured plants not only affect herbivores and pathogens but may also signal alarm to neighbouring plants by triggering defense responses. This is called *inter-plant priming*.

Many of these compounds have been referred to as "secondary metabolites" to distinguish them from the "primary metabolites" required for the growth of plants. These secondary metabolites however, are likely to be essential for successful competition or reproduction.

2.2.2 *Chemical Compounds Classes in Plant Volatiles*

The volatile compounds emitted by plants are generally lipophilic and belong to several different classes but are united by their low molecular weight (from 30 to 300 amu) and vapour pressure sufficient to be released and dispersed into the air under normal pressure and temperature. In aromatic and scented plants, they originate from four categories of chemicals: terpenes derivatives, aromatic derivatives, fatty acid derivatives and amino-acid derivatives. Other groups seem to be more sporadic [3, 4, 10–13].

Terpenes Derivatives

Terpenes derivatives, also called isoprenoids, are defined as materials derived from the head-to-tail linkage of the isoprene moiety (2-methylbutane). The isopropyl part of 2-methylbutane is defined as the head, and the ethyl residue as the tail. Depending on the number of isoprene subunits one differentiates between hemi- (C_5), mono- (C_{10}), sesqui- (C_{15}), di- (C_{20}), sester- (C_{25}), tri- (C_{30}), tetraterpenes (C_{40}) and polyterpenes $(C_5)_n$ with $n > 8$. In mono-, sesqui-, di- and sesterterpenes, the isoprene units are linked to each other from head-to-tail; tri- and tetraterpenes contain one tail-to-tail connection in the centre (Fig. 2.1).

Sometimes skeletal rearrangements occur and fragmentation or degradation reactions can reduce the number of carbon atoms so that the empirical formula does not contain a simple multiple of five carbons, thus providing irregular terpenes. Nonetheless, the natural product chemist quickly recognizes the characteristic terpene framework of the structure.

Terpenes encountered among the volatile compounds from plants are exclusively mono-, sesqui- and diterpenes, as well as irregular ones.

Monoterpenes. These substances can be further divided into three groups depending on whether they are acyclic (e.g. myrcene), monocyclic (e.g. limonene) or bicyclic (e.g. α-pinene). Within each group, the monoterpenes may be simple unsaturated hydrocarbons (e.g. limonene) or may have functional groups and be alcohols (e.g. α-terpineol), aldehydes (e.g. citronellal), ketones (e.g. carvone), esters (e.g. linalyl acetate) (Fig. 2.2).

Sesquiterpenes. Like monoterpenes, the sesquiterpenes fall chemically into groups according to the basic carbon skeleton; the common ones are either acyclic (e.g. α-farnesene), monocyclic (e.g. γ-bisabolene) or bicyclic (e.g. α-guaiene) (Fig. 2.3). However, within each group there are many different compounds known. Today, several thousands sesquiterpenoids with well-defined structures, belonging to some 200 skeletal types, are listed.

Diterpenes. Very few diterpenes are reported in floral scents; this may be due to their general low volatility (Fig. 2.4).

Irregular terpenes. The irregular terpenes include compounds varying in the number of carbon atoms from 8 to 18. Among these are apocarotenoids, which are biodegradation products of carotenoid compounds (C_{40}) like β-carotene. Ionones

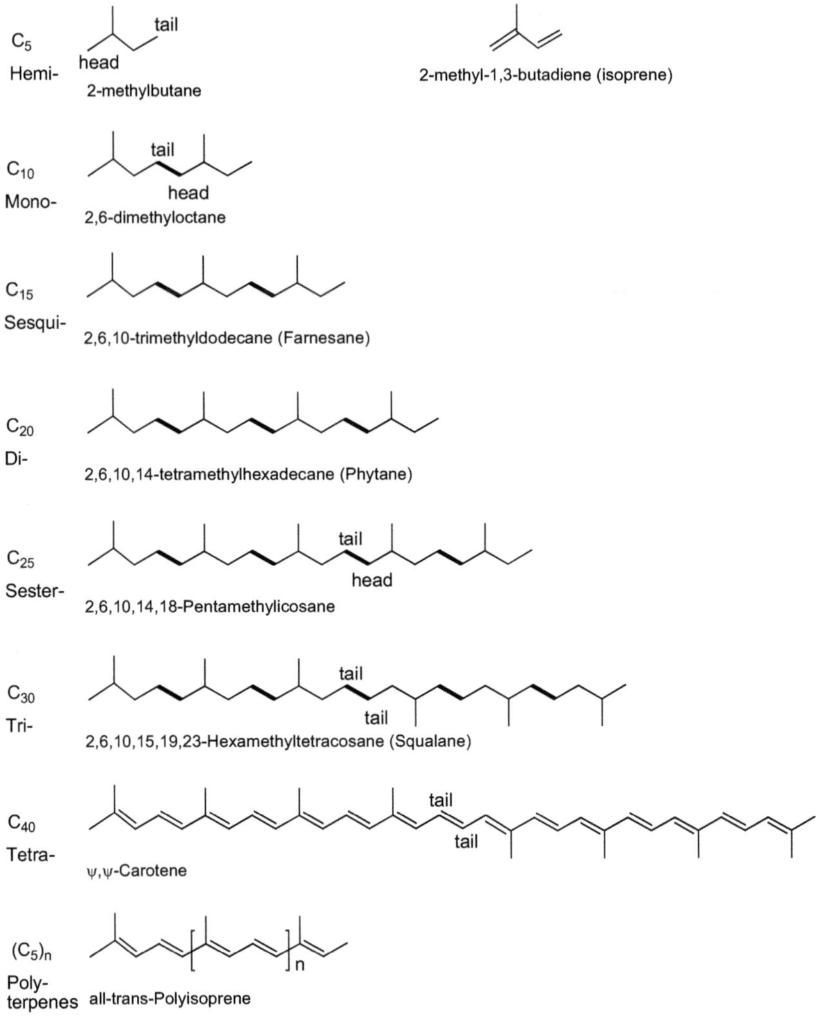

Fig. 2.1 Parent hydrocarbons of terpenes (isoprenoids)

such as dihydro-β-ionone, 6-methyl-5-hepten-2-one and geranyl acetone are the compounds the most often cited (Fig. 2.5).

Aromatic Derivatives

The second category of volatile organic compounds, aromatic derivatives also named as phenolic compounds or benzenoids, are mainly synthesized *via* the shikimate pathway. This pathway got its name from shikimic acid, which is the key step in the formation of the aromatic compounds. Examples of these include: acetophenone, *ortho*-vanillin, cinnamyl alcohol (Fig. 2.6).

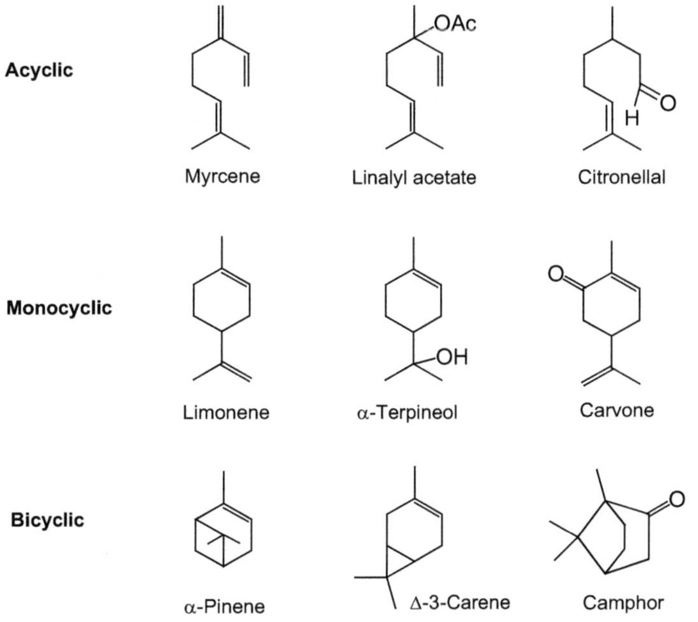

Fig. 2.2 Chemical structures of some monoterpenic compounds

Fatty Acid Derivatives

Fatty acid derivatives are often associated with green leaf odour emitted immediately following the breakdown and lipoxygenation of lipid membranes after mechanical damage. However, these green leaf volatiles are sometimes also produced by flowers. Among the fatty acid derivatives, both saturated and unsaturated hydrocarbons are fairly common, the majority having between 2 and 17 carbon atoms (Fig. 2.7). Aldehydes, alcohols and ketones are also common. Free acids are less common, whereas esters encompass the largest number of different chemical structures. Special mention should be made of the six carbon-compounds known as "green-leaf" volatiles like (*Z*)-3-hexenyl acetate found in vegetative as well as floral scents of numerous plants. This compound probably plays a role in plant defense [1, 7, 8].

Amino Acids Derivatives

Many plant volatiles including aldehydes, alcohols, esters, acids and nitrogen- and sulfur containing compounds are derived from amino acids such as alanine, valine, leucine, isoleucine and methionine, which play an important role in plant defense by recruiting the natural enemies of the attacking herbivore. Amino acids on de-amination form α-keto acid, which in turn forms formaldehyde, acids, alcohols

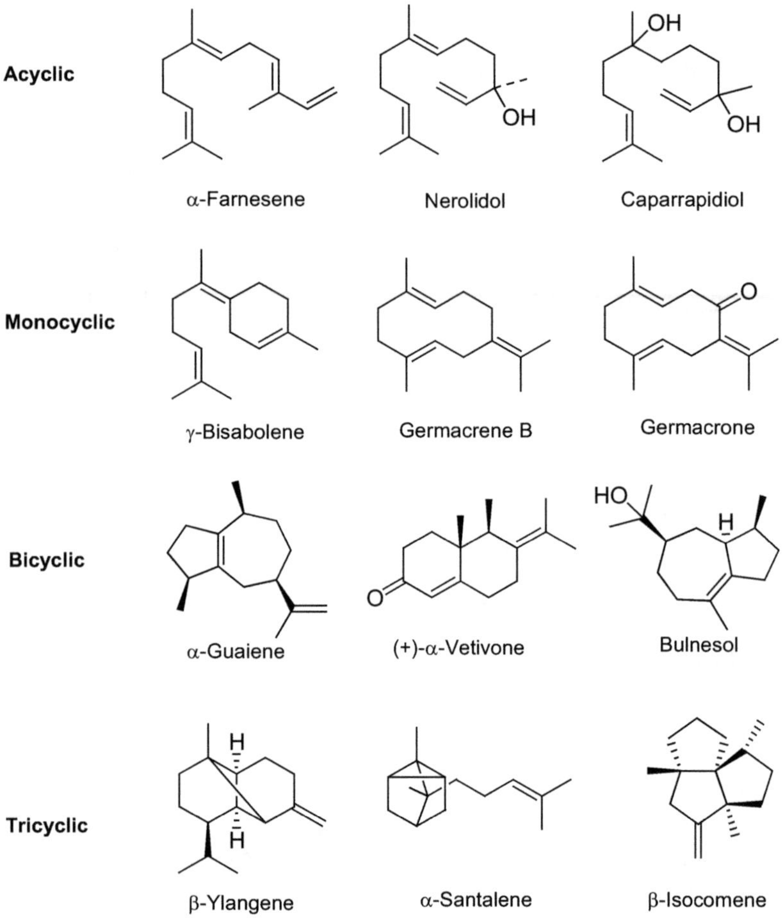

Fig. 2.3 Chemical structures of some sesquiterpenic compounds

and esters on decarboxylation, reduction, oxidation and esterification. Methionine and cysteine have been found to be the precursor of sulfur containing volatiles such as methanethiol, dimethyl disulfide and thioesters responsible for the odour of garlic, onions and boiled potatoes (Fig. 2.8).

2.2.3 Variation in Plant Volatiles

The presence, yield and composition of secondary metabolites in plants, in particular volatile compounds, can be affected in a number of ways, from their formation in the plant to their final isolation. Factors affecting volatile compounds production

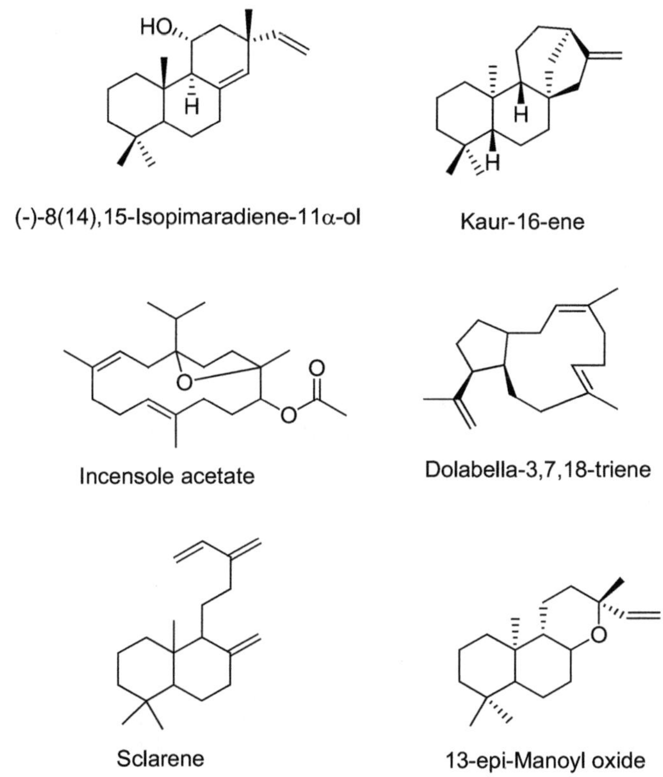

Fig. 2.4 Chemical structures of some diterpenic compounds

include: (1) *physiological variations* such as organ development, pollinator activity cycle, type of plant material (leaves, flowers, etc.), type of secretory structure, seasonal variation, mechanical and chemical injuries; (2) *environmental conditions* like climate, pollution, diseases and pests, edaphic factors; (3) *geographic variation*; (4) *genetic factors and evolution*; (5) *storage* [14].

2.3 Why Investigate Plant Volatiles?

Knowledge of the identity and relative amounts of the volatile substances emitted by plants is of great importance to several fields of basic and applied research mainly in chemistry and biology. So, they are studied for different purposes.

The first one is purely *economic* as plant volatiles can be used in a wide variety of consumer goods such as detergents, soaps, toilet products, cosmetics, pharmaceuticals, perfumes, confectionery food products, soft drinks, distilled alcoholic beverages (hard drinks) and insecticides.

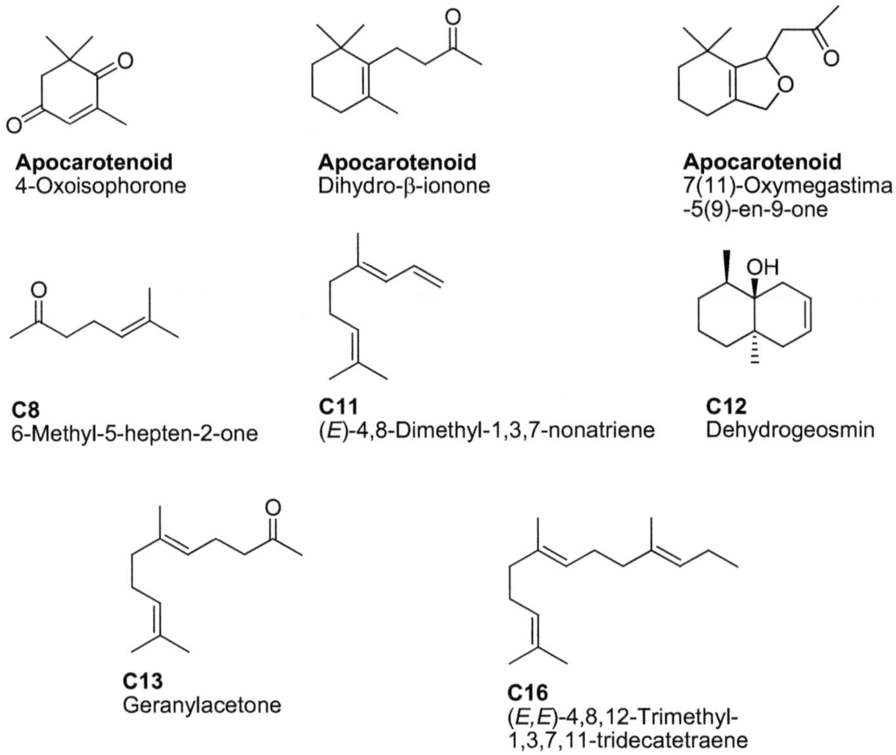

Fig. 2.5 Chemical structures of some irregular terpenes

The second one concerns more specifically *ecology* and focuses for example on plant-plant communication, plant-insect interaction, plant pollination and defense, thermo-tolerance and other environmental stress adaptation.

Studying volatile compounds from plants may also help to understand the *phylogeny or systematic* of some plants through chemotaxonomy. The chemotaxonomy (from chemistry and taxonomy), also called chemosystematics, is the attempt to classify and identify organisms (such as plants), according to demonstrable differences and similarities in their chemical compositions.

At last, volatile compounds are also studied for their *biosynthesis*. Although many of the volatile constituents of plants have been identified, many of the enzymes and genes involved in their biosynthesis are indeed still not known. Such investigation could be interesting for biotechnological process. Indeed, for some years now the demand for natural aroma chemicals is growing fast, in response to both consumers, who are asking for a return to nature, as well as perfumers and flavorists looking for novel creative ingredients. However, the quality and supply of traditional natural flavour and fragrance chemicals are often limited. So, in addition to extraction from natural sources, viable alternative and innovative ways to provide flavour and fragrance chemicals include today

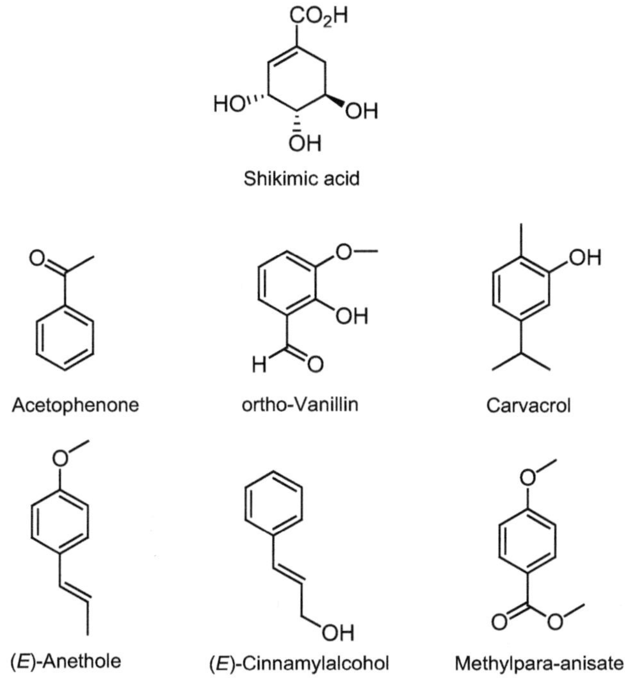

Fig. 2.6 Chemical structures of some aromatic compounds

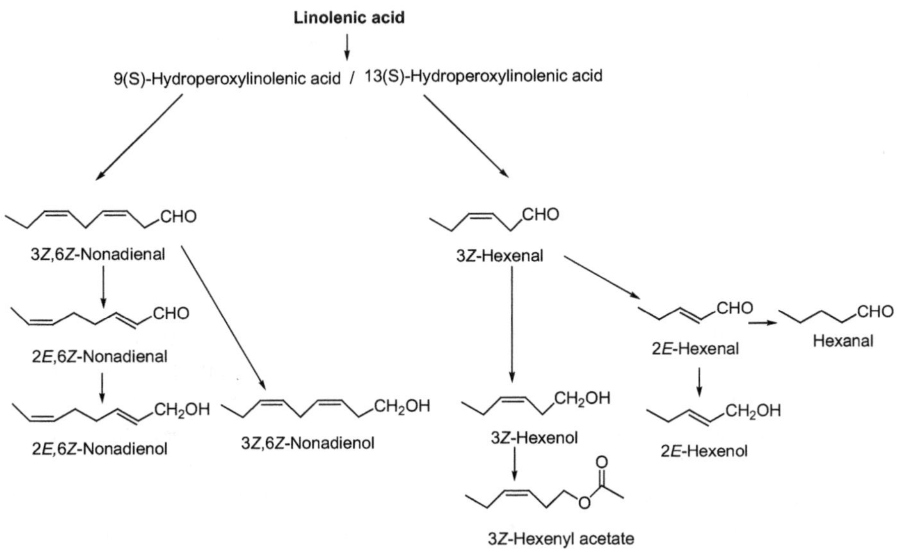

Fig. 2.7 Chemical structures of some fatty acid derivatives

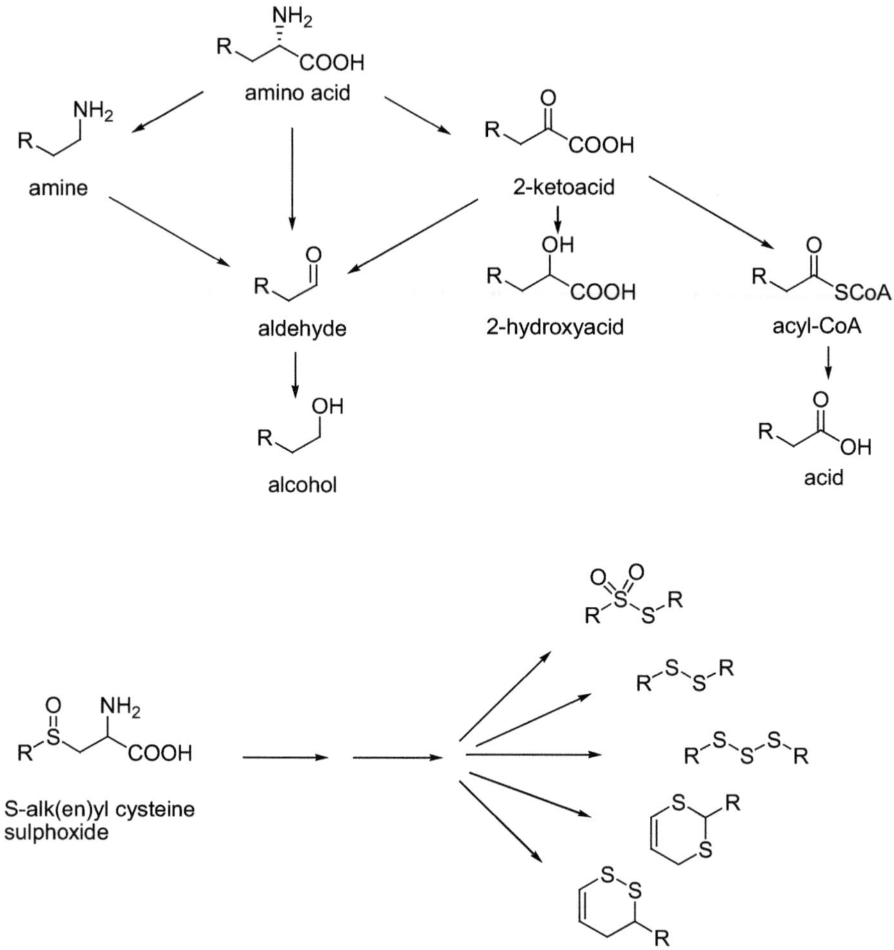

Fig. 2.8 Chemical structures of some amino acid derivatives

biotechnological routes, *i.e.* microbial fermentation, biotransformation using whole cells, biocatalysis using enzymes, plant tissue culture and transgenic plants. So, production or modification of flavour by genetic engineering is thus dependent on the knowledge and availability of genes that encode enzymes of key reactions that influence or divert the biosynthetic pathways of plant-derived volatiles.

This increasing scientific interest in plant volatile compounds has led to the development of a variety of systems for the collection and analysis of volatiles. The choice of which system to use in a particular experiment for collection and analysis obviously depends on the objective which is set. One must consider all the advantages and disadvantages of each technique.

2.4 How to Investigate Plant Volatiles?

2.4.1 Sampling Techniques

Extracting and trapping methods for volatile compounds stored in plants are very numerous. The more commonly used methods will be described hereafter.

Extracting Methods

Extracting methods [15–17] can be divided into two groups: (1) traditional and (2) innovative methods. In relation to the technique used, the qualitative and quantitative composition of essential oils extracted from a same part of a plant is quite different.

Traditional Methods

Distillation. This is the most popular, widely used and cost-effective method for producing essential oils throughout the world. Distillation simply implies vaporizing or liberating the volatile compounds from the plant cellular membranes in the presence of moisture, by applying high temperature and then cooling the vapour mixture to separate the oil from the water on the basis of the immiscibility and density of the essential oil with respect to water. There are different techniques of distillation: hydrodistillation, water and steam distillation, direct steam distillation, distillation with cohobation and hydrodiffusion.

Hydrodistillation. Hydrodistillation is the simplest and oldest process available for obtaining essential oils from plants. Hydrodistillation differs from steam distillation mainly in that the plant material is almost entirely covered with water in the still which is placed on a furnace. An important factor to consider in water distillation is that the water present in the tank must always be enough to last throughout the distillation process; otherwise the plant material may overheat and char. In this method, water is made to boil and the essential oil is carried over to the condenser with the steam which is formed. Water-distilled oil is slightly darker in colour and has much stronger still notes than oils produced by other methods. Hydrodistillation is extensively used by small-scale producers of essential oil.

Water and steam distillation. To eliminate some of the drawbacks of water distillation, some modifications were made to the distillation units. A perforated grid was introduced in the still to support the plant material and to avoid its direct contact with the hot furnace bottom. When the water level is kept below the grid, the essential oil is distilled by the rising steam from the boiling water.

Direct steam distillation. In direct steam distillation, the plant material is distilled with steam generated outside the tank in a steam generator or boiler. As water and steam distillation, the plant material is supported on a perforated grid above the steam inlet. In water and steam distillation, the steam is at atmospheric pressure and hence its maximum temperature is 100 °C, whereas steam in a modern pressure boiler operating at, for example, 50 psi pressure will have a temperature correspondingly higher. Moreover, there is no limitation to the steam generation when an external boiler is used as a source of steam.

Distillation with cohobation. Cohobation is a technique that can be used for hydrodistillation or for water and steam distillation. It uses the process of returning the distillate water to the still after the oil has been separated from it so that it can be re-boiled. This method has been developed for oils which have partial solubility in water. Indeed, although most of the essential oils have finite solubility in water, some oils like those of rose, lavender and geranium have comparatively higher solubility. In such extractions, the loss with the outgoing water of distillation can become alarmingly high. This problem can be solved by returning the condensate water from the separator back to the still: this is known as cohobation. It is evident that this cannot be done with steam distillation as the water level in the still will keep building up to continuous steam injection.

Hydrodiffusion. This system was first described in 1983. Unlike traditional steam distillation, hydrodiffusion works on the diffusion principle of allowing steam to enter the top of the plant charge and diffuse oil from the oil glands. The system is connected to a steam source, and low pressure steam is passed into the plant material from a boiler. The condenser, which is directly under the basket within the still, is of the tube type. The oil and water are collected below the condenser in a typical oil separator (such as a Florentin flask). The yield of oil is generally high and the process is advantageous because of reduced steam consumption, shorter distillation time and absence of hydrolysis, as the raw material does not come in contact with boiling water.

Cold expression. This technique is used for the very delicate type of *Citrus* family oils which are easily destroyed even by moderate heat and damaged by the steam because of their terpene and aldehyde composition. The peels of ripe *Citrus* fruits are first squeezed by hand or by using special devices that press at room temperature. Secondly, the oil is rinsed off in cold running water. The essence is then collected after decantation.

Maceration in organic solvents. For delicate flowers such as Rose, Jasmine, or Violet, volatile compounds can also be extracted by maceration in organic solvents. This technique permits recovery of many volatile compounds that are lost during distillation due to high temperature. The solvent dissolves all extractable matter from the plant which includes waxes, pigments and highly volatile compounds. The extraction efficiency depends on organic solvents (e.g. pentane, or petroleum ether), the use of agitation and choice of temperature (between 25 °C and 30 °C) to increase the solubility. The solution containing solvent and dissolvable plant material is then filtered and the filtrate subjected to low pressure distillation to recover the solvent for further use. The remaining waxy mass is called *concrete*.

The concentrated concretes are processed further to remove the waxy materials which dilute the volatile compounds. To achieve this, the waxy concrete is warmed and stirred with alcohol (absolute ethanol). During the heating and stirring process the concrete breaks up into minute globules. Since the aromatic molecules are more soluble in alcohol than is wax, an efficient separation of the two takes place. But along with volatile molecules a certain amount of wax also becomes dissolved and this can only be removed by agitating and freezing the solution at very low temperatures. In this way most of the wax precipitates out. As a final precaution the purified solution is cold filtered leaving only the wax-free material which after solvent evaporation is called *absolute*.

Enfleurage. This is a method of extraction involving the absorption of the volatile compounds into a fatty substance. It is usually used to extract volatile compounds from fairly soft plant materials, e.g. flowers, buds, petals and so on. In the "Enfleurage" method the flowers or other plant materials are spread in a glass dish which contains a thin layer of scentless and purified fat. In less modern times, purified lard was often used for this purpose. The plant material is then left for 24, 48 or 72 h, and then replaced by fresh plant material at 24, 48 or 72 h intervals. This process is repeated time and time again until the fat medium is "saturated" with volatile compounds. The resulting fat is called a *pomade*.

Like concretes, this pomade is dissolved in a solvent (often ethanol). The fatty substance itself does not dissolve, but sinks to the bottom of the vessel, while the volatile components are dissolved in the alcohol. The alcohol is then gently heated to give the absolute. Volatile compounds from Rose, Tuberose, Jasmine and Neroli are collected in this way.

Innovative Methods

Important progress has been made in the development of novel separation techniques with shortened extraction times, reduced solvent consumption, and enhanced prevention of oxidation and isomerization, especially for thermolabile and chemically highly active constituents.

Microwave extraction. Microwaves are increasingly being used as the heat source to assist the extraction of volatile compounds. The developed microwave extraction methods include the Solvent-Free Microwave extraction (SFME) and the Microwave Steam Diffusion (MSD). The Solvent-Free Microwave Extraction in particular is the most commonly used. This method involves placing the fresh plant material in the microwave reactor without addition of water or any other solvent. The internal heating of the water within the sample distends its cells and leads to rupture of the glands and oleiferous receptacles. This process thus frees essential oil, which is evaporated by the *in-situ* water of the plant material. A cooling system outside the microwave oven continuously condenses the vapours which are collected on specific glassware. The excess of water is refluxed back to the extraction vessel in order to restore the *in-situ* water to the sample [18–20].

Supercritical fluid extraction. This is a solvent-free extraction method, usually carried out using CO_2 due to its advantages as a solvent for volatile compounds. Under high pressure, CO_2 turns into a liquid and acts as a solvent that can be used to extract volatile compounds from plant material. CO_2 is forced into a stainless steel tank containing plant material, then the pressure is released. As pressure decreases, CO_2 returns to a gaseous state and only the plant extract remains. One of the main drawbacks of solvent-free extraction is its limitation to non-polar and medium-polar substances, since it is mostly applied with CO_2.

Ultrasound-assisted extraction. This extraction method can be used for the extraction of volatile compounds localized in both surface glands, where a mild ultrasonic treatment is enough, and inside the cells, where stronger treatment is needed. Ultrasound-assisted extraction increases the performance of solvents and is performed at lower temperature, which is less likely to result in losses of thermally unstable compounds, but isomerisation and decomposition may occur for chemically unstable compounds. Ultrasound-assisted extraction provides smaller extraction yields than most classical methods such as hydrodistillation or steam distillation and some recent extraction methods.

Trapping Methods

Several fields of study include analyses of floral volatiles in relation to pollination biology, measurements of volatiles such as isoprene released from phytosynthetic tissues in response to changes in light and temperature, and volatile emissions induced by herbivore damage. In most cases, the emitted volatiles have to be sampled and concentrated prior to subsequent analysis. Such investigation is often referred to as *headspace* analysis [21–23]. Headspace sampling is a non-destructive method for collecting volatiles in the airspace (or headspace) surrounding above-ground plant parts. Compared with solvent extractions of volatiles from plant tissues, headspace analysis gives a more realistic picture of the volatile profile emitted by plants and detected by insects that respond to plant volatiles, making this method most suitable for many ecologically relevant applications.

There are different headspace sampling techniques: the *static headspace sampling* (no air circulation) with the use of Solid Phase Micro-Extraction (SPME); and the *dynamic headspace sampling* which includes Close-Loop Stripping (air is continuously recycled), Pull and Push-Pull Systems (air is constantly taken up from the outside, passed over the plant sample and through an absorbent trap). In all these volatile collection methods, the used chamber for headspace collection should be free of material that retains volatiles or cause bleeding of compounds that may contaminate the system. Good choices for materials include glass, Teflon and metal which are easy to clean and do not show bleeding, whereas materials such as rubber, plastic, glues, adhesives and wood should be avoided.

Concerning more specifically static headspace analysis, the plant or its parts are enclosed in a container and emitted volatiles are trapped onto an adsorbent. The air surrounding the plant remains "static" which means it is not circulated in the chamber. Volatiles are enriched on the adsorbing matrix without sampling impurities of a continuous air stream that may obscure the detection of low-abundant volatile compounds. Thus, this method is advantageous for sampling volatile compounds from low emitting plants.

An important advance in static headspace analysis was the development of SPME which is a true "green revolution" in sample preparation techniques. Cheap, fast, simple and solvent-free, this method enables the collection of volatiles at detection limits in the ppbv (parts per billion by volume) range. Solid phase micro-extraction is based on ad/absorption and desorption of volatiles from an inert fiber coated with different types of ad/absorbents. The fiber is attached within the needle of a modified syringe and volatiles can be sampled by inserting the needle through a septum of a headspace collection container and pushing the plunger to expose the fiber. Following equilibration between the fiber and the volatile sample (a few minutes to half an hour), the fiber is retracted into the needle and can be transferred to a gas chromatograph injector for direct thermal desorption. Solid phase microextraction fibers can be reused approximately 100 times. By carefully selecting the polarity and thickness of the fiber coating, compounds of different polarity and volatility ranging from high-boiling or semi-volatile to volatile compounds can be sampled. Solid micro-extraction does not provide directly quantitative information since the adsorbed amount depends on the fiber-coating affinity for the compound in addition to its concentration in the headspace. Thus, the results obtained may misrepresent some volatiles and over-represent others. However, quantification of volatiles by SPME may be possible by the application of internal or external calibration.

2.5 Analysis Techniques

Plant volatile compounds extracted or trapped on ad/absorbing matrices are routinely analysed by the standard technique of gas chromatography [24, 25]. A chromatographic system includes four fundamental blocks: (1) An injector; (2) a column placed in an oven; (3) a detector and (4) a data system.

Injector. Samples are either injected as solvent extracts into the heated injector in a split or splitless mode or desorbed from the adsorbent such as the SPME fiber by placing it directly in a thermal desorption tube, heated to 250–300 °C. The injection system then transfers the mixture of volatile compounds into the chromatographic column. Ideally, this must be done in a quantitative manner, without discrimination due to molecular weight or component volatility and without chemical alteration of any constituent substance. So, an injector is a sort of entrance door into the column where component separation is achieved.

Column. Volatiles from plants are commonly separated on fused capillary columns with different stationary phases, such as the non-polar dimethyl polysiloxanes (e.g. OV-1, DB-1, DB-5, CPSil 5) and the more polar polyethylene glycol polymers, including Carbowax 20 M, DB-Wax and HP-20 M. Columns are usually 30–60 m long (and have a stationary phase film thickness of 0.2–0.3 μm and an internal diameter of 0.25 mm or 0.32 mm).

Detector. Following separation on a GC column, volatile compounds can be analysed by two different types of detectors. The first type, for example, a *flame ionization detector* (FID) provides only information on retention times, while detectors of the second type, such as *Mass Spectrometry* allow additional structure evaluation.

Flame ionization detectors. With such a detector, organic compounds are ionized in a hydrogen/air flame, producing a signal proportional to the mass flow of carbon. They are commonly used for quantitative analysis because of their wide linear dynamic range, their very stable response and their high sensitivity with detection limits of the order of picograms and nanograms per compound.

Mass Spectrometry detectors. These are the most popular type of detector for routine plant volatile GC analysis. In the mass spectrometers of most standard GC-MS bench-top instruments, compounds exiting the GC column are ionized by electron impact (EI) and the resulting positively charged molecules and molecule fragments are selected according to their mass-to-charge (m/z) ratio by entering a quadrupole mass filter. Total ion chromatograms are obtained, which provide information on the retention time of each compound and its mass spectrum consisting of a characteristic ion fragmentation pattern.

Data systems. For identification of compounds in GC-MS analysis, suggestions can be obtained from popular computerized mass spectral libraries such as Wiley [26], NIST [27] or MassFinder [28] and other databases not computerized developed by Adams [29] or Joulain and König [30].

Concerning the computerized MS library search, it is usually performed using PBM (probability – based matching) algorithm, a library-search routine that uses a reverse search to verify that peaks in the reference spectrum are present in the unknown spectrum. Extra peaks in the unknown are ignored, thus allowing the analysis of a spectrum resulting from a mixture of compounds. The PBM search results displays the list of the best 20 matches that resulted from the library search showing the name of each compound, the molecular weight and the correlation coefficient or quality. The correlation coefficient is used as the first identification criterion. However, mass spectra libraries cannot be used as unique and absolute criteria for the identification of chromatogram peaks. This process is not always straightforward but complicated by several issues.

First of all, even under highly reproducible mass spectrometry conditions, two mass spectra of the same compound are not absolutely identical, but only similar to each other. Measured on the same mass spectrometer under identical experimental conditions, the similarity is usually excellent. However, in most cases, the reference spectrum contained in the library has been recorded with another mass spectrometer than the unknown mass spectrum, and the spectra may differ to a certain degree. Using several types of spectrometer (quadrupole, ion trap...) result in many cases in notable differences. Impurities of co-eluting peaks or air may also cause significantly different spectra.

Furthermore, the essential feature of mass spectrometry for essential oils is that mass spectra are not particularly unique in many cases. Within the broad class of monoterpenes and sesquiterpenes found in essential oils, a large number of isomers of the same molecular formula but with different structure exist. Many times, their mass spectra are very similar making the peak identification somewhat difficult and sometimes impossible. The issue of substances with almost identical mass spectra can only be resolved by using additional experimental data that are able to distinguish between them. The gas chromatography is indisputably the complementary technique used to differentiate such compounds. The first solution consists of identifying the components by co-chromatography with standards. If authentic material is not available, the second solution is to combine the experimental values for retention times and mass spectra that are two independent parameters: substances with same mass spectra usually exhibit different retention times, or in other words, co-eluting peaks could only pure chance give rise to identical mass spectra. However, the retention time of the separated compounds cannot be utilised as such owing to their dependence on many factors such as column length, column polarity, carrier gas velocity and oven temperature program. All these elements prevent comparison of the retention times obtained with those of other scientists. A superior approach is then to use a derived value from the retention time called *Retention Index* or *Kovats Index*. These have the advantage of being fairly insensitive to experimental conditions and can therefore be replicated for a given stationary phase. They represent a common language among chromatographers.

The Retention Index (RI) or Kovats Index (KI) is a relative value determined by comparing the retention time of the compounds with retention times of two standard compounds. Several homologous series of organic compounds can be used as standards, but *n*-alkanes have been used exclusively. The formula for RI calculation depends on temperature conditions.

For isothermal conditions:

$$\mathrm{RI(x)} = 100\frac{\log\left(t_{R(x)}\right) - \log\left(t_{R(z)}\right)}{\log\left(t_{R(z+1)}\right) - \log\left(t_{R(z)}\right)} + 100z$$

For linear temperature program conditions:

$$\mathrm{RI(x)} = 100\frac{t_{R(x)} - t_{R(z)}}{t_{R(z+1)} - t_{R(z)}} + 100z$$

where:

RI(x) is the retention index of compound x
z is the number of carbon atoms in alkane z
$t_{R(x)}$ is the retention time of compound x
$t_{R(z)}$ is the retention time of alkane z
$t_{R(z+1)}$ is the retention time of alkane z + 1

To use the retention index system, a measure of gas chromatograms of all *n*-alkanes covering the desired retention time range needs to be done; e.g. to investigate monoterpenic and sesquiterpenic compounds a range of C_9–C_{26} is needed, so all terpene peaks will elute between known alkanes. In order to measure the complete alkane pattern, with only a single injection, mixtures of *n*-alkanes are usually prepared from pure chemicals at 5 % concentration in pentane. The measurement of the alkane pattern obviously has to be carried out under exactly the same experimental conditions as the future analyses. Different GC-MS systems will exhibit slightly different alkane patterns, so for each system the alkane pattern has to be determined separately. Few compilations of retention indices are currently available. Those which should be mentioned are:

Davies [31], a compilation of RI of some 400 monoterpenes and sesquiterpenes on either or both types of stationary phase (non-polar and polar).

Kondjoyan and Berdagué [32], a compilation of RI of more than 2,000 volatile compounds on non-polar (type DB-1 or DB-5) and polar (type Carbowax) stationary phases.

Adams [29] and Joulain and König [30] list the retention indices of compounds on non-polar stationary phases (DB-5 and CPSil-5) and their mass spectra.

Massfinder [28] also represents another excellent two dimensional search algorithm that takes both retention index on non-polar stationary phase (DB-5) and mass spectrum similarity into account. The library of this program covers approximately 2,000 compounds commonly found in essential oils, particularly monoterpenic and sesquiterpenic compounds, diterpenes and related aromatic and aliphatic constituents like esters or lactones. Most spectra acquired on quadrupole spectrometers will be very similar and fully compatible with Massfinder mass spectral collection. Mass spectra acquired on ion traps may exhibit close similarity, but depending on the measurement conditions may differ slightly.

To sum up, a correct identification of a volatile compound requires at least the determination of Retention Indices on two columns with different polarities and a good match quality of the mass spectrum of the compound of interest with that of authentic standard.

2.6 Investigation on Volatile Compounds by LCSNSA

2.6.1 Environment Protection and Biodiversity Conservation

It is recognized that tropical oceanic islands contribute disproportionably for their area to global biodiversity. However, these islands are also specifically vulnerable to many threats including habitat loss, habitat degradation and fragmentation, and invasion of exotic species. Tropical islands have therefore been considered biodiversity hotspots containing many plant and animal species that are not found elsewhere but that are under severe threat. A series of islands scattered in the western Indian Ocean along the southeast coast of Africa has been recognized as a global biodiversity hotspot. Dominated by the nation of Madagascar, the fourth largest island on earth, this hotspot also includes the independent nations: Seychelles (including Aldabra), the Comoros, Mauritius (including Rodrigues), and the French overseas departments: Reunion, Mayotte (one of the Comoros) and the Iles Eparses. Three of them (Reunion, Mauritius and Rodrigues) are grouped together and make up the Mascarene Islands. Reunion Island (2,512 km^2) is the biggest of the three and also contains the largest proportion of intact habitat types (ca 30 %) because of its rugged topography which has precluded agriculture and large-scale urbanization. Today, it appears obvious that all initiatives that can ensure biodiversity protection on this volcanic island will enable to preserve representative tracts of Mascarene ecosystems. Systematic conservation planning is widely considered the most effective approach for designing protected area and other ecological network. However, effective systematic conservation planning requires expert knowledge on the distribution of plant and animal species, patterns of species richness and composition, and the way they are interrelated. Thus, in this context, biologists, botanists and chemists from the University of Reunion Island decided to collaborate and contribute to a better knowledge of the fauna and flora of Reunion Island.

Among the several ongoing scientific projects, the phytochemical study about species of the genera *Psiadia* and *Melicope* was envisaged by the LCSNSA for chemotaxonomic purpose [33–36].

The genus *Psiadia* (Asteraceae) comprising aromatic plants is widely distributed in tropical and subtropical regions. The taxonomy of this genus is rather complex. Thus, in the Mascarenes, the taxonomic scheme established by Scott is built on the segregation of the genus into five groups mostly based on morphological characters. Nine species endemic to Reunion Island and Mauritius were investigated.

The genus *Melicope* (Rutaceae) encompasses more than 200 species occurring from the Malagasy and Indo-Himalayan regions, east to Hawaiian and Marquesas Islands and south to New Zealand. In the revision by Hartley, all *Euodia* species from the Malagasy region, including the Mascarene Islands were transferred to the *Melicope* species. Six of these species are endemic to Reunion Island. The six species including sub-species and varieties were investigated.

For all these species, the chemical composition of the essential oil obtained from their leaves by hydrodistillation has been examined. The chromatographic analyses by GC-FID and GC-MS allowed the quantification and identification of their volatile constituents. However, some of them could not be identified by computer matching with MS libraries (laboratory-made and commercial) and Retention Index. So, their complete structure was elucidated by means of mono- and bidimensional NMR spectroscopic techniques after isolation and purification by repeated column chromatography. Thus, several new molecules were identified. This is the case of an acetylated monoterpene (1) and a sesquiterpene alcohol (2), both detected at a high percentage in the essential oils of *Psiadia anchusifolia*, *Psiadia argentea*, *Psiadia boivinii* and *Psiadia salaziana*. The occurrence of these unusual terpenoids in these four species supposes that the latter have the same biosynthetic pathway and it may also suggest that the two new molecules could be used as a chemotaxonomic tool for the characterization of some *Psiadia* species. By NMR spectroscopy, we have also identified a new oxygenated sesquiterpene (3), major constituent of *Melicope obscura* essential oil (Fig. 2.9).

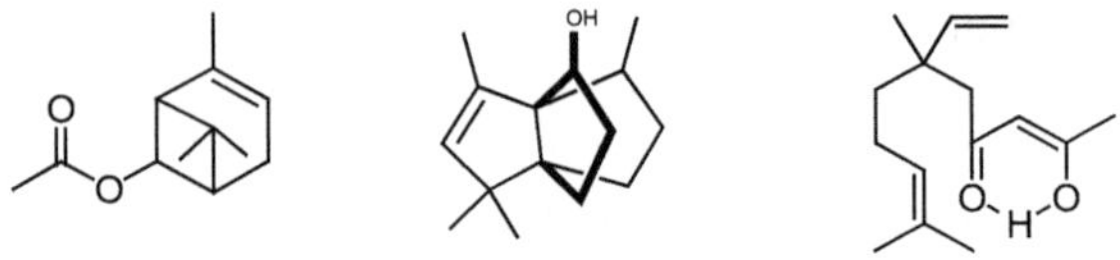

Fig. 2.9 New compounds isolated from *Psiadia* and *Melicope* species

Environment protection and biodiversity conservation requires also a better knowledge of inter-specific interactions between plants and their pollinators. No other plant family shows as wide a range of pollinator-linked floral forms as Orchidaceae, which exhibit pollination systems among most diverse, specialized and complex of all angiosperms. Orchid pollination mechanisms have primarily involved the insect orders Hymenoptera (bees, wasps and ants; these pollinate roughly 60 % of orchid species), Diptera (flies and mosquitoes), Lepdoptera (moths, hawkmoths and butterflies) and Coleoptera (beetles). Approximately 3 % of orchid species are estimated to be pollinated by birds and 5–20 % of species are thought to be self-pollinating. However, in spite of the large size of this model family and a long history orchid pollination biology, the identity and specificity of most orchid pollinators remain inadequately studied, especially in the tropics where the family has undergone extensive diversification. Thus, in order to better understand the reproductive biology of the tropical orchids on Reunion Island, biologists and chemists from The University of Reunion investigated the pollination syndromes i.e. floral morphology, breeding system, pollinator diversity, floral scent profile and fruity success of orchid species of the genera *Angraecum* [37, 38], *Jumellea* [39] and *Bulbophyllum* [40]. Floral scents were studied using headspace solid phase micro-extraction combined with GC-MS analyses. On the basis of this work, it has been demonstrated that one species of *Angraecum* (*A. cadetii*) is pollinated by a highly unexpected pollinator, an undescribed species of raspy cricket (*Glomeremus orchidophilus*, Gryllacrididae). Although Orthoptera are well known for herbivory, this represents the first clearly supported case of orthopteran-pollination.

2.6.2 *Commercial Goals*

The second objective of the investigation of the LCSNSA on aromatic plants is more commercial. It concerns not only essential oils but also the composition of headspace from fragrant flowers.

Essential Oils

Essential oils are valuable natural products used as raw materials in many fields, including perfumes, cosmetics, aromatherapy, spices and nutrition, etc. Aromatherapy is the therapeutic use of fragrances or volatile compounds to cure, mitigate or prevent diseases and infections by means of inhalation. This has attracted our attention and encouraged us to screen plants to study the biological activities of their oils from chemical and pharmacological investigations to therapeutic aspects. Hopefully, this will lead to new information on plant applications and new perspective on the potential use of these natural products. Of course, it is clear that only a detailed knowledge of the constituents of the essential oil will allow for a better and specially directed application. Among the aromatic plants investigated as potential biological and pharmacological resources, there are:

Cedrelopsis grevei native to Madagascar is a small tree, known under various common names such as Katafray, Matahora, Bemafaitra and particularly exploited not only for its wood but for its medicinal properties as well. The essential oils extracted from its bark and its leaves and commercially produced by traditional distilleries in Madagascar, are in particular commonly used in folk medicine as fortifying, tonic, relaxing and postnatal medication. The bark essential oil is also used to cure rheumatism and muscular pains and is known to exert antifungal and antibiotic activities. The essential oils from the bark and leaves commercially available have been examined separately [41]. A total of 55 compounds have been identified constituting around 77 % and 92 % of the volatile compounds for respectively bark and leaf essential oils. Both oils were found to have a similar composition rich in terpenes. However the relative percentages of some compounds notably differed. The bark essential oil contained β-pinene, *cis*-sesquisabinene hydrate and caryophyllene oxide as the main components whereas the leaf essential oil was largely dominated by *trans*-β-farnesene, β-pinene, *cis*-sesquisabinene hydrate and *ar*-curcumene.

Ayapana triplinervis commonly known as Ayapana in Hindi, is native to South America and can be found in the Amazon region of Brazil, Ecuador, Peru, the three Guianas, Puerto Rico, Hawaii but is also well represented in other countries such as India, Vietnam and the Mascarene Islands (Reunion, Mauritius, Rodrigues). Since it is widely used in folk medicine, the plant has been extensively investigated for its biological and pharmacological properties. However, till now the information about its chemical composition has remained poor. Concerning more specifically volatile compounds from *A. triplinervis*, few investigations exclusively devoted to essential oil samples of Brazilian, Indian and Vietnamese origins have been carried out. Thus, we decided to investigate the chemical composition of the volatile oil of *A. triplinervis* from Reunion Island, where this plant is locally really appreciated for its healing virtues in particular its digestive properties. Our study [42] includes the analysis of the leaf essential oil of three specimens of *A. triplinervis* collected at two distant locations (North of the island for samples 1 and 2, South of the island for sample 3), in different growth phases (flowering for samples 1 and 3, vegetative for sample 2) in order to elucidate the chemical character of this species on the island and to investigate the relationship between essential oil composition, developmental stage and geographic locations. Analysis by GC-FID and GC-MS enabled us to identify and quantify a total of 39 constituents accounting for 97.1–98.0 % of the oils. The three essential oil samples, all obtained by hydrodistillation, showed a high percentage of the aromatic compound thymohydroquinone dimethyl ether (89.9–92.8 %). All other minor components remained more or less unchanged both qualitatively and quantitatively with respect to the stage of growth. On the contrary, variations were observed with geographic distribution on the island.

The composition of the oils isolated from the three *A. triplinervis* specimens investigated here by us was then compared to those obtained from the same species collected in different geographical sites in the world (Brazil and India). For comparison purposes, only the main components were considered. It appears then a clear geographical diversity in *A. triplinervis* with respect to the chemical composition and the main components of the essential oils. Two chemotaxonomic groups may be established:

Group 1: Species that mainly synthesize the aromatic ether thymohydroquinone dimethyl ether. This was described for species from Brazil and India (Saugar).
Group 2: Species producing principally the oxygenated sesquiterpene selina-4 (15),7(11)-dien-8-one. This is the case of *A. triplinervis* from India (Lucknow).

On this basis, our *A. triplinervis* samples from Reunion Island considerably dominated by thymohydroquinone dimethyl ether (average: 91.2 %) may be included in the group 1. However, in this group, by looking at the other main components, three chemotypes may tentatively be considered: chemotype thymohydroquinone dimethyl ether/α-phellandrene/borneol (India/Saugar); chemotype thymohydroquinone dimethyl ether/β-caryophyllene (Brazil); chemotype thymohydroquinone dimethyl ether (Reunion Island).

Toddalia asiatica (Rutaceae) is an evergreen woody liana native of tropical Asia from India and Sri Lanka to Malaysia; it is also found in temperate Asia, tropical Africa, Madagascar and Mascarene Islands. This species is widely used in folk medicine as remedies for a human variety of ailments such as malaria, stomach complaints and coughs. As a consequence of the ethnobotanical uses, this species has been investigated chemically. So far, alkaloids, coumarins, benzopyranones, terpenoids and cyclohexylamides are the most characteristic compounds. Most of them as well as crude extracts and essential oils of the plant were studied for their pharmacological activities revealing important biological properties: anti-HIV, antimalarial, antiplatelet aggregation, antipyretic, anti-inflammatory, analgesic, antiviral, antimicrobial, spasmolytic, antifeedant, anticancer and skin whitening. Despite this extensive investigation little is known about its essential oil composition. The objective of our study was then to characterize *Toddalia asiatica*, growing wild in Reunion Island, through the composition of its essential oil in order to observe the homogeneity of the composition or, conversely, to evidence a chemical variability among specimens from different geographical regions of the world [43].

Thus, comparing the results obtained for our plant material with those reported for the same species from India, profound differences were revealed in the composition of essential oils: monoterpenes such as linalool (12.4 %), α-borneol (9.4 %), *p*-cymene (8.8 %), α-terpineol (8.6 %) were reported as the characteristic components of Indian specimen, while the two aldehydes, prenal (24.6 %) and 2-hydroxy-3-methyl-but-2-enal (22.8 %) were the most important volatiles of Reunion Island specimen. Moreover, it is well known that aldehydes usually play a significant role in odor composition. So, these two major constituents could be putatively considered as key odorants of *Toddalia asiatica* essential oil principally characterized by choking, pungent, powerful, aldehydic and fruity notes.

Headspace

Floral perfumes continue to inspire perfumers and the headspace technology offers perfumers the option to clone the essence of flowers from which no oils can be extracted such as orchids and has led to a surge of new scents. So, in order to discover new fragrances, which may become a source of inspiration for a new generation of perfumes or home fragrances, we decided to investigate by micro-extraction on solid phase, the headspace of around 100 flowers *in vivo*, from wild or ornamental, endemic or indigenous plants of La Réunion. The results of 60 flowers were selected and given in a book for perfumers (Book ongoing).

2.7 Conclusions

Several practical approaches to plant volatiles analysis are described (isolation techniques, separation and detection techniques, compound identification procedure). More precisely, a selection of sampling methods for volatile compounds that range from conventional, inexpensive, solvent-free, quick sampling methods to innovative methods, as well as an overview of detection and identification methods of volatiles including GC-FID and GC-MS are presented.

Some examples of research projects conducted in LCSNSA are highlighted. In particular, research of bioactive molecules and chemotaxonomic studies concerning volatile compounds generated by essential oils extracted from endemic or indigenous plants of Reunion, Mauritius and Madagascar islands are developed as well as applications in perfume industry and plant-insect interactions dedicated to volatiles found in the airspace (headspace) surrounding flowers.

Such research topics underline the engagement of the LCSNSA to enhance volatile compounds (essential oils and headspace) and to fulfil environment protection and biodiversity conservation.

Acknowledgements This work was supported by the Regional Council of La Réunion. The authors wish to express their gratefulness to Hermann Thomas and Max Félicité for their help in plant material collection.

References

1. Pichersky E, Gershenzon J (2002) The formation and function of plant volatiles: perfumes for pollinator attraction and defense. Curr Opin Plant Biol 5:237–243
2. Knudsen JT, Eriksson R, Gershenzon J, Stahl B (2006) Diversity and distribution of floral scent. Bot Rev 72:1–120
3. Dudareva N, Pichersky E, Gershenzon J (2004) Biochemistry of plant volatiles. Plant Physiol 135:1893–1902

4. Dudareva N, Negre F, Nagegowda DA, Orlova I (2006) Plant volatiles: recent advances and future perspectives. Crit Rev Plant Sci 24:417–440
5. Caissard JC, Joly C, Bergougnoux V, Hugueney P, Mauriat M, Baudino S (2004) Secretion mechanisms of volatile organic compounds in specialized cells of aromatic plants. Rec Res Dev Cell Biol 2:1–15
6. Langenheim JH (1994) Higher-plant terpenoids: a phytocentric overview of their ecological roles. J Chem Ecol 20:1223–1280
7. War AR, Sharma HC, Paulraj MG, War MY, Ignacimuthu S (2011) Herbivore induced plant volatiles – their role in plant defense for pest management. Plant Signal Behav 6:1973–1978
8. Paré PW, Tumlinson JH (1999) Plant volatiles as a defense against insect herbivores. Plant Physiol 121:325–331
9. Qualley AV, Dudareva N (2008) Aromatic volatiles and their involvement in plant defense. In: Schaller A (ed) Induced plant resistance to herbivory. Springer Science + Business Media B.V, The Netherlands, pp 409–432
10. Spinelli F, Cellini A, Marchetti L, Nagesh KM, Piovene C (2011) Emission and function of volatile organic compounds in response to abiotic stress. In: Shanker A, Venkateswarlu B (eds) Abiotic stress in plants – mechanisms and adaptations. InTech, Rijeka, pp 409–432
11. Knudsen JT, Tollsten L, Bergström G (1993) Floral scents – a checklist of volatile compounds isolated by headspace techniques. Phytochemistry 33:253–280
12. Knudsen JT, Gershenzon J (2006) The chemical diversity of floral scent. In: Dudareva N, Pichensky E (eds) Biology of floral scent. CRC Press, Boca Raton, pp 27–52
13. Breitmaier E (2006) Terpenes: flavors, fragrances, pharmaca, pheromones. Wiley-VCH Verlag GmbH & Co. KGaA, Weinheim, p 214
14. Figueiredo AC, Barroso JG, Pedro LG, Scheffer JJC (2008) Factors affecting secondary metabolites production in plants: volatile components and essential oils. Flavour Fragr J 23:213–226
15. Tandon S (2008) Distillation technology for essential oils. In: Handa SS, Khanuja SPS, Longo G, Rakesh DD (eds) Extraction technologies for medicinal and aromatic plants. ICS UNIDO, Trieste, pp 115–143
16. Ormeno E, Goldstein A, Niinemets U (2011) Extracting and trapping biogenic volatile organic compounds stored in plant species. Trends Anal Chem 30:978–989
17. Smadja J (2009) Chemical composition and localization of essential oils. In: Chemat F (ed) Essential oils and aromas: green extractions and applications. Har Krishan Bhalla & Sons, Dehradun, pp 122–146
18. Chemat F, Lucchesi M, Smadja J (2004) Extraction par microondes sans utilisation de solvant, de composés naturels volatils. European Patent EP 1439218 (A1)
19. Chemat F, Lucchesi M, Smadja J (2004) Solvent-free microwave extraction of volatile natural substances. United States Patent US2004187340 (A1)
20. Visinoni F, Chemat F, Smadja J (2006) Extraction par microondes de composés naturels volatils. European patent EP 1629725 (A1)
21. Tholl D, Röse USR (2006) Detection and identification of floral scent compounds. In: Dudareva N, Pichensky E (eds) Biology of floral scent. CRC Press, Boca Raton, pp 3–52
22. Tholl D, Boland W, Hansel A, Loreto F, Röse USR, Schnitzler JP (2006) Practical approaches to plant volatile analysis. Plant J 45:540–560
23. Harlalka R (2008) Solid phase micro-extraction and headspace trapping extraction. In: Handa SS, Khanuja SPS, Longo G, Rakesh DD (eds) Extraction technologies for medicinal and aromatic plants. ICS UNIDO, Trieste, pp 145–163
24. Gauvin A (2006) Hyphenated methods. 8-7-01 analysis and characterisation of essential oils using GC-MS. In: Gross M, Caprioli R (eds) Encyclopedia of mass spectrometry, 8th edn. Elsevier, Oxford, pp 737–749
25. Stashenko EE, Martinez JR (2012) GC-MS analysis of volatile plant secondary metabolites. Gas chromatography in plant science, wine technology, toxicology and some specific applications. InTech, Rijeka, pp 247–270

26. Software: McLafferty FW (2009) Wiley registry of mass spectral data, 9th edn. Wiley, Hoboken
27. Software: National Institute of Standards and Technology (2011) Mass spectral database for windows (data version: NIST11)
28. Software: Koenig WA, Joulain D, Hochmuth DH (2008) Terpenoids and related constituents of essential oils. MassFinder 4. In: Hochmuth DH (ed) Convenient and rapid analysis of GCMS, Hamburg, Germany
29. Adams RP (2001) Identification of essential oil components by gas chromatography/quadrupole mass spectroscopy. Allured Publishing, Carol Stream
30. Joulain D, König WA (1998) The atlas of spectral data of sesquiterpene hydrocarbons. E.B-Verlag, Hamburg
31. Davies NW (1990) Gas chromatographic retention indices of monoterpenes and sesquiterpenes on methyl silicone and carbowax 20M phases. J Chromatogr 503:1–24
32. Kondjoyan N, Berdagué JL (1996) A compilation of relative retention indices for the analysis of aromatic compounds, 1st edn. Laboratoire Flaveur, France
33. Gurib Fakim A, Gauvin A, Smadja J, Govinden-Soulange J, Kodja H (2000) Composition of the essential oil of the endemic *Psiadia arguta* Pers (Voigt) from Mauritius. J Essent Oil Res 12:556–558
34. Gauvin A, Susperregui A, Barth P, Rémy L, Déléris G, Smadja J (2004) An acetylated monoterpene and a sesquiterpene alcohol from *Psiadia anchusifolia*. Phytochemistry 65:897–901
35. Govinden-Soulange J, Magan N, Gurib-Fakim A, Gauvin A, Smadja J, Kodja H (2004) Chemical composition and in vitro activities of the essential oils from endemic *Psiadia* species growing in Mauritius. Biol Pharm Bull 27:1814–1818
36. Smadja J, Strasberg D, Le Goff G, Gauvin-Bialecki A (2010) Volatile compounds from *Melicope obscura*. Chem Biodivers 7:467–475
37. Micheneau C, Fournel J, Warren BH, Hugel S, Gauvin-Bialecki A, Pailler T, Strasberg D, Chase M (2010) Orthoptera, a new order of pollinator. Ann Bot 105:355–364
38. Hugel S, Micheneau C, Fournel J, Warren BH, Gauvin-Bialecki A, Pailler T, Chase MW, Strasberg D (2010) *Glomeremus* species from the Mascarene islands (Orthoptera, Gryllacridae) with the description of the pollinator of an endemic orchid from the island of Reunion. Zootaxa 2545:58–68
39. Micheneau C, Fournel J, Gauvin A, Pailler T (2008) Auto-pollination in a long-spurred endemic orchid (*Jumellea stenophylla*) on Reunion Island (Mascarene Archipelago, Indian Ocean). Plant Syst Evol 272:11–22
40. Humeau L, Micheneau C, Gauvin-Bialecki A, Fournel J, Pailler T (2011) Sapromyiophily in the native orchid, *Bulbophyllum variegatum* on Reunion (Mascarene Archipelago, Indian Ocean). J Trop Ecol 27:591–599
41. Gauvin A, Ravaomanarivo H, Smadja J (2004) Comparative analysis by gas chromatography–mass spectrometry of the essential oils from bark and leaves of *Cedrelopsis grevei* Baill, an aromatic and medicinal plant from Madagascar. J Chromatogr A 1029:279–282
42. Gauvin-Bialecki A, Marodon C (2008) Essential oil of *Ayapana triplinervis* from Réunion island: a good natural source of thymohydroquinone dimethyl ether. Biochem Syst Ecol 36:853–856
43. Smadja J, Laurençon L, Thomas H, Bialecki A (2001) Medicinal plant potentialities from Réunion Island. The particular case of *Toddalia asiatica* (Rutaceae). In VIII international ethnobotany symposium proceedings, Lisbon, pp 625–635

Chapter 3
An Investigation into the Use of Concept Cartoons in the Teaching of "Metals and the Reactivity Series" at the Secondary Level

Hiteyeshi Lallbeeharry and Fawzia B. Narod

Abstract The present study was designed in an attempt to promote argumentation, and to enhance students' engagement and understanding during the teaching of "Metals and the Reactivity Series" by making use of concept cartoons. The research was carried out in a Girls' State Secondary School situated in the North of Mauritius. The sample comprised 35 Form III students (age 13–14 years), who were in the third year of secondary schooling. The study was based on an action research, during which concept cartoons were used in the teaching of "Metals and the Reactivity Series". Four different data-collection tools were used during the study, namely, observation checklists, students' achievement tests, students' questionnaire, and a semi-structured interview. Our findings have revealed that the use of concept cartoons in the teaching of "Metals and the Reactivity Series" has promoted both teacher-student and student-student interactions. Moreover, concept cartoons have proved to be very effective in eliciting students' interest, and in motivating them to participate during the lessons through discussions and argumentations. Most importantly, our findings have clearly indicated that concept cartoons can serve as stepping stones to improve students' understanding and performance by helping teachers to identify their problems and misconceptions. It has also been found that concept cartoons depicting different viewpoints, or those including both correct and incorrect statements, have not only helped in engaging learners in argumentation, but have also promoted thinking. The study has also brought to light that it is important for teachers to include a gamut of statements, ranging from simple to more challenging ones, while designing concept cartoons in order to cater for students of different abilities. We also report here that concept cartoons can be used as teaching tools in conjunction with various strategies, like hands-on practical, demonstration, video-viewing and group work.

H. Lallbeeharry
Sugar Industry Labour Welfare Fund, Port-Louis, Mauritius

F.B. Narod (✉)
Department of Science Education, Mauritius Institute of Education, Réduit, Mauritius
e-mail: zia373@eudoramail.com

M. Gupta Bhowon et al. (eds.), *Chemistry: The Key to our Sustainable Future*,
DOI 10.1007/978-94-007-7389-9_3, © Springer Science+Business Media Dordrecht 2014

3.1 Introduction

Science (including Chemistry as a branch of Science) is defined as knowledge based on observed facts and tested truths arranged in an orderly system [1]. It has also been defined as "*a careful, disciplined, logical search for knowledge about any and all aspects of the universe, obtained by examination of the best available evidence and always subject to correction and improvement upon discovery of better evidence*" [2]. Not surprisingly, throughout history, scientists have worked by reviewing theories, models and explanations based on new evidence by using the scientific method [3]. In the scientific method, the validity of scientific claims is settled by referring to observations of phenomena (data). These data are obtained by observations and measurements taken in situations that range from natural settings to completely contrived ones, such as the laboratory [4]. Thus, it has been rightly advocated by Driver et al. [5] that learning Science should not be viewed only as the acquisition of facts, but should also initiate learners into the scientific ways of thinking and making sense of the practices of the scientific community. Further, Wellington and Osborne [6] pointed out that learning to reason in Science involves learning how to construct arguments, which link evidence with ideas and theories, while La Velle and Erduran [7] claimed that argumentation can help Science students to clarify their ideas. The role of argumentation in Science Education has also been emphasized by Newton et al. [8] who claimed that "*pedagogies which foster argument lie at the heart of an effective education in science*". This close relationship between argumentation and Science Education posits that argumentation should be an important part of Science Education, as supported by rising evidence that argumentation is a powerful strategy for teaching and learning in Science [9]. Moreover, Erduran [10] also argued that chemical education theory and practice would benefit from argumentation.

On the other hand, according to the constructivist model of learning [11], teachers need to recognize that learners have some prior conceptions or knowledge, and that they can only make sense of new situations in terms of their existing understanding. The constructivist approach thus requires teachers to find out students' current level of understanding or prior knowledge [11], and to confront them with situations and problems, which are slightly beyond this level of understanding. This would lead to two important features of the constructivist classroom situation, namely, negotiation and interaction [12]. Gray [12] also emphasised that genuine student-student and student-teacher dialogue are very important in a constructivist classroom; the author also claimed that purposeful talk by students would lead to meaningful discussion and to the emergence of meanings. In light of these deliberations, it is clear that constructivists also support that argumentation and dialogue are indispensable in Science classrooms.

Nonetheless, it is paradoxical that debate, discussion and argumentation are not common features in Science classrooms [13]; the teaching of Science (including Chemistry) is generally done through the didactic approach [14] which is passive, and does not promote scientific discourse and argumentation in classroom situation.

The latter report [14] also points out that Science teaching in schools is uninteresting and unexciting, and recommends that more child-centred active learning approaches to Science teaching be employed to develop inquisitiveness, reasoning and problem-solving skills amongst learners. In light of the above discussions, the present study has been specifically designed to use concept cartoons to teach "Metals and the Reactivity Series" at the Secondary Level, in an attempt to create an active classroom learning situation and to encourage argumentation and students' engagement during the lessons.

The topic "Metals and the Reactivity Series" has been selected, as many of the concepts involved figure out as pre-requisites for learning the "Periodic Table" in the Cambridge International Examinations (CIE) 'O'- Level syllabus. It has indeed been highlighted in recent CIE 'O' level Examiners' reports that many students failed to realize some prior concepts of the "Periodic Table" leading to misconceptions about the topic. Further, the examiners also reported that many students gave general answers in the examinations, showing that they have learnt by rote [15]. According to Chuckowree [16] and Sobron [17], one of the possible factors responsible for the misconceptions and problems encountered by students might be attributed to the way Chemistry is taught; most teachers use the traditional chalk and board method (the expository method). Indeed, as mentioned above, based on a survey carried out [14], it has been recommended that active learner-centred approaches be used in the teaching and learning of Science in Mauritius. Thus, this paper is based on an investigation on the use of concept cartoons in the teaching of "Metals and the Reactivity Series" in an attempt to promote argumentation and to shift to an active learner-centred classroom situation.

3.2 Theoretical Framework: Concept Cartoons

Concepts are reported to be "*the basic parts of knowledge, and they help human beings to organize and categorize the learning outcomes*" [18]. According to the constructivist theory of learning, in order to learn a new concept, students should adapt their prior knowledge with the new learning through interactions with their peers and teacher. Thus, learning is considered to be an active process which allows students to construct meaning and to interpret situations based on their previous knowledge and experiences [19]. In other words, students should be allowed to talk, debate, and investigate their own ideas to be able to learn new concepts [20]. This posed two main problems for teachers, firstly classroom management problems [20], and secondly, research has shown that though teachers were able to tap on students' prior ideas, it was a challenge for them to restructure these ideas [21]. In view of these problems, the first concept cartoons were created by Keogh and Naylor in 1991 [22, 23] while they were helping primary teachers to develop knowledge and understanding of Science. These concept cartoons were created in an attempt to bring out and challenge students' prior ideas and to allow them to develop the correct scientific ideas; Keogh and Naylor were indeed trying to look for alternative approaches to engage students in thinking about science concepts.

Concept cartoons are cartoon-style drawings featuring different cartoon characters putting forward a range of viewpoints about the science involved in everyday life situations [22]. Since they offer different ways of looking at a situation, concept cartoons provide a stimulus for probing and developing ideas. The typical features of concept cartoons include: application of scientific ideas in everyday situations (rather than purely theoretical), visual representation, minimal text and alternative viewpoints of the scientific ideas (all given equal status); they are not necessarily designed to give only one single correct answer. One of the most important aspects of a concept cartoon is that by presenting diverse perspectives around a scientific idea, it not only motivates learners to voice out their ideas, but also allows them to encounter uncertainty and cognitive conflict which underpin the constructivist learning theory. Further, several features of concept cartoons have been identified which make them potentially relevant to promoting argumentation and interaction in science classrooms [13, 23]. The authors [13, 23] claim that the conversation between the cartoon characters invite learners into the debate, while the visual stimulus and everyday situation engage learners in non-threatening discussions with peers and teacher to clarify or justify ideas.

Taking into consideration the importance of argumentation in Science education, and the claims that concept cartoons [22] not only provide a focus and a context, but also a purpose for discussions and debate in classroom situation, the present study has been undertaken to use concept cartoons in the teaching of "Metals and the Reactivity Series" at the Form III Level (students of age 13–14 years). To our knowledge, no research has been carried out in the local context on the use of concept cartoons in the teaching of "Metals and the Reactivity Series" at the Form III Level.

3.3 Methodology

3.3.1 Sample Description

The study was conducted in a Girls' State Secondary School situated in the northern zone of Mauritius, during the second term of 2010, in the months of June and July. The sample consisted of 35 students (age 13–14 years) of Form III, the third year of secondary schooling in Mauritius. Upon successful completion of Form III, the students will be eligible for the 2-year preparatory classes to sit for the CIE 'O'-Level Examinations in October/November 2012.

3.3.2 Research Design

The study was carried out through an action research consisting of 3 cycles, each cycle including 3 lessons of 40 min each. All the lessons were planned and implemented, by making use of concept cartoons in the teaching of "Metals and

Table 3.1 An overview of the three cycles of the action research and the use of concept cartoons

Lesson	Aims	Use of concept cartoons
Cycle 1		
1	To introduce students to "*ores*" as sources of metals To consolidate on the properties of metals To relate the uses of metals to their properties	Use of four concept cartoons (CC 1–4)
2	To explain about alloys and their uses	Use of two concept cartoons (CC 5 and 6)
3	To develop understanding of: • *The uses of iron in everyday life* • *The process of rusting*	Use of three concept cartoons (CC 7–9)
Cycle 2		
4	To develop understanding of the reactivity of some metals with water	Use of three concept cartoons (CC 10–12)
5	To discuss about the reactions of some metals with acids	Use of four concept cartoons (CC 13–16)
6	To introduce students to the reactivity series, and to the reactions of metals with oxygen	Use of three concept cartoons (CC 17–19)
Cycle 3		
7	To consolidate about the reactivity series and its importance To develop understanding of "displacement" reactions	Use of two concept cartoons (CC 20 and 21). Students were also required to draw/prepare concept cartoons to explain their reasoning following a minds-on activity
8	To consolidate about displacement reactions	Use of two concept cartoons (CC 22 and 23). Students were also required to draw/prepare concept cartoons following a practical activity
9	To introduce students to the action of heat on metal carbonates	Use of two concept cartoons (CC 24 and 25). Students were also required to draw concept cartoons based on a practical activity

the Reactivity Series" in line with the curriculum requirements. Table 3.1 gives details about the concepts addressed in each lesson, and the use of concept cartoons. As given in Table 3.1, a total of 25 concept cartoons were used during the study.

3.3.3 Use of Concept Cartoons in the Present Study

How Were Concept Cartoons Used During the Lessons?

As depicted in Table 3.1, concept cartoons have been used in the teaching of different concepts pertaining to the topic "Metals and the Reactivity Series". Concept cartoons were prepared and used in various ways during the lessons in the present action research. A total of 25 concept cartoons were used during the study; all the concept cartoons have been prepared by the teacher-researcher. The cartoon characters were either hand-drawn or prepared by making use of Paint.

The concept cartoons included from 3 to 6 cartoon characters, discussing about concepts pertaining to "Metals and the Reactivity Series" (Table 3.1). They included speech bubbles with ***statements representing different viewpoints about the concepts***. In some cases, the statements were ***complete*** and students were required to analyse and discuss about them. In other cases, the statements were ***incomplete*** and students were required to complete them by identifying the missing words.

Further in Cycle 2 of the action research, the concept cartoons included ***both correct and incorrect statements in the speech bubbles***; students were required to study the statements, identify the incorrect ones and explain their reasoning. In some lessons of Cycle 2, the concept cartoons were used in conjunction with other strategies, like teacher demonstration, practical activity or video viewing. Students were required to use information gathered through these strategies to identify and rectify incorrect statements in the concept cartoons.

In Cycle 3 of the action research, some concept cartoons presented included ***empty speech bubbles***, which students had to complete during the lessons. Most importantly, in this last cycle, students were also ***required to draw and prepare their own concept cartoons***, following minds-on and practical activities during the lessons.

During the lessons, the concept cartoons were presented to students, who were required to analyse and/or complete the concept cartoons, individually, in pairs, or in groups. This was usually followed by teacher-questioning and whole-class discussions.

Some concept cartoons prepared and used during the present study are given in Figs. 3.1, 3.2, 3.3, 3.4, 3.5 and 3.6.

When Were Concept Cartoons Used During the Lessons?

Concept cartoons were used at different stages during the lessons in the present study. Concept cartoons were used in some cases:

- At the start of the lessons, during the introduction to arouse students' interest and/or to help them recall their prior knowledge.
- During the teaching activities, to help students develop understanding of the concepts to be learnt and to attain the lesson objectives.
- As formative assessment tools to assess students' understanding of the concepts being addressed.
- At the end of the lessons, to summarise the lessons by engaging the students in analysing and/or completing concept cartoons.

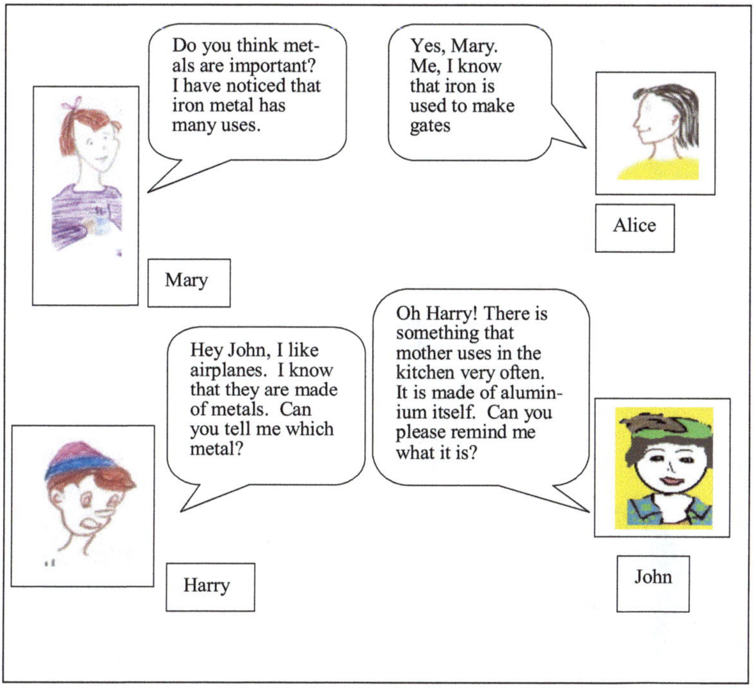

Fig. 3.1 Concept cartoon based on the uses of aluminium and iron (CC 1 used in Lesson 1 of Cycle 1)

Data Collection Tools

Both qualitative and quantitative approaches have been used during the present study. The following data-collection tools were used during the study to allow for triangulation:

(a) Observation Checklists
Throughout the action research, students were observed during all the lessons; observational data were recorded by making use of carefully designed observation checklists in line with the aims and objectives of the research. The checklists were filled during the implementation of the lessons using concept cartoons. The observation checklists were based on the following criteria:

- Students' participation, motivation and interest;
- Students' talk and argumentation;
- Classroom interactions (teacher-students and students-students);
- Students' understanding of the concepts; and
- Students' responses to the use of concept cartoons and the various ways of using them during the lessons.

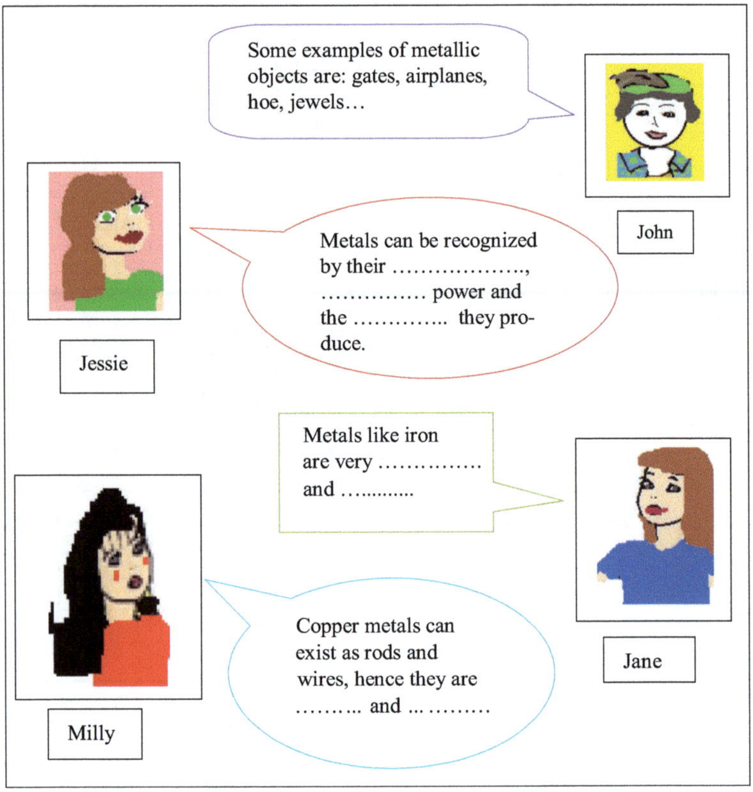

Fig. 3.2 Concept cartoon based on the properties of metals (CC 4 used to summarize Lesson 1 of Cycle 1). Students were required to complete the statements by making use of learning cards

(b) Students' Achievement Tests
After each cycle of the action research, a Students' Achievement Test was administered in order to gather information about students' understanding of the concepts being taught through use of concept cartoons. The Tests were designed, taking into consideration the learning outcomes of the lessons. Thus, 3 Tests were prepared, namely Tests 1, 2, and 3, and were administered after completing Cycles 1, 2 and 3 respectively. All three Tests carried 25 marks each, and students were allocated 30 min to complete each one.

(c) Students' Questionnaire
After completing the three cycles of the action research, a questionnaire was prepared and administered to the students in view of gathering information about their attitudes and opinions regarding use of concept cartoons during the teaching of "Metals and the Reactivity Series". The questionnaire included both close-ended and open-ended questions; the open-ended questions ensured qualitative dimensions to the students' views and opinions, while the close-ended ones helped to achieve objectivity. Before administering the questionnaire to the students, consent was gained from the participants who were assured confidentiality and anonymity.

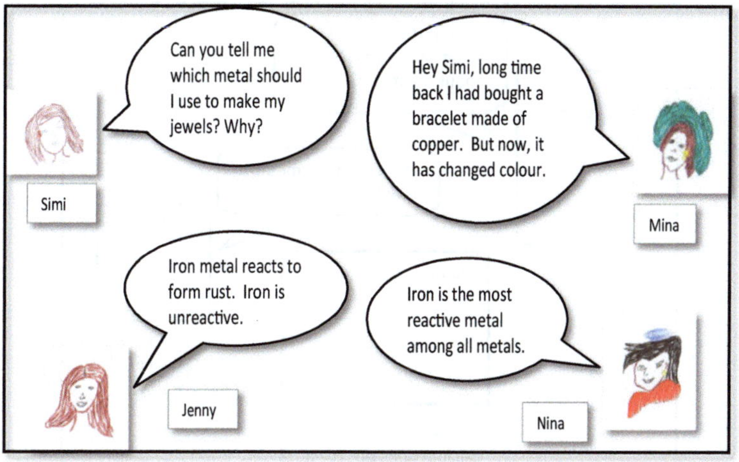

Fig. 3.3 Concept cartoon based on the reactivity of different metals, including incorrect statements which students were required to identify (CC 10 used in Lesson 4, Cycle 2)

(d) Interview
A semi–structured interview was conducted with a group of 5 students, after completing all the three cycles, in order to allow students to express their thoughts freely [24] about the use of concept cartoons during the lessons. A list of 11 questions focusing on students' opinions about the use of concept cartoons in classroom teaching was drawn to help them voice out their views and attitudes. The interview also helped the researchers to find out information, which could not be observed and/or obtained from the questionnaire, and also to triangulate the data collected through the other tools.

3.4 Results and Discussion

Both qualitative and quantitative approaches have been used in the present study by making use of observation checklists, students' achievement tests, students' questionnaire and interview. The data gathered from the different tools are presented and discussed in this section.

3.4.1 Data Gathered from Observation Checklists

In Cycle 1 of the action research, concept cartoons prepared by the teacher-researcher were used during the lessons to allow participants to discuss about the

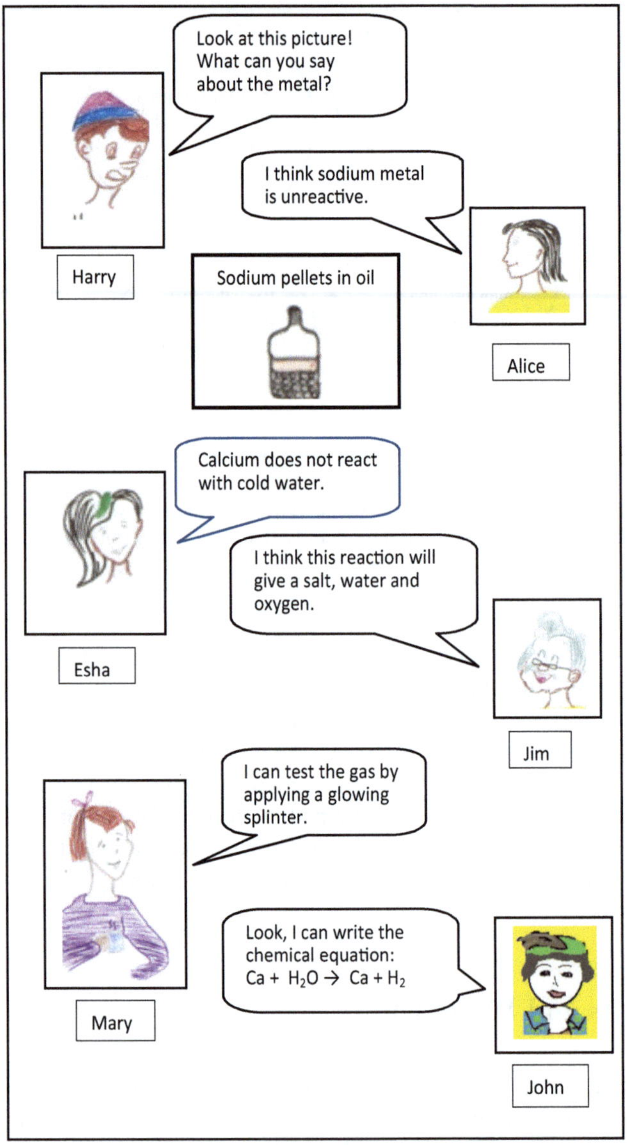

Fig. 3.4 Concept cartoon based on reactions of metals with water. Students were required to analyse the concept cartoon and to identify the incorrect statements (CC 11 used in Lesson 4, Cycle 2)

concepts under study; concept cartoons were also used as tools for assessment and lesson summary. In Lesson 1, when students were first exposed to concept cartoons, it was observed that most of them displayed amazement and surprise. It was also noted that though a few students were participating during the discussions, some

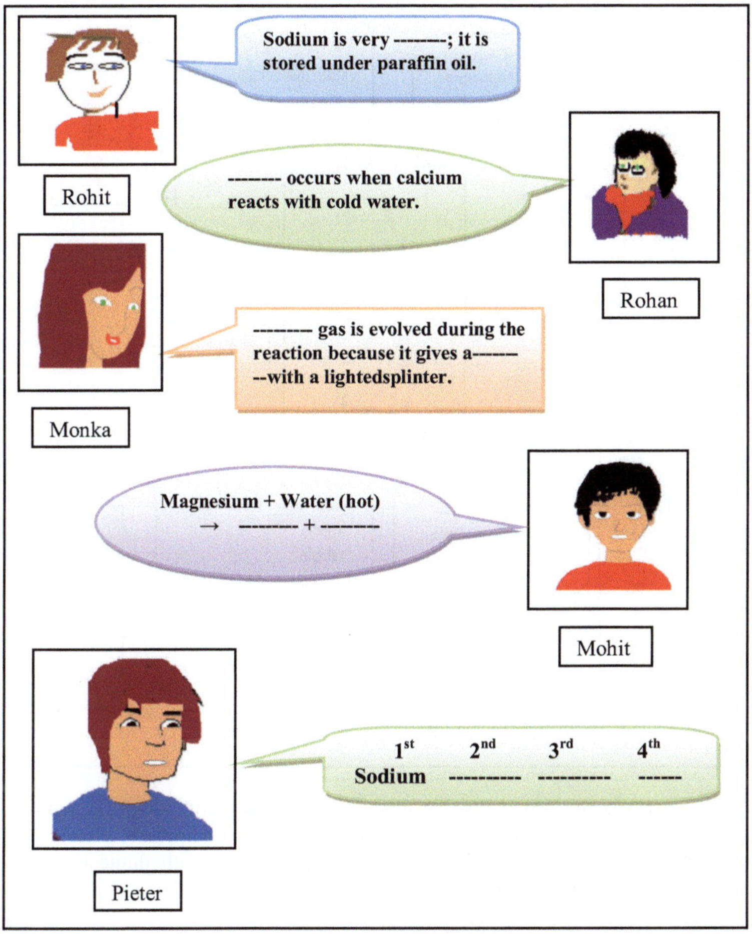

Fig. 3.5 Concept cartoon (CC 12) used to summarize Lesson 4, Cycle 2. Students were required to complete the statements

were initially shy to share their opinions. Students were also found to be enjoying the lessons when concept cartoons were used; for instance, they were heard sharing their ideas regarding jewellery, gold, silver and alloys in Lesson 2. As the action research moved to Lesson 3 of Cycle 1, it was found that students' participation increased both in discussions, and in answering teacher's questions. Surprisingly, the timid students were also found to be participating during the discussions, and were even sharing their opinions. An unexpected observation during Cycle 1 was that some students were found to get bored when statements presented in the concept cartoons were straightforward and low order.

Following reflection after Cycle 1, it was decided to make use of concept cartoons in a more challenging manner in the following cycle. Thus, the concept cartoons prepared and used during Cycle 2 included both correct and incorrect statements;

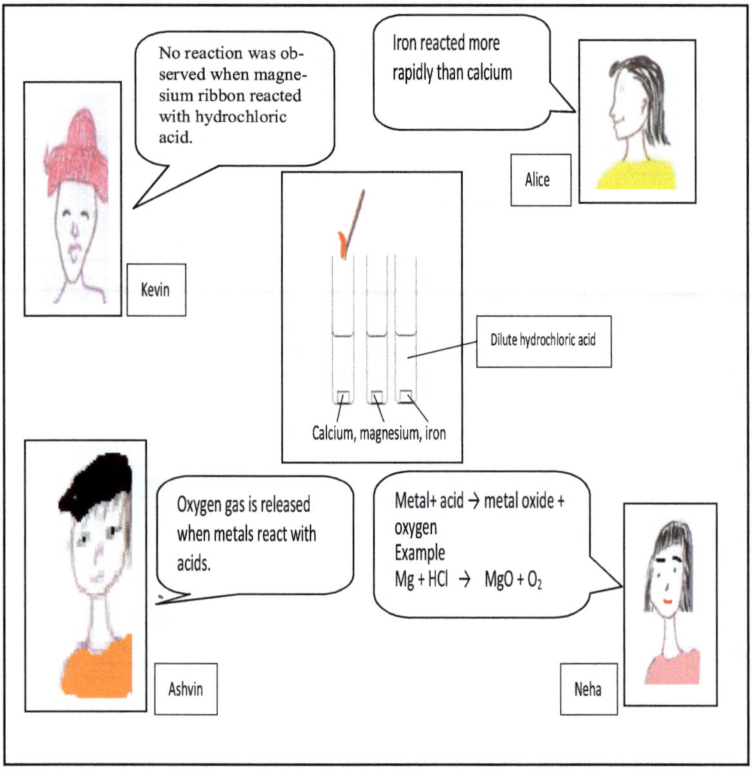

Fig. 3.6 Concept cartoon based on the reaction of metals with acids including incorrect statements. Students were required to identify the incorrect statement(s) after carrying out a practical on the reaction of magnesium, zinc, copper and iron with dilute hydrochloric acid (CC 14 used in Lesson 5, Cycle 2)

students were required to analyse the concept cartoons, and identify the incorrect statements following demonstrations, practical activities or video viewing. They were also required to justify their answers. It was observed that this change has helped to enhance students' attention and to further encourage thinking. All students were found to be participating during the class discussions; even the shy ones were comfortable to share their views with peers and the teacher. The concept cartoons succeeded in arousing interest of all students; they were fully engaged in sharing their own experiences about the concepts with peers and the teacher. An important observation made during the lessons of Cycle 2 was the increased interactions, both students-students and students-teacher as compared to the lessons of Cycle 1. It was also noted that students' engagement and involvement increased when dealing with real-life situations. Moreover, our observations have also revealed that use of concept cartoons has allowed teacher to access students' ideas, prior knowledge and also misconceptions about certain concepts.

In Cycle 3, it was decided to use concept cartoons in an attempt to further enhance students' thinking skills, and to improve their writing skills during the lessons. Thus, students were exposed to concept cartoons with empty speech bubbles, which they had to complete during the lessons. In some cases, students were also required to draw and prepare their own concept cartoons in groups based on practical activities or minds-on tasks. Like in Cycles 1 and 2, it was again observed that students were actively participating during classroom discussions; there were both students-students and teacher-students interactions. On the other hand, some students found it challenging to identify and write appropriate statements in the speech bubbles; however, it was interesting to observe that they interacted and discussed with teacher while trying to write statements to fill the speech bubbles. Thus, this further provided opportunities for teacher-students' interactions. Other important observations during Cycle 3 were that using demonstrations and hands-on activities together with concept cartoons further enhanced students' involvement and engagement. Students were fully engaged in observing the changes occurring (for example the displacement reactions between magnesium & copper(II) sulfate, between zinc & lead(II) nitrate and finally between copper & silver nitrate), following which they were reflecting and discussing about these changes in an attempt to complete the concept cartoons by writing the observations, chemical equations of the displacement reactions as well as justification based on the reactivity series. Some students even asked the teacher if they could include illustrations of the experiments while constructing the concept cartoons. They were also arguing and discussing among themselves in groups while trying to fill in empty speech bubbles and/or to prepare their own concept cartoons. After preparing the concept cartoons, each group had to present them to their peers, further encouraging them to think critically and to share their ideas.

3.4.2 Concept Cartoons Prepared by Students

This section discusses the findings gathered from some concept cartoons prepared by participant students during the study. It was noted that when asked to prepare concept cartoons about specific displacement reactions, some students preferred to write the chemical equation of the displacement reaction to be considered at the centre of the concept cartoon and presented different statements in the speech bubbles. On the other hand, some students have used drawings at the centre of the concept cartoons to depict their ideas about the displacement reaction between magnesium and zinc oxide. Analysis of students-prepared concept cartoons has revealed that the participants were able to use their knowledge of reactivity series appropriately to prepare the concept cartoons. They were able to observe reactions carried out through practical work, and to write statements in the speech bubbles of the concept cartoons. Hence, constructing concept cartoons in conjunction with hands-on experiments allowed students to enhance their conceptual understanding of the reactivity series and displacement reactions. Students' creativity was also encouraged while drawing the cartoon characters during the preparation of the concept cartoons.

Furthermore, preparation of concept cartoons by students allowed the teacher to identify their difficulties, for example, it was noted that some students failed to balance the chemical equations in the concept cartoons. Our findings thus support those of Dabell [25] who earlier reported that use of concept cartoons in the classroom can reveal misconceptions and uncertainties. Indeed, during the present study it has been found that a few students encountered difficulties in writing the formulae of compounds and balancing chemical equations while preparing their concept cartoons.

Thus said, our observations have clearly revealed that concept cartoons were able to create a classroom atmosphere which was friendly and enjoyable, and where students felt comfortable to share their prior experiences, voice out their opinions, discuss with peers and teacher, and also ask questions to teacher. These observations concur well with findings from Kabapinar [26] who stated that concept cartoons provide a non-threatening atmosphere which makes students more confident to share their views. Students were not only found to be enjoying the lessons, but they were also more cognitively involved and engaged during the lessons using concept cartoons as earlier pointed out by Keogh and Naylor [21].

3.4.3 Data Gathered from Students' Achievement Tests

Three students' achievement tests, namely, Tests 1, 2 and 3 were administered at the end of Cycles 1, 2 and 3 respectively to gather information about students' understanding of concepts taught during each cycle. The percentage scores obtained by each student in the 3 achievement tests are given in Table 3.2, while Table 3.3 gives a comparison of the minimum, maximum and average percentage scores in the three tests (Students have been assigned code numbers X1 to X35 for ethical reasons).

Data gathered from the students' achievement tests clearly indicate that students performed quite well or very well in all three tests; no student failed in any of the tests. Out of the three tests, the maximum score was 96 %, while the minimum score was 46 %. It needs to be highlighted that in all the three tests, the maximum score was above 80 %. Nevertheless, it is also clearly evident from Table 3.2 that the performance of most students was better in Test 1 as compared to Tests 2 and 3. This could be attributed to the fact that concepts addressed in Cycle 1 were simpler and easier, and related to the properties and uses of metals as well as rusting which students have learnt in lower classes. However, the lessons in Cycles 2 and 3 dealt with the reactivity series, reactions of metals with water, acids and oxygen, as well as displacement reactions and decomposition of metal carbonates; these concepts not only involved description of the reactions occurring but also involved writing of formulae, and writing and balancing of chemical equations. Indeed, during Cycles 2 and 3, observational data have revealed that some students encountered difficulties in writing formulae and balancing chemical equations. Thus, in Cycles 2 and 3, the concepts addressed were more challenging than in Cycle 1; this might explain the slight decrease in students' performance in Tests 2 and 3 as compared to Test 1. Yet, it needs to be acknowledged that most students were able to score well in Tests 2 and 3, even though they were exposed to more difficult concepts.

Table 3.2 Students' percentage scores obtained in Tests 1, 2 and 3

	Percentage scores (%)		
Students' code	Test 1	Test 2	Test 3
X1	76	66	86
X2	92	68	94
X3	92	66	82
X4	92	66	90
X5	76	66	74
X6	70	64	90
X7	76	66	86
X8	80	66	86
X9	88	74	70
X10	90	80	70
X11	84	82	74
X12	66	76	70
X13	80	72	70
X14	76	76	70
X15	72	72	60
X16	86	66	60
X17	92	56	70
X18	72	56	52
X19	88	74	66
X20	96	64	80
X21	88	56	80
X22	76	56	84
X23	84	60	66
X24	90	72	84
X25	92	84	80
X26	82	80	76
X27	82	74	76
X28	88	76	80
X29	92	68	72
X30	84	76	46
X31	88	72	54
X32	92	64	52
X33	84	64	56
X34	90	56	66
X35	80	64	66

Table 3.3 Minimum, maximum and average percentage scores of students in Tests 1, 2 and 3

	Test 1	Test 2	Test 3
Minimum percentage score (%)	66	56	46
Maximum percentage score (%)	96	84	94
Average percentage score (%)	84	69	72

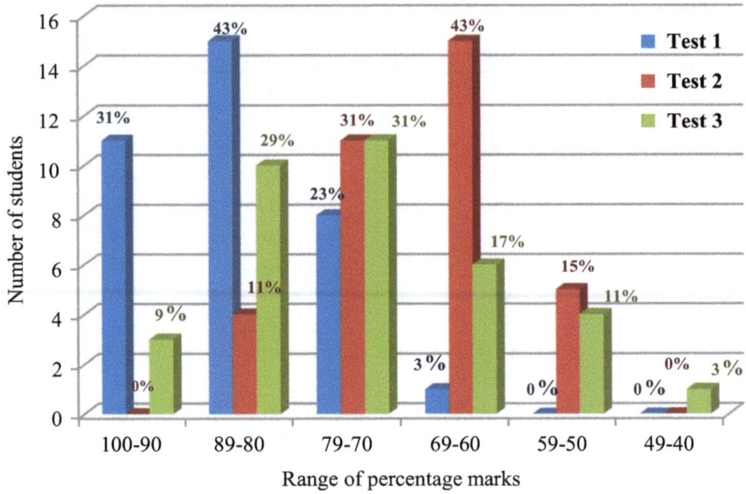

Fig. 3.7 Comparison of percentage marks obtained by students in Tests 1, 2 and 3

In order to have a more detailed view about students' performance in the three tests, it was decided to evaluate marks scored by the students in specific percentage ranges over the three tests. Figure 3.7 shows the variation in the number of students scoring marks in specific percentage ranges over the three tests. It is interesting to note from Fig. 3.7, that whatever the test, the majority of the students had a percentage mark superior to 49 %. Only 1 student scored less than 50 % in Test 3.

In Test 1, all students scored marks above 60 %, with 23 % of students scoring in the range 70–79 %, 43 % students scoring in the range 80–89 % and 31 % students scoring in the range 90–100 % (Fig. 3.7). In Test 2, most students scored in the range 60–69 % (43 % of students) and the range 70–79 % (31 % of students). In Test 3, the majority of the students are represented in the range 70–79 % (31 % of students) and 80–89 % (29 % of students). As aforementioned, going from Test 1 to Tests 2 and 3, there has been a slight decline in the students' performances due to the increasingly difficult concepts addressed. In spite of this, it is clear from Fig. 3.7 that the majority of students have scored in the range of 60–100 in the three tests, and that no student failed in any of the tests, showing that they have developed proper understanding of the concepts learnt in the topic "Metals and the Reactivity Series" through use of concept cartoons. Thus, findings from the students' achievement tests clearly point out to the fact that use of concept cartoons has contributed positively to students' conceptual understanding of the topic "Metals and the Reactivity Series".

3.4.4 Findings from Students' Questionnaire

As mentioned in section "Data Collection Tools", a questionnaire was administered to all participating students after completing the action research to gather

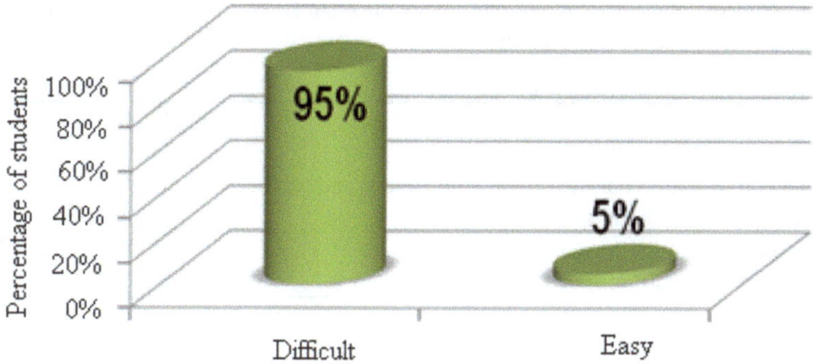

Fig. 3.8 Students' perceptions of the topic "Metals and Non-Metals" learnt in Form II (second year of secondary schooling in Mauritius)

Fig. 3.9 Students' responses to whether they liked the use of concept cartoons in the teaching and learning of the topic "Metals and the Reactivity Series"

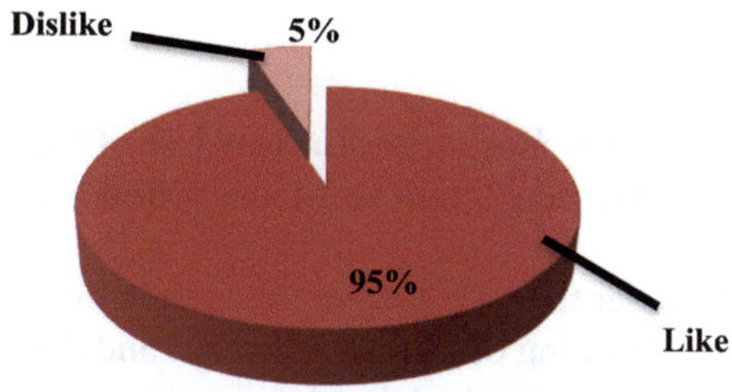

information about their attitudes and opinions regarding use of concept cartoons in the teaching and learning of "Metals and the Reactivity Series". The major findings gathered through the students' questionnaire are presented and discussed in this section.

It was deemed necessary to find out about students' perceptions of the topic "Metals and Non-Metals" learnt in Form II (second year of secondary schooling in Mauritius), as many of the concepts learnt are prerequisites for the topic "Metals and the Reactivity Series" in Form III. As illustrated in Fig. 3.8, it is obvious that most students (95 %) found the topic "Metals and Non-Metals" at Form II level difficult, while only a minority of 5 % found it easy. This finding is quite interesting because though most participants claimed that the topic "Metals and Non-Metals" in Form II was difficult, all of them performed well in Test 1 and scored above 60 %. Indeed, the lessons in Cycle 1 of the present study dealt a lot with prerequisites acquired in "Metals and Non-Metals" by making use of concept cartoons. Thus, it can be seen that use of the concept cartoons has greatly enhanced understanding of these concepts resulting in a good performance in Test 1.

Findings from the questionnaire have revealed that out of 35 students, 34 (95 %) liked the use of concept cartoons during the lessons (Fig. 3.9). Most of the students stated that the lessons were *more interesting* and *enjoyable*; some mentioned that the concept cartoons were very attractive and *helped to reduce boredom* during the lessons. This finding is in line with those of Allibacus [27] who earlier reported that

Table 3.4 Students' responses to questions about the effectiveness of concept cartoons in helping them to learn "Metals and the Reactivity Series"

Did the use of concept cartoons help you to:-	YES	NO
(i) Understand the concepts better?	35 (100 %)	–
(ii) Encourage you to think?	34 (97 %)	1 (3 %)
(iii) Enable you to share your ideas with your friends and teacher?	35 (100 %)	–
(iv) Discuss and argue about the concepts?	32 (91 %)	3 (9 %)
(v) Participate more in the lessons by talking?	31 (89 %)	4 (11 %)
(vi) Motivate you to learn more about chemistry?	29 (83 %)	6 (17 %)

use of concept cartoons rendered lessons more interesting and exciting. Some students also stated that they were *more motivated* during the lessons including use of concept cartoons, thereby confirming the motivating nature of concept cartoons as earlier highlighted by researchers [20].

Students' Responses to How Use of Concept Cartoons During the Lessons Has Helped Them to Learn "Metals and the Reactivity Series"

In order to probe further about the effectiveness of concept cartoons in the teaching and learning of the topic "Metals and the Reactivity Series", students were asked six questions (Table 3.4) related to the use of concept cartoons. Students were required to answer by YES or NO and also to give reasons for their answer. The findings are presented in Table 3.4.

From Table 3.4, it is found that all participating students unanimously agreed that use of concept cartoons during the lessons has:

- Helped them to understand the concepts better and
- Enabled them to share ideas with their friends and teacher.

Further, it is also evident from Table 3.4 that the majority of students claimed that use of concept cartoons has:

- Encouraged them to think more (97 %);
- Helped them to discuss and argue about the concepts (91 %);
- Helped them to participate more in the lessons by talking (89 %); and
- Motivated them to learn more about Chemistry (83 %).

Some students stated that they were happy to share their ideas with the teacher and their friends, and that they had enough time to talk about any misconceptions and doubts. Students claimed that sharing and discussing their opinions/ideas about the concepts has helped to enhance their understanding of these concepts. Most students stated that the colorful pictures in the concept cartoons were not only interesting, but were also informative. This has helped to arouse their curiosity and to increase their motivation. Some students also mentioned that concept cartoons with correct and incorrect statements encouraged them to argue and discuss about

Table 3.5 Students' perceptions about the different ways in which concept cartoons were used during the lessons

How did you find using concept cartoons in the following ways?	Difficult	Easy	Boring	Interesting
At the start of the lesson	3 (9 %)	14 (40 %)	–	18 (51 %)
To encourage discussions about the concepts	–	19 (54 %)	–	16 (46 %)
With correct/incorrect statements	4 (12 %)	8 (23 %)	5 (14 %)	18 (51 %)
With partly-filled speech bubbles	1 (3 %)	14 (40 %)	4 (12 %)	16 (45 %)
Drawing your own concept cartoons	12 (34 %)	5 (14 %)	11 (32 %)	7 (20 %)
With empty speech bubbles	5 (14 %)	11 (31 %)	3 (9 %)	16 (46 %)

the concepts and have helped to clear their doubts and misconceptions. A few students (3 representing 9 %) claimed that use of concept cartoons did not help them to discuss and argue about the concepts, and stated that the statements in the concept cartoons were not challenging enough.

Students' Perceptions About the Different Ways in Which Concept Cartoons Were Used During the Lessons

In order to gather students' views regarding the various ways in which concept cartoons were used during the research, they were asked to complete Table 3.5 by ticking their choice in the appropriate column.

From Table 3.5, it is noted that most students found using concept cartoons interesting (51 %) and easy (40 %) at the start of the lessons; though 9 % claimed it to be difficult, no one found it boring. Further, all students agreed that use of concept cartoons to encourage discussions was either easy (54 %) or interesting (46 %); no one found it difficult or boring.

When asked about concept cartoons with correct and incorrect statements, 51 % of the students found them interesting, 23 % found them easy, while the others stated that they were boring (14 %) and difficult (12 %). Our findings also show that regarding use of concept cartoons with partly-filled speech bubbles, the majority of the students found them easy (40 %) and interesting (45 %), while one student (3 %) stated that they were difficult and four students (12 %) stated that they were boring. The same pattern is more or less observed for concept cartoons with empty speech bubbles which students had to fill – the majority found them easy (31 %) or interesting (46 %) while the remaining few claimed that they were difficult (14 %) or boring (9 %). When asked about drawing their own concept cartoons, a greater majority claimed that they found it difficult (34 %) or boring (31 %), while fewer students found them easy (14 %) or interesting (20 %) as shown in Table 3.5.

Findings from Table 3.5 clearly point out to the fact that the students were more in favour of using complete teacher-prepared concept cartoons as most of them claimed that use of concept cartoons at the start of the lessons was easy and

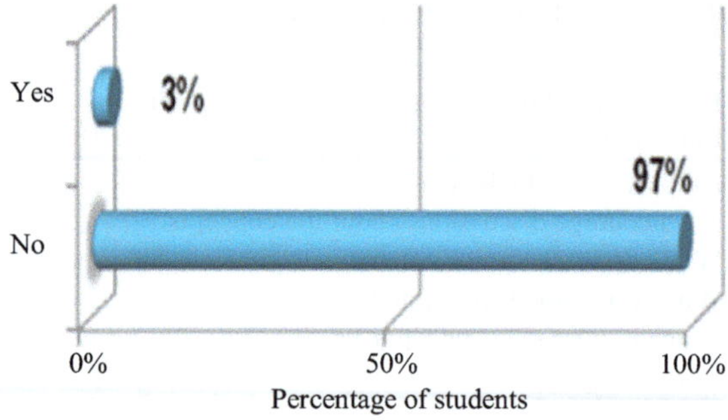

Fig. 3.10 Students' responses to whether they encountered difficulties when concept cartoons were used

interesting. Indeed, at the start of the lessons fully prepared concept cartoons were used with complete statements. However, when it comes to concept cartoons with correct/incorrect statements, or with incomplete speech bubbles, or with empty speech bubbles to be filled, our findings have clearly revealed that a few students found them to be difficult. Students also claimed that using concept cartoons in these latter ways requires more thinking from the part of the students and is more challenging; it also demands proper and concise writing skills to communicate ideas in one sentence in a speech bubble. Those students who found these concept cartoons difficult and boring commented that they preferred to use them during group activities rather than individually, as they feel more motivated and comfortable to discuss with peers in order to complete the statements or to fill empty speech bubbles. Students who found use of concept cartoons with empty speech bubbles easy and interesting stated that these concept cartoons encouraged them to think and enhanced understanding of the concepts.

As evidenced from Fig. 3.10, out of the 35 participants only one (3 %) stated that she encountered difficulties when concept cartoons were used, while the majority (97 %) claimed that they did not encounter difficulties. This might be attributed to the fact that concept cartoons use minimal texts, and are based on real life situations which students are familiar with; this tends to increase students' confidence allowing them to participate more and to be mentally engaged in the lessons as recently highlighted by Naylor and Keogh [22].

Students' Responses to Whether Use of Concept Cartoons is Time-Consuming

Most students (71 %) affirmed that use of concept cartoons is time-consuming as portrayed in Fig. 3.11. This finding is in line with data gathered during observations; additional time was needed to complete the lessons. This can be explained by the fact that more time was needed to allow students to think, to share their ideas, to interact with the teacher and also to put forward appropriate arguments. This is in clear contrast with passive expository-type lessons where students just listen to teacher without interactions and argumentations.

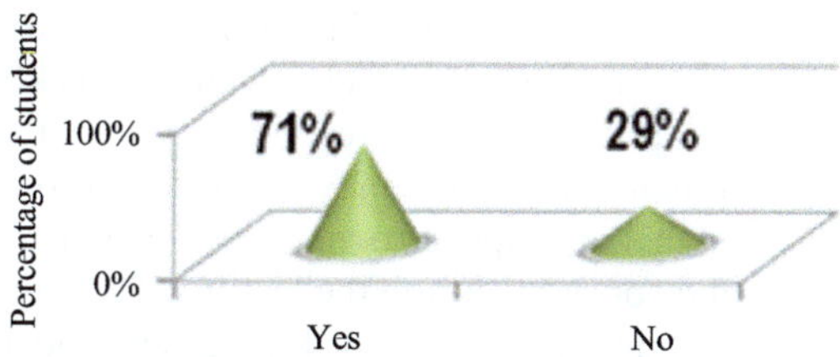

Fig. 3.11 Students' responses to whether use of concept cartoons is time-consuming

Students' Views Regarding Advantages and Disadvantages of Using Concept Cartoons During the Lessons

In the last part of the questionnaire, students were invited to give their views about the advantages and disadvantages of using concept cartoons during the lessons. Some of the advantages mentioned by students are that use of concept cartoons:

- Is enjoyable and interesting;
- Helps to understand and retain information;
- Allows more sharing of ideas/opinions during the lessons;
- Motivates them to participate in the lessons;
- Reduces boredom;
- Promotes discussions; and
- Encourages teacher-students and students-students interactions.

These statements concur well with observational data, and with findings from previous questions of the questionnaire. Further, these findings are also compatible with those from other researchers, who reported that concept cartoons promote students' understanding [28], make lessons enjoyable for students, and help them to develop better understanding of concepts [29]. It has also been reported that use of concept cartoons helps to generate motivation, engagement and participation, leading to lively discussions and making the lessons more interactive [21, 22, 29].

According to some students, disadvantages of using concept cartoons are that, it can be time-consuming, and can lead to much noise during class discussions. Both of these might be attributed to increased class discussions and interactions. At this stage, it needs to be highlighted that though use of concept cartoons is time-consuming, teachers should decide when and how to use them during the lessons in view of enhancing conceptual understanding, and of promoting argumentation. Nevertheless, for successful use of concept cartoons in classroom, teachers must pay particular attention to time-management, as well as managing the discussions and argumentation going on to avoid noise problems. A few students also admitted that use of concept cartoons can be boring if the statements given are too simple, straightforward and non-challenging.

3.4.5 Findings Obtained from Interview

After completing the three cycles of the action research, an interview was conducted with a random group of five students, in view of triangulating the data gathered through the other tools. All interviewed students stated that they found use of concept cartoons interesting, fun, user-friendly and enjoyable, thereby confirming findings from observation checklists and questionnaire. They found the concept cartoons interesting and attractive, and claimed that the classroom environment was more enjoyable with use of concept cartoons. The students also asserted that concept cartoons helped them to discuss freely and to share their ideas about the concepts during the lessons. Two students even admitted that at first they were feeling shy to talk to the teacher, and to share their ideas even with their friends; they were not feeling at ease because they feared their answers would be wrong. However, with use of concept cartoons during the lessons, they were able to overcome their shyness and to talk to the teacher without fear. This fully substantiates findings from earlier research, which claim that "*not only concept cartoons' motivating nature but also non-threatening atmosphere make the strategy effective particularly for students who require confidence in putting forward their views in teaching context*" [26]. The interviewed students also stated that use of concept cartoons has helped them to understand concepts better, and in a more meaningful way, as they were based on real life experiences and situations. Furthermore, out of the five students, three pointed out that challenging statements need to be employed in the concept cartoons in order to further foster thinking.

3.5 Conclusions

The present paper was based on an investigation into the use of concept cartoons in the teaching of the topic "Metals and the Reactivity Series" at Form III level (third year of secondary schooling in Mauritius) in response to the needs to shift from teacher-centred classroom situation to learner-centred situation, and to promote students' argumentation and engagement. It is indeed widely felt that there is a mismatch between the way Scientists work and the way Science is taught in actual classroom situation, which is usually passive, and does not promote scientific discourse amongst learners.

Our findings have clearly pointed out that all the students were actively engaged in discussions both with their peers and with the teacher, when concept cartoons were used, showing that concept cartoons promoted both student-student and student-teacher interactions. The present study has also brought to light that using concept cartoons with empty speech bubbles, which students need to complete, or allowing students to prepare their own concept cartoons during the lessons also contributed to enhanced interactions between students and teacher. It is also evident from our findings that use of concept cartoons has created a conducive and enjoyable classroom environment that was non-threatening so that all students (even the shy or less confident ones) were able to share their ideas/opinions, clearly supporting results from earlier research [26].

Results from the study have also revealed that use of concept cartoons during the lessons has encouraged students to share their ideas about the concepts under study, and most importantly to easily voice out their prior conceptions and experiences.

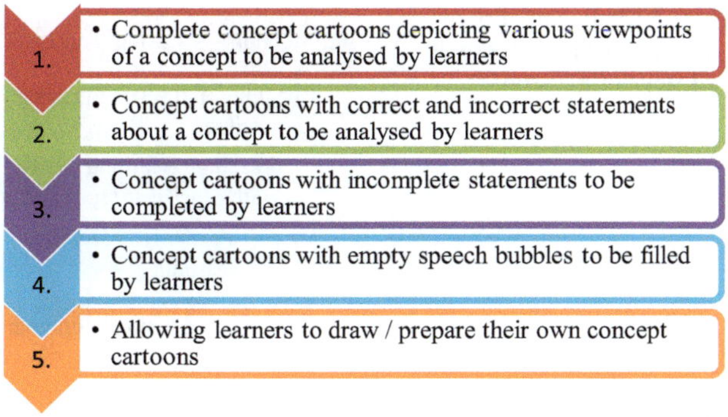

Fig. 3.12 Various ways of using concept cartoons during classroom teaching

This is an essential feature of the use of concept cartoons in relation to the constructivist approach to teaching and learning, which is based on the premise that learners can only construct new knowledge and make sense of new information by making use of their prior knowledge and understanding. Accordingly, teachers must take cognizance of learners' prior knowledge during the teaching/learning process. Concept cartoons can thus prove to be very useful in the constructivist classroom, as they provide a context for learners to share their prior knowledge about the concept(s) and for teachers to identify learners' prior conceptions and use these meaningfully during the lessons.

Further, our study has shown that use of concept cartoons had a positive impact on students, not only in terms of enhanced motivation, participation and interest during the lessons, but also on their conceptual understanding and performance. We found that use of concept cartoons has helped the participants to develop proper understanding of concepts involved in "Metals and the Reactivity Series" as evidenced by their performances in the students' achievement tests. Indeed, two noteworthy aspects of concept cartoons in promoting conceptual understanding include firstly, their ability to engage students in the teaching/learning process, and secondly, they allow teachers to detect students' misconceptions which can be addressed during the lessons. We have indeed found that students were actively engaged in the lessons when concept cartoons were used – they were engaged in analysing or completing the statements, in filling empty speech bubbles or in preparing their own concept cartoons. It is indeed reported in literature that learners' engagement during lessons contributes to learning and development [30]. Our findings also corroborate with those of other authors, who earlier pointed out the effectiveness of concept cartoons in allowing teachers to identify and address students' misconceptions thereby promoting conceptual change [25, 26].

Based on the present action research, we would like to emphasize that concept cartoons have been successfully used in the teaching and learning of concepts involved in "Metals and the Reactivity Series", and thus confirm the practicability of using concept cartoons in classroom teaching. We would also like to report here three key aspects of the use of concept cartoons in classroom teaching. Firstly, we have clearly shown in this study that concept cartoons can be used in a variety of ways during the lessons as depicted in Fig. 3.12.

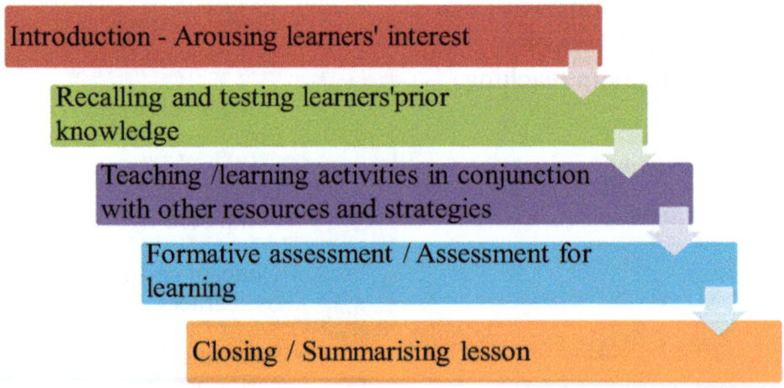

Fig. 3.13 Potentiality of use of concept cartoons at all steps of a lesson

It needs to be highlighted that using concept cartoons in various ways during the lessons endows additional benefits to learners, like *promoting thinking*, *enhancing writing skills* (filling empty speech bubbles and preparing their own concept cartoons); *promoting creativity* (drawing/preparing their own concept cartoons), as well as *enhancing observation skills* (using concept cartoons in conjunction with practical work), in addition to promoting classroom interactions and increasing motivation and engagement.

Secondly, the present study has also brought to light that concept cartoons can be effectively used during all the steps of a typical lesson as portrayed in Fig. 3.13.

The third key aspect of concept cartoons that we would like to report here is that its use in classroom teaching is not rigid and exclusive, that is, it can easily accommodate use of other strategies like group work, minds-on activities, teacher-demonstration, hands-on practical work as well as video-viewing. In the present study, we have been able to use concept cartoons as teaching tools to complement the afore-mentioned strategies, thereby adding variety to the lessons and rendering the classroom environment richer and more conducive and interesting for students.

Findings from the present study have also revealed that using concept cartoons in classroom situation takes additional teaching time, and also demands adequate preparation from the part of the teacher. Teachers should not only have a good mastery of the concept, but should also take into consideration its application in real life situations to be able to prepare relevant concept cartoons for teaching purposes. Another important recommendation based on the present study is that during preparation of concept cartoons, teachers should be careful to include statements of varying levels to cater for learners of all abilities; otherwise learners can get bored/unmotivated if the statements are too straightforward and low-order as these would not challenge them to think. Thus, for successfully incorporating use of concept cartoons in lessons, it is important for teachers to take into consideration the preparation of the concept cartoons, as well as proper planning and implementation of the lessons, in order to promote conceptual understanding, and at the same time ensuring proper time-management for completion of lessons within the allocated time.

In conclusion, the present study has clearly brought into perspective how concept cartoons can be used in the teaching of "Metals and the Reactivity Series".

The study thus confirms the practicability of using concept cartoons in Chemistry Education. Further, the research has also revealed that use of concept cartoons has had a positive impact on students, in terms of enhanced interest, motivation, engagement and increased classroom interactions (teacher-student and student-student), as well as enhanced conceptual understanding and improved performance. Thus, concept cartoons can provide teachers with a means to shift to a learner-centred classroom situation as earlier recommended [14]. We also report here that use of concept cartoons has encouraged learners to voice out their prior conceptions, thereby lending support to claims made by other authors [18]. Last but not least, the present research has also brought to light how concept cartoons can be successfully used in various ways in classroom teaching, in conjunction with other teaching strategies, and also at various steps during a lesson.

References

1. Barnhart CL, Barnhar RK (1992) The World book dictionary, vol 2. World Book, Inc., Chicago
2. Randi J, cited in: Wudka J (1998) From Antiquity to Einstein. http://physics.ucr.edu/~wudka/Physics7/Notes_www/node5.html. Accessed 19 Apr 2012
3. Wudka J (1998) From Antiquity to Einstein. http://physics.ucr.edu/~wudka/Physics7/Notes_www/node6.html#SECTION02121000000000000000. Accessed 19 Apr 2012
4. Rutherford FJ, Ahlgren A (1990) Science for all Americans. Oxford. Cited in: Backer PR (1998) What is the scientific method? http://www.engr.sjsu.edu/pabacker/scientific_method.htm. Accessed 19 Apr 2012
5. Driver R, Squires A, Rushworth P, Wood-Robinson V (1994) Making sense of secondary science: Research into children's ideas. http://edutech.csun.edu/eduwiki/index.php/Driver_et_al,_1994. Accessed 19 Apr 2012
6. Wellington JJ, Osborne J (2001) Language and literacy in science education. Open University Press, Buckingham, Philadelphia
7. La Velle BL, Erduran S (2007) Argument and developments in the science curriculum. Sch Sci Rev 88:31–40
8. Newton P, Driver R, Osborne J (1999) The place of argumentation in the pedagogy of school science. Int J Sci Educ 21:553–576
9. Erduran S, Jimenez-Aleixandre MP (2008) Argumentation in science education: perspectives from classroom-based research. Springer, Dordrecht. Cited in: Erduran S, Villamanan R (2009) Cool argument: engineering students' written arguments about thermodynamics in the context of the Peltier effect in refrigeration. Educación Química 20:119–125
10. Erduran S (2007) Breaking the law: promoting domain-specificity in chemical education in the context of arguing about the periodic law. Found Chem 9:247–263
11. Bybee RW, Powell JC, Trowbridge LW (2008) Teaching secondary school science. Strategies for developing scientific literacy, 9th edn. Pearson Merrill Prentice Hall, Upper Saddle River, New Jersey, Columbus, Ohio
12. Gray A (1997) Constructivist teaching and learning; SSTA research centre report #97–07: http://saskschoolboards.ca/research/instruction/97-07.htm#What%20is%20Constructivism. Accessed 13 Mar 2012
13. General Teaching Council for England (2011) Teaching & Learning Academy. Case study 4: using concept cartoons in science. http://www.gtce.org.uk/tla/rft/sciencetalk0411/case_studies/cs4/. Accessed 13 Mar 2012

14. Mauritius Research Council (2004) Teaching and learning of science in schools (Republic of Mauritius), vol 1. Recommendations and Action Plan (2004–2006). Republic of Mauritius
15. Luckpoteea M, Narod F (2012) An investigation into the use of the concept attainment model in teaching the "Periodic Table" at 'O'-Level through an action research. In: Bhowon MG, Jhaumeer-Laulloo S, Li Kam Wah H, Ramasami P (eds) Chemistry for sustainable development. Springer, Dordrecht, Heidelberg, London, New York, pp 265–300
16. Chuckowree V (2005) The use of pedagogical aids in the teaching of periodic table at form five level, an action research. PGCE Dissertation, Mauritius Institute of Education, Réduit
17. Sobron S (2008) Use of games in the learning of periodic table at form four level, an action research. B.Ed. Dissertation, University of Mauritius and Mauritius Institute of Education, Réduit
18. Birisçi S, Metin M (2010) Developing an instructional material using a concept cartoon adapted to the 5E model: a sample of teaching erosion. Asia-Pac Forum Sci Learn Teach 11:1–16
19. Driver R, Bell B (1986) Students' thinking and the learning of science: a constructivist view. Sch Sci Rev 67:443–456
20. Kabapinar F (2005) Effectiveness of teaching *via* concept cartoons from the point of view of constructivist approach. Educ Sci Theory Pract 5:135–146
21. Keogh B, Naylor S (1996) Teaching and learning in science: a new perspective. Paper presented at the BERA Conference, Lancaster. http://www.leeds.ac.uk/educol/documents/000000115.htm. Accessed 13 Jan 2012
22. Naylor S, Keogh B (2012) Concept cartoons: what have we learnt? Paper presented at the Fibonacci Project European Conference, Inquiry-Based science and mathematics education: bridging the gap between education research and practice. Leicester. http://search.conduit.com/results.aspx?q=Keogh+%26+Naylor%2C+1993&Suggest=&stype=Results&FollowOn=True&SelfSearch=1&SearchType=SearchWeb&SearchSource=1&ctid=ct1060933&octid=CT1060933. Accessed 21 June 2012
23. Naylor S, Keogh B (2000) Concept cartoons in science education (The ConCISE Project). Milligate House Publishers, Sandbach, UK
24. Bogdan R, Biklen SK (1998) Qualitative research for education: an introduction to theory and methods, 3rd edn. Pearson, Boston
25. Dabell J (2004) The maths coordinator's file-using concept cartoons. PFP Publishing. Cited in: Birisçi S, Metin M (2010) Developing an instructional material using a concept cartoon adapted to the 5E model: a sample of teaching erosion. Asia-Pac Forum Sci Learn Teach 11:1–16
26. Kabapinar F (2009) What makes concept cartoons more effective? Using research to inform practice. Educ Sci 34:104–118
27. Allibacus A (2008) Use of concept cartoons as a strategy to teach acids and bases at Form II level. An action research. PGCE Dissertation. Mauritius Institute of Education, Réduit
28. Keogh B, Naylor S (1999) Concept cartoons, teaching and learning in science: an evaluation. Int J Sci Educ 21:431–446. Cited in: Kabapinar F (2009) What makes concept cartoons more effective? Using research to inform practice. Education and Science 34:104–118
29. Sing Teach (2009) Cartoons in the classroom. Sing Teach 18:6–7. National Institute of Education, Nanyang Technological University. http://singteach.nie.edu.sg/science-ed/100.html. Accessed 1 Aug 2012
30. Furrer C, Skinner E (2003) Sense of relatedness as a factor in children's academic engagement and performance. J Educ Psychol 95:148–162. Cited in: Toshalis E, Nakkula MJ (2012) Motivation, engagement, and student voice: the students at the centre series. Jobs for the future. http://www.studentsatthecenter.org/sites/scl.dl-dev.com/files/Motivation%20Engagement%20Student%20Voice_0.pdf. Accessed 22 May 2012

Chapter 4
Electron Correlation Energy in the Ground State of the Helium Sequence

Khalil H.A. AL-Bayati and Nada I.I. AL-Zubaidi

Abstract The purpose of this work is to examine five trial wave functions for the helium atom and the helium-like ions (He, Li^+, Be^{2+},. . . . .Ne^{8+}). The comparisons between the trial wave function have been done in view of the calculated energy (E) and the electron–electron repulsion, $\left\langle \frac{1}{r_{12}} \right\rangle$, for the atom and all the ions considered. The calculations show that as the wave function becomes more correlated, E_{corr} increases and the percentage error in energy calculation decreases, whereas this percentage, for all systems under study, decreases as Z increases. The best trial wave function in this work gives results closer to the experimental findings due to a reduction in electron–electron repulsion.

4.1 Introduction

The concept of electron correlation energy (E_{corr}) was first proposed by Wigner in his classic study of free electrons in metal [1], and later Lowdin defined the correlation energy as the difference [2]:

$$E_{corr} = E_{exact} - E_{HF} \tag{4.1}$$

where E_{exact} is the exact, non-relativistic energy, and E_{HF} is the Hartree-Fock energy. The calculations require accurate evaluation of the correlation energy as the latter is important in quantum chemistry [4]. The correlation problem is usually treated using

K.H.A. AL-Bayati (✉)
Department of Physics, College of Science for Women, Baghdad University, Baghdad, Iraq
e-mail: drkhalilhadi@yahoo.com

N.I.I. AL-Zubaidi
Department of Physics, College of Science, Diyala University, Diyala, Iraq

M. Gupta Bhowon et al. (eds.), *Chemistry: The Key to our Sustainable Future*, DOI 10.1007/978-94-007-7389-9_4, either configuration interaction [5], or Variation method [6] or Perturbation Theory. Electron correlation results from the Pauli exclusion principle, and by mutual repulsion as governed by understanding the factors that determine the correlation energy are complicated by the famine of parameters that can be conveniently varied to measure their effect [7]. The correlation energy can be divided into the radial correlation energy $(E_{rad,\ corr})$ and the angular correlation energy $(E_{ang,\ corr})$. Radial correlation accounts for the tendency of the two electrons to be at different distance from the nucleus while angular correlation accounts for the tendency of the two electrons to occupy positions on opposite sides of the nucleus [8].

Consider a system of two electrons such as, neutral helium atom He (Z = 2) or ions with two electrons such as H^- (Z = 1) and Li^+ (Z = 3) [9].

The Hamiltonian operator for the helium sequence can be expressed in atomic units [10, 11]:

$$H = -\frac{1}{2}\nabla_1^2 - \frac{1}{2}\nabla_2^2 - \frac{Z}{r_1} - \frac{Z}{r_2} + \frac{1}{r_{12}} \tag{4.2}$$

The Schrödinger equation (S.E.) cannot be solved exactly, because of the electron–electron repulsion term and this is why the approximation method such as the Variation Method [9] became very important to solve the S.E.

4.2 Theory

The Hartree-Fock wave function for a helium-like ion is [12]:

$$\psi_{HF}(r_1, r_2) = \varphi(r_1)\varphi(r_2) \tag{4.3}$$

where $\varphi(r_i)$ is spherically symmetric spatial orbital, where $i = 1$ or 2.

The orbital function is defined as [13]:

$$\varphi(r) = \left(\frac{\alpha^3}{\pi}\right)^{\frac{1}{2}} \exp(-\alpha r) \tag{4.4}$$

The first and simplest trial function for a helium-like ion is written by Kellner [13]:

$$\psi_1(r_1, r_2) = \varphi(r_1)\varphi(r_2) = \frac{\alpha^3}{\pi} \exp(-\alpha(r_1 + r_2)) \tag{4.5}$$

This wave function is the Hartree-Fock solution where the orbital is a Slater-type function with the optimal exponent [13] $\alpha = Z - \frac{5}{16}$ which is calculated by variation treatment. For helium atom $\alpha = \frac{27}{16}$, this means that the other electron partially screens the nucleus, reducing its effective charge from 2 down to about 1.6875 [9].

Physically this parameter may be interpreted as the effective nuclear charge felt by each of the electrons [9]. It is based on the orbital concept where electrons have wave functions that are independent of the coordinates of other electrons. Since it does not provide for electron correlation it will serve as the standard in the subsequent series of calculations.

Some electron correlation can be built into the second type of trial wave function by assuming that each electron is in an orbital which is a linear combination of two scaled hydrogen 1 s orbitals.

$$\varphi(r) = \exp(-\alpha r) + \exp(-\beta r) \tag{4.6}$$

Under the orbital approximation, this assumption gives a trial wave function of the form [14]:

$$\begin{aligned}\psi_2(r_1, r_2) = \varphi(r_1)\varphi(r_2) = &\exp(-\alpha(r_1 + r_2)) + \exp(-(\alpha r_1 + \beta r_2)) \\ &+ \exp(-(\beta r_1 + \alpha r_2)) + \exp(-\beta(r_1 + r_2))\end{aligned} \tag{4.7}$$

Inspection of this wave function indicates that 50 % of the time the electrons are in different orbitals, while for the first trial wave function the electrons were always in the same orbital. The energy calculation in this case is more complex than that using the previous one and that it is being minimized simultaneously with respect to two parameters rather than just one. It is also clear that introducing some electron correlation into the trial wave function has improved the agreement between theory and experiment, but still more improvements are needed.

Electron correlation can be increased further by eliminating the first and last term in the second type wave function. This yields the third trial type wave function of the form [9, 10]:

$$\psi_3(r_1, r_2) = \exp(-(\alpha r_1 + \beta r_2)) + \exp(-(\beta r_1 + \alpha r_2)) \tag{4.8}$$

This trial wave function is called Eckart wave function. It places the electron in different scaled hydrogen 1 s orbitals 100 % of the time and it allowed two different shielding factors, suggesting that one electron is relatively close to the nucleus, and the other is further away [9]. This wave function allows for (in-out) correlation since the two electrons may be supposed to be in distinct orbitals [15]. This equation is then minimized simultaneously with respect to the adjustable two parameters, α and β.

The third type of trial wave function, however, still rests on the orbital approximation and, therefore, does not treat electron correlation appropriately. The extent of electron correlation in the fourth type of trial function can be increased further by introducing electron correlation directly by adding a term, r_{12} involving the inter electron separation [16]:

$$\psi_4(r_1, r_2) = N \cdot \exp[-\alpha(r_1 + r_2)] \cdot [1 + b \times r_{12}] \quad (4.9)$$

This wave function is called Hylleraas type wave function [3]. It is a linear expansion of basis functions that are constructed with Slater orbitals. It contains two variational parameters α and b and these coefficients are determined variationally. Physically, α is once again the effective nuclear charge while b may be interpreted as a measure of electronic correlation [6]. The term $1 + b \times r_{12}$ makes the wave function larger for large values of r_{12}; this is as it should be, because the repulsion between the electrons makes it energetically more favorable for the electrons to avoid each other [3].

The fifth type of trial wave function brought about further improvement by adding Hylleraas r_{12} term to the third trial wave function as given by Eq. (4.10):

$$\psi_5(r_1, r_2) = N \cdot [\exp(-(\alpha r_1 + \beta r_2)) + \exp(-(\beta r_1 + \alpha r_2))] \cdot [1 + b r_{12}] \quad (4.10)$$

This wave function is called Chandrasekhar wave function [17], and it consists of three parameters which give rise to fairly complicated variational expression for the ground state energy.

4.3 Results and Discussion

Table 4.1 shows the calculated and the experimental energy values for helium sequence.

Table 4.2 shows the expectation value of electron–electron repulsion $\langle V_{ee} \rangle = \left\langle \frac{1}{r_{12}} \right\rangle$ using the five trial wave functions and the relation between $\langle V_{ee} \rangle$ versus Z is represented in Fig. 4.1. The percentage error of energy is shown in Table 4.3.

Table 4.4 gives $E_{corr}\%$ (the percentage correlation energy) defined as [3, 8]:

$$E_{corr}\% = \frac{E_{corr}}{E} \times 100, \text{ where E is the calculated energy in a.u.}$$

The calculation with the first trial wave function $\psi_1(r_1, r_2)$ gives energy for helium atom E = −2.8477 a.u. and an error of 1.93 % with $\alpha = 1.6875$. This error is due to this type of function, because it includes one parameter (α), which may be interpreted as the effective nuclear charge felt by each of the electrons. The error in this calculation is attributed to the fact that the wave function is based on the orbital approximation and, therefore, does not appropriately take electron correlation into account. In other words, this wave function gives the electrons too much independence, given that they have like charges and tend to avoid one another.

Table 4.1 The calculated energy (E) and experimental energy (E_{exp}) for helium sequence using five trial wave functions (results are in a.u.)

		$-E$					
Species	Z	ψ_1	ψ_2	ψ_3	ψ_4	ψ_5	$-E_{exp}$ [18]
H^-	1	0.4727	0.4870	0.5133	0.5088	0.5255	0.5277
He	2	2.8477	2.8603	2.8757	2.8911	2.9014	2.9037
Li^+	3	7.2227	7.2350	7.2487	7.2682	7.2772	7.2799
Be^{2+}	4	13.5977	13.6098	13.6230	13.6441	13.6525	13.6556
B^{3+}	5	21.9727	21.9847	21.9970	22.0200	22.0278	22.0309
C^{4+}	6	32.3477	32.3596	32.37227	32.3950	32.4029	32.4062
N^{5+}	7	44.7227	44.7346	44.7471	44.7700	44.0778	44.7814
O^{6+}	8	59.0977	59.1096	59.1220	59.1452	59.1532	59.1566
F^{7+}	9	75.4727	75.4846	75.4970	75.5200	75.5281	75.5317
Ne^{8+}	10	93.8477	93.8595	93.8720	93.8954	93.9032	93.9068

Table 4.2 The expectation value of electron–electron repulsion $\left\langle \frac{1}{r_{12}} \right\rangle$, for helium sequence using five trial wave functions

		$\langle V_{ee} \rangle = \left\langle \frac{1}{r_{12}} \right\rangle$				
Species	Z	ψ_1	ψ_2	ψ_3	ψ_4	ψ_5
H^-	1	0.4297	0.3965	0.2958	0.3731	0.3208
He	2	1.0547	1.0281	0.9921	0.9743	0.9492
Li^+	3	1.6797	1.6544	1.6242	1.5924	1.5721
Be^{2+}	4	2.3047	2.2799	2.2519	2.2144	2.1960
B^{3+}	5	2.9297	2.9052	2.8698	2.8378	2.8203
C^{4+}	6	3.5547	3.5304	3.5041	3.46179	3.4449
N^{5+}	7	4.1797	4.1555	4.1297	4.08611	4.0696
O^{6+}	8	4.8047	4.7806	4.7552	4.71061	4.6945
F^{7+}	9	5.4297	5.4057	5.3805	5.33525	5.3192
Ne^{8+}	10	6.0547	6.0308	6.0057	5.95996	5.9439

From the second trial wave function, the value of the energy for He atom is E $= -2.8603$ a.u. obtained for $\alpha = 2.1603$ and $\beta = 1.2141$ and energy error % is 1.49. The result of this wave function shows that the value of energy and energy error % is less than that in $\psi_1(r_1, r_2)$ and this means that $\psi_2(r_1, r_2)$ is better than $\psi_1(r_1, r_2)$ because it includes two parameters α, β and this wave function indicates that 50 % of the time, the electrons are in different orbitals, while for the first trial wave function the electrons are in the same orbital 100 % of the time. Therefore $\psi_2(r_1, r_2)$ is introducing some electron correlation into the trial wave function with $E_{corr}\% = 0.441$ and this means it becomes a correlated wave function.

Third trial wave function $\psi_3(r_1, r_2)$ shows that the value of the energy for He atom is -2.8757 a.u. with $\alpha = 2.18317$ and $\beta = 1.18853$, and energy error % is 0.9656. This result indicates $\psi_3(r_1, r_2)$ is better than $\psi_1(r_1, r_2)$ and $\psi_2(r_1, r_2)$, and it becomes more correlated. From them $E_{corr}\% = 0.974$, because it is assumed that one electron is closer to the nucleus than the other; such an effect may be called radial correlation.

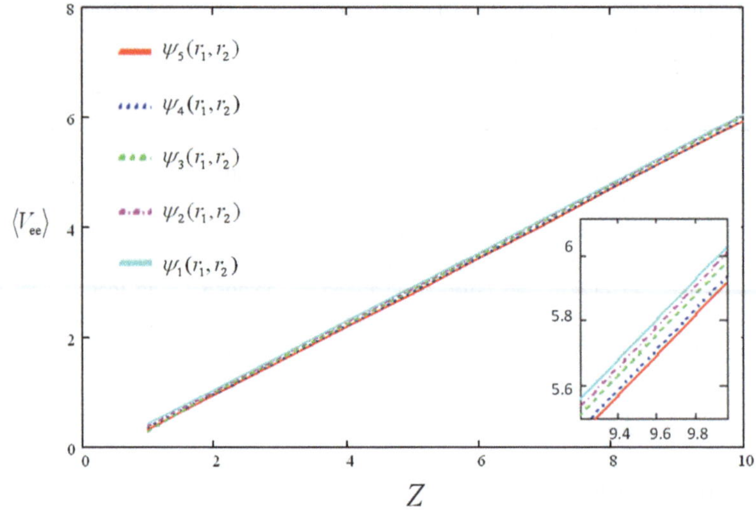

Fig. 4.1 $\langle V_{ee}\rangle$ versus Z

Table 4.3 The energy error % for helium sequence using five trial wave functions

Species	Z	Energy error % ψ_1	ψ_2	ψ_3	ψ_4	ψ_5
H^-	1	10.40	7.709	2.7283	3.585	0.410
He	2	1.930	1.490	0.9656	0.433	0.078
Li^+	3	0.790	0.617	0.428	0.161	0.037
Be^{2+}	4	0.400	0.335	0.239	0.085	0.022
B^{3+}	5	0.260	0.210	0.154	0.0515	0.014
C^{4+}	6	0.180	0.144	0.105	0.035	0.010
N^{5+}	7	0.130	0.104	0.077	0.025	0.0076
O^{6+}	8	0.100	0.079	0.059	0.019	0.0057
F^{7+}	9	0.078	0.062	0.046	0.015	0.0047
Ne^{8+}	10	0.063	0.050	0.037	0.012	0.0038

Table 4.4 The percentage correlation energy (E_{corr}%) for helium sequence using four correlated trial wave functions

Species	E_{corr}% ψ_2	ψ_3	ψ_4	ψ_5
H^-	2.936	7.910	7.095	10.048
He	0.441	0.974	1.501	1.851
Li^+	0.170	0.359	0.626	0.749
Be^{2+}	0.089	0.186	0.340	0.401
B^{3+}	0.055	0.110	0.215	0.250
C^{4+}	0.037	0.076	0.146	0.170
N^{5+}	0.027	0.055	0.106	0.123
O^{6+}	0.020	0.041	0.080	0.094
F^{7+}	0.016	0.032	0.063	0.073
Ne^{8+}	0.013	0.026	0.051	0.059

The value of the energy for He atom using fourth trial wave function $\psi_4(r_1, r_2)$ is -2.8911 a.u. The energy error % with $\alpha = 1.8497$ and $b = 0.3658$ is 0.433. Comparing the result of this function with the previous equations, it is clear that $\psi_4(r_1, r_2)$ is a better trial wave function because it takes into account the instantaneous electron correlation. Therefore $E_{corr}\%$ increases to 1.501 , because it introduces the inter electronic distance; such a correlation is called angular correlation.

The fifth type trial wave function $\psi_5(r_1, r_2)$ shows that the value of the energy for He atom is -2.9014 a.u. with $\alpha = 2.2088$, $\beta = 1.4364$ and $b = 0.2934$ and energy error % is 0.078. On comparing the results of this wave function with the previous equations, it is clear that $\psi_5(r_1, r_2)$ is a much better trial wave function because it includes three parameters α, β and b where the parameter b adds weight to the r_{12} term which most directly represents electron correlation in the wave function. Thus, it becomes more correlated from previous wave functions and therefore $E_{corr}\%$ increases to 1.851.

The discussion above is nearly the same for all the helium sequence except that the hydride ion is a particularly interesting case because the first two trial wave functions do not predict stable ion where the starting point for H^- should be the third trial wave function. This indicates that electron correlation is an especially important issue for atoms and ions with small nuclear charge.

4.4 Conclusions

Some conclusions from this research are:

(a) The calculations show that as wave function becomes more correlated, $E_{corr}\%$ increases and percentage energy error decreases.
(b) The three parameters wave function becomes more like the two and one parameter wave function as Z increases.
(c) Except for the hydride ion results for $\psi_4(r_1, r_2)$, the improved agreement with experimental results (the lower total energy) is due to a reduction in electron–electron repulsion as shown in Tables 4.1 and 4.2.
(d) As shown in Table 4.3, the percentage of energy error for atom and ions decreases as Z increases, because the attraction of the nucleus is increased with $\langle V_{ee} \rangle$.
(e) Fifth type trial wave function is better than all the wave functions used in this work, because it gives results closer to the experimental results.

References

1. Wigner E (1934) On the interaction of electrons in metals. Phys Rev 46:1002–1011
2. Löwdin PO (1959) Correlation problem in many-electron quantum mechanics I. Review of different approaches and discussion of some current idea. Adv Chem Phys 2:207–322

3. Levine IN (2000) Quantum chemistry, 5th edn. Prentice-Hall Inc, New Jersey, p 739
4. Hollett JW, Gill PMW (2011) The two faces of static correlation. J Chem Phys 134:114111–114116
5. Weiss AW (1961) Configuration interaction in simple atomic systems. Phys Rev 122:1826–1836
6. Li WK (1988) A lesser known one-parameter wave function for the helium sequence and the virial theorem. J Chem Educ 65:963–964
7. Slamet M, Sahni V (1995) Coulomb holes and correlation potentials in the helium atom. J Phys Rev A 51:2815–2825
8. Wilson CL, Montgomery HE, Sen KD, Thompson DC (2010) Electron correlation energy in confined two-electron systems. J Phys Lett A 374:4415–4419
9. Griffiths DJ (2005) Introduction to quantum mechanics, 2nd edn. Pearson Education Inc., New Jersey, p 468
10. Bressanini D, Morosi G (2008) A compact boundary-condition – determined wavefunction for two-electron atomic systems. J Phys B At Mol Opt Phys 41:145001–145006
11. Hameka HF (2004) Quantum mechanics. Wiley, New Jersey, p 194
12. Engel T (2006) Quantum chemistry and spectroscopy. Pearson Education Inc., New York, p 489
13. Gill PMW, Besley NA, O'Neill DP (2004) Wigner intracule for the Kellner helium-like ions. Int J Quantum Chem 100:166–171
14. Maysoon AM (2011) Self consistent field calculation using two parameter wave function for helium atom and helium-like ions. J Baghdad Sci 8:527–531
15. Coulson CA, Neilson AH (1961) Electron correlation in the ground state of helium. Proc Phys Soc 78:831–837
16. Scully MO, Allen RE, Dou Y, Kapale KT, Kim M, Chen G, Svidzinsky A (2004) Molecular calculations with two-center correlated orbitals. Chem Phys Lett 389:385–392
17. Chandrasekhar S (1944) Some remarks on the negative hydrogen ion and its absorption coefficient. Astro Phys J 100:176–180
18. Pekeris CL (1958) Ground state of two-electron atoms. Phys Rev 112:1649–1658

Chapter 5
Hydrocarbon Generating Potentials of Benue Trough Coals

Aliyu Jauro, Brian Horsfield, Heinz Wilkes, and Muhammad B. Abubakar

Abstract Benue trough is an inland coal bearing basin that has been only lightly explored for petroleum. An organic geochemical analysis of coal samples from Kumo and Lamza in the upper part of the trough and from Lafia-Obi in the lower part was undertaken to assess their hydrocarbon generative potentials. The highest $\%R_o$ and T_{max} of 0.9–1.1 % and 449–455 °C was found in Lafia-Obi samples, next was Lamza with 0.5–0.7 % and 434–438 °C and Kumo with 0.4–0.6 % and 415–429 °C. The HI of Lafia-Obi, Lamza and Kumo are within the ranges of 136–246 mg HC/g TOC, 184–269 mg HC/g TOC and 167–322 mg HC/g TOC respectively. The Odd-over-Even-Predominance (OEP) values and pristane/phytane (Pr/Ph) ratios of the samples suggest that they are dominantly terrestrial in nature and were deposited in an oxic to suboxic environment. Ratios of 20*S*/(20*S* + 20*R*)$\alpha\alpha C_{29}$ cholestanes, 20*S*/(20*S* + 20*R*)$\beta\beta C_{29}$ cholestanes, $\beta\alpha/(\beta\alpha + \alpha\beta)$ C_{30} hopanes, Ts/Ts + Tm C_{27} hopanes and other maturity parameters suggest that Lafia-Obi is highest in maturity (peak oil window), followed by Lamza (early oil window) and then Kumo (immature).

5.1 Introduction

The potentials of terrestrial organic matter, coals and coaly shale for generating oil have been established [1, 2]. The importance of this potential source of hydrocarbons has been highlighted by the discovery of a number of oil and gas

A. Jauro (✉) • M.B. Abubakar
National Centre for Petroleum Research and Development (Energy Commission of Nigeria), Abubakar Tafawa Balewa University, PMB 0248, 740004 Bauchi, Nigeria
e-mail: alijauro@daad-alumni.de; ambello2002@yahoo.com

B. Horsfield • H. Wilkes
Organic Geochemistry Section, GeoForschungsZentrum, Telegrafenberg, 14467 Potsdam, Germany
e-mail: brian.horsfield@gfz-potsdam.de; heinz.wilkes@gfz-potsdam.de

M. Gupta Bhowon et al. (eds.), *Chemistry: The Key to our Sustainable Future*, DOI 10.1007/978-94-007-7389-9_5, © Springer Science+Business Media Dordrecht 2014 fields around the world. Some typical examples include the Junggar and Tarim basins in the north-western China, the Herald and Lulita fields in the Danish Central Graben of the North Sea, the Mahakam Delta in Indonesia and the Niger Delta in Nigeria [3, 4].

Benue trough is a lightly explored sedimentary basin and has had some small discoveries. One of the three wells drilled in the inland basin established the existence of an effective petroleum system. Kolmani-River 1 drilled by Shell Nigeria Exploration Company, to a depth of 3,000 m in 1999, encountered some 934 billion standard cubic meter of gas and a little oil, but the well was plugged and suspended due to the sub-commercial inferred reserve [5]. The second well, Nasara-1, was found to be dry, but the occurrence of migrated oil was reported [5, 6]. Although many coals generate significant amounts of liquid hydrocarbons that are similar to naturally occurring oils, poor expulsion efficiency limits their source rock potential [7]. Significant amounts of liquid products are assimilated by vitrinite matrix of most coals prior to expulsion. The coals from Benue trough are Cretaceous and majority of humic coals capable of generating and expelling non-volatile are Cretaceous to Tertiary in age [4]. The presence of petroleum source rock is a major prerequisite to petroleum accumulation. Previous studies have indicated that some coal beds constitute one of the potential source rocks in the Benue trough [8, 9]. Assessing the source rock potentials of the coals would reduce petroleum exploration risks by determining whether the coal bearing basins could be explored further for petroleum.

Benue trough originated from early Cretaceous rifting of the Central West African basement uplift. It forms a regional structure 150 km wide which is exposed from the northern frame of the Niger Delta running northeastwards for about 1,000 km terminating underneath Lake Chad (Fig. 5.1). The trough contains as much as 6,000 m of Cretaceous-Tertiary sediments, including those predating the middle Santonian which have been compressed, deformed, folded, faulted, and uplifted in several places, producing more than 100 anticlines and synclines [10]. The Benue trough is subdivided into lower, middle and upper parts (Fig. 5.1). The upper part of the trough bifurcates at its northeastern end into the Gongola and the Yola basins (Fig. 5.1).

The aims of this study are to use some vitrinite reflectance, Rock-Eval pyrolysis and biomarkers in assessing the maturity, organic matter type and hence the hydrocarbon source rock potentials of Lafia-Obi, Lamza and Kumo coal samples.

5.2 Materials and Methods

5.2.1 Samples

Four coal samples were collected from each of the three locations, making a total number of twelve (12). Samples were collected at the depth of 27.5 m from Kumo (KM) in Maiganga coal mines of the Gongola basin and from outcrops at

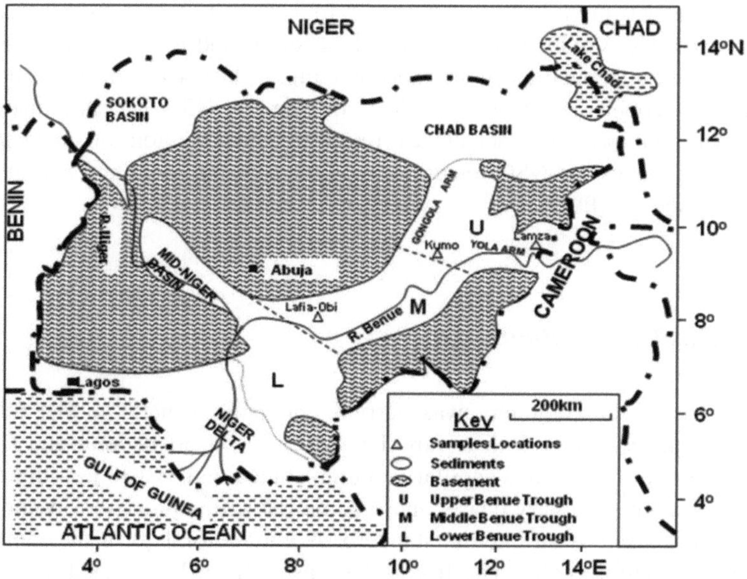

Fig. 5.1 Geological map of Nigeria, showing the major sedimentary basins and samples location (Modified from Jauro et al. [9])

Lamza (LM) in the Yola basin of the upper part of the trough (Fig. 5.1). Samples were also collected at the depth of 20.7 m from Lafia-Obi in the middle part of the trough (Fig. 5.1).

5.2.2 Rock-Eval Pyrolysis and Total Organic Carbon (TOC)

The TOC and Rock-Eval analyses were conducted by Applied Petroleum Technology AS, Kjeller, Norway, using a Rock-Eval 6 instrument and based on the Norwegian Industry Guide to Organic Geochemical Analysis [11].

5.2.3 Solvent Extraction of Coal Samples

The samples were pulverized and passed through a 250 μm sieve size. The powdered sample (5 g) was soxhlet extracted with an azeotropic mixture of dichloromethane and methanol (250 ml), in the volume ratio of 99:1 for 24 h. Solvent was recovered with the aid of a turbo vaporizer. Asphaltenes removal was achieved by the addition of 60-fold *n*-hexane. The deasphalted extract (maltenes) and residue (asphaltenes) were dried to a constant weight in a nitrogen atmosphere. The deasphalted extract was dissolved in 1 ml *n*-hexane. 5.0 mg/ml of 5α-androstane in hexane and 2.0 mg/ml of ethylpyrene in toluene were added to the mixture as internal standards. Volumes of the internal standards added were based on extract weight-to-internal standard volume ratio of 1 mg:1 μl. A 40 μl of the standard oil (Norwegian Geochemical Standard (NGS) oil sample, NSO-1) was prepared in a similar way but without asphaltenes removal. Fractionation of maltenes into aliphatics, aromatics and NSO fraction was achieved using Medium Pressure Liquid Chromatography (MPLC).

5.2.4 Gas Chromatography

Capillary gas chromatography was carried out using Agilent model 6890 gas chromatograph equipped with a vaporizing injector, a 50 m long and 0.20 μm internal diameter fused silica capillary column coated with a 0.33 μm film of Ultra 1 methylsiloxane stationary phase, and a flame ionization detector. 5α-androstane in hexane was used as an internal standard. A volume of 1 μl of the sample was injected by an autosampler. The front inlet was a HP PTV operated in a splitless mode. Helium was used as the carrier gas with an average velocity of 24 cm/s. Saturated hydrocarbon fractions were analyzed using an initial oven temperature of 40 °C for 2 min, then increase at the rate of 5 °C min^{-1} up to a temperature of 300 °C where it was held for 60 min.

5.2.5 Gas Chromatography–Mass Spectrometry

Biomarker analysis of the aliphatic hydrocarbon fraction was carried out using an Agilent 6890 series gas chromatograph interfaced with a Finnigan MAT 95 XL series mass selective detector operated in Metastable Reaction Monitoring (MRM) mode. Sample (1 μl aliquot) dissolved in *n*-hexane was injected using an automatic sampler. The front inlet was a HP PVT injection system operated in a splitless mode equipped with a fused silica capillary column (SGE BPX5; 50 m length, inner diameter of 0.22 μm, film thickness of 0.25 μm). Helium was used as the carrier gas with a flow rate of 20 cm^3/min. The GC oven temperature was held at 50 °C for 1 min, and then raised to 310 °C at 3 °C/min where it was held isothermally for 20 min, making a total run time of 107.7 min. The electron impact mass spectra were obtained at 70 eV in acquired mode. Peaks on mass chromatogram were identified based on comparison with known standards and published elution time. Both peak heights and areas of selected compounds were measured from the selected ion mass chromatograms, and were used in calculating the various compounds ratios.

5.3 Results

5.3.1 Vitrinite Reflectance, Rocks-Eval Data and Total Organic Carbon

The vitrinite reflectance values, rock-eval pyrolysis and total organic carbon (TOC) of the coals are shown in Table 5.1. The samples' vitrinite reflectance (%R_o) falls within 0.4–1.1 %. Lafia-Obi has the highest %R_o values of 0.9–1.1, followed by Lamza with %R_o values of 0.5–0.7. Kumo has the least %R_o values of 0.4–0.6. The TOC of the samples varies from 23.8 to 75.8 wt%. In Lafia-Obi, TOC ranges from 23.8 to 66.7 wt%, while in Lamza and Kumo, TOC contents are in between 50.7–75.8 wt% and 40.6–61.1 %, respectively. Rock Eval data indicate that the maximum temperature for hydrocarbon generation (T_{max}) of Lafia-Obi coal samples falls within the range of 449–455 °C, while those of Lamza and Kumo are within the ranges 434–438 °C and 415–429 °C, respectively. The hydrogen index (HI) of the samples varies from 136 to 322 mg HC/g TOC. The HI of Lafia-Obi, Lamza and Kumo are within the ranges of 136–246 mg HC/g TOC, 184–269 mg HC/g TOC and 167–322 mg HC/g TOC respectively.

5.3.2 Extract Composition and Distribution of n-Alkanes

The concentrations of extracts from the coal samples and ratios of Pr/Ph, OEP, Pr/*n*-C_{17}, and Ph/*n*-C_{18} are shown in Table 5.2. Bitumen concentrations vary from 17.1 to 146.5 mg/g TOC. The lowest and highest extracts of 100.5–146.5 mg/g TOC and 17.1–55.5 mg/g TOC were obtained from Lamza and Lafia-Obi coal samples, respectively. Bitumen from Kumo are within the range 26.7–72.3 mg/g TOC.

Table 5.1 Vitrinite reflectance and rock-eval pyrolysis data of the Benue trough coal samples

			TOC	S_1	S_2	S_3	T_{max}	OI	HI
Location	Identity	%R_o	wt%	mgHC/g	mgHC/g	mgHC/g	°C	$mgCO_2$/gTOC	mgHC/g
Lafia-Obi	**LF-2**	1.0	66.7	4.4	164.3	1.3	452	2	246
	LF-4	1.1	23.8	0.7	39.6	1.2	455	5	166
	LF-6	1.0	61.1	1.9	83.1	13.6	449	22	136
	LF-9	0.9	27.0	3.9	41.2	1.7	452	6	153
Kumo	**KM-2**	0.5	56.0	2.8	150.3	22.9	415	41	268
	KM-3	0.6	61.1	2.1	124.8	24.6	421	40	204
	KM-5	0.4	57.7	1.7	96.3	28.7	415	50	167
	KM-6	0.4	40.6	1.1	130.6	13.9	429	34	322
Lamza	**LM-1**	0.5	72.2	7.5	194.0	7.6	436	10	269
	LM-3	0.5	73.3	4.3	171.0	11.3	435	15	233
	LM-4	0.6	75.8	4.7	174.5	10.2	434	13	230
	LM-5	0.7	50.7	2.2	93.3	12.6	438	25	184

Table 5.2 Composition of the coal extracts from Benue trough

Location	Identity	Bitumen mg/g TOC	Maltenes mg/g TOC	Asphaltenes mg/TOC	Pr/Ph	Pr/nC_{17}	Ph/nC_{18}	OEP
Lafia-Obi	**LF-2**	38.8	21.3	13.5	3.4	8.9	2.2	1.0
	LF-4	17.1	12.9	2.0	3.6	1.2	0.2	1.0
	LF-6	54.1	5.8	21.9	3.2	0.5	0.1	1.0
	LF-9	55.5	29.3	15.9	5.1	0.7	0.1	1.0
Kumo	**KM-2**	49.6	14.1	29.0	4.5	4.0	0.7	2.1
	KM-3	51.3	19.1	30.9	3.1	2.5	0.4	5.9
	KM-5	26.7	9.1	16.7	5.7	4.3	0.6	3.4
	KM-6	72.3	16.9	51.2	3.1	0.8	0.2	3.1
Lamza	**LM-1**	146.5	33.2	6.3	10.9	12.1	0.9	1.2
	LM-3	100.5	30.9	45.8	11.5	12.2	0.9	1.3
	LM-4	118.7	27.7	6.2	12.0	11.9	0.9	1.2
	LM-5	121.5	38.7	36.7	9.8	7.6	0.6	1.2

Pr pristine, *Ph* phytane, *OEP* odd-over–even predominance = $(C_{25} + 6C_{27} + C_{29})/(4C_{26} + 4C_{28})$

The concentrations of deasphalted extracts (maltenes) in Lafia-Obi and Kumo are 5.8–29.3 mg/g TOC and 9.1–19.1 mg/g TOC respectively. The distribution of the maltenes in the samples (Table 5.2) revealed a similar pattern as that of the whole extract.

Figure 5.2 shows the gas chromatograms of the aliphatic fractions of coal extracts. The chromatograms generally exhibit a unimodal *n*-alkane distribution within the nC_{15}-nC_{33} range. There is absence of n-C_{27}, n-C_{29} and n-C_{31} dominance (OEP ~ 1), except in Kumo, where OEP is greater than 2 (Fig. 5.2, Table 5.2). The samples from Kumo are poor in low molecular weight *n*-alkanes, with high Pr/Ph ratios above 3. The highest Pr/Ph ratios of 9.8–12.0 were observed in Lamza coals.

5.3.3 *Terpanes*

The total ion current (TIC) chromatograms of the aliphatic fractions of representative samples showing the triterpanes distribution are shown in Fig. 5.3. Identities of the annotated peaks on the mass chromatograms are given in Table 5.3 and biomarker ratios are shown in Table 5.4.

Lamza and Lafia-Obi samples have abundant C_{27} to C_{35} hopanes, while Kumo samples are rich only in the C_{27} to C_{33} hopanes. The concentration of 17α(H)-22,29,30-trisnorhopane (Tm) is generally higher than the 18α(H)-22,29,30-trisnorneohopane (Ts) in all the samples, with the exception of Lafia-Obi where the reverse is the case. 17β(H),21β(H)-Hopane is either absent or its concentration is below the detection limit in all the coal samples. Oleanane is present in lower concentration in Lamza and Lafia-Obi samples. Gammacerane was not detected in all the samples (Fig. 5.3, Table 5.3).

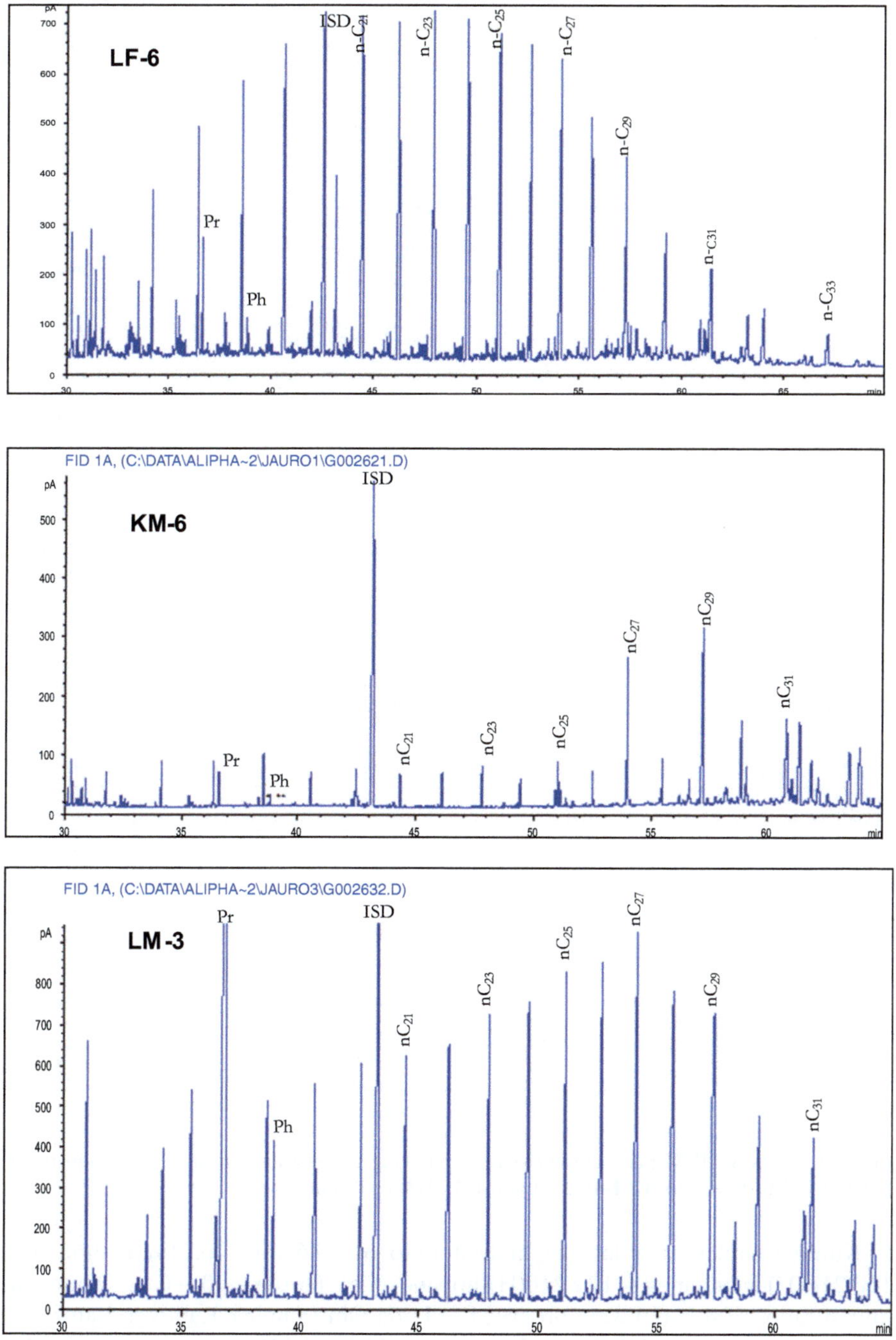

Fig. 5.2 Representative gas chromatograms showing *n*-alkanes distribution in Lafia-Obi (**LF-6**), Kumo (**KM-6**) and Lamza (**LM-3**) coal samples

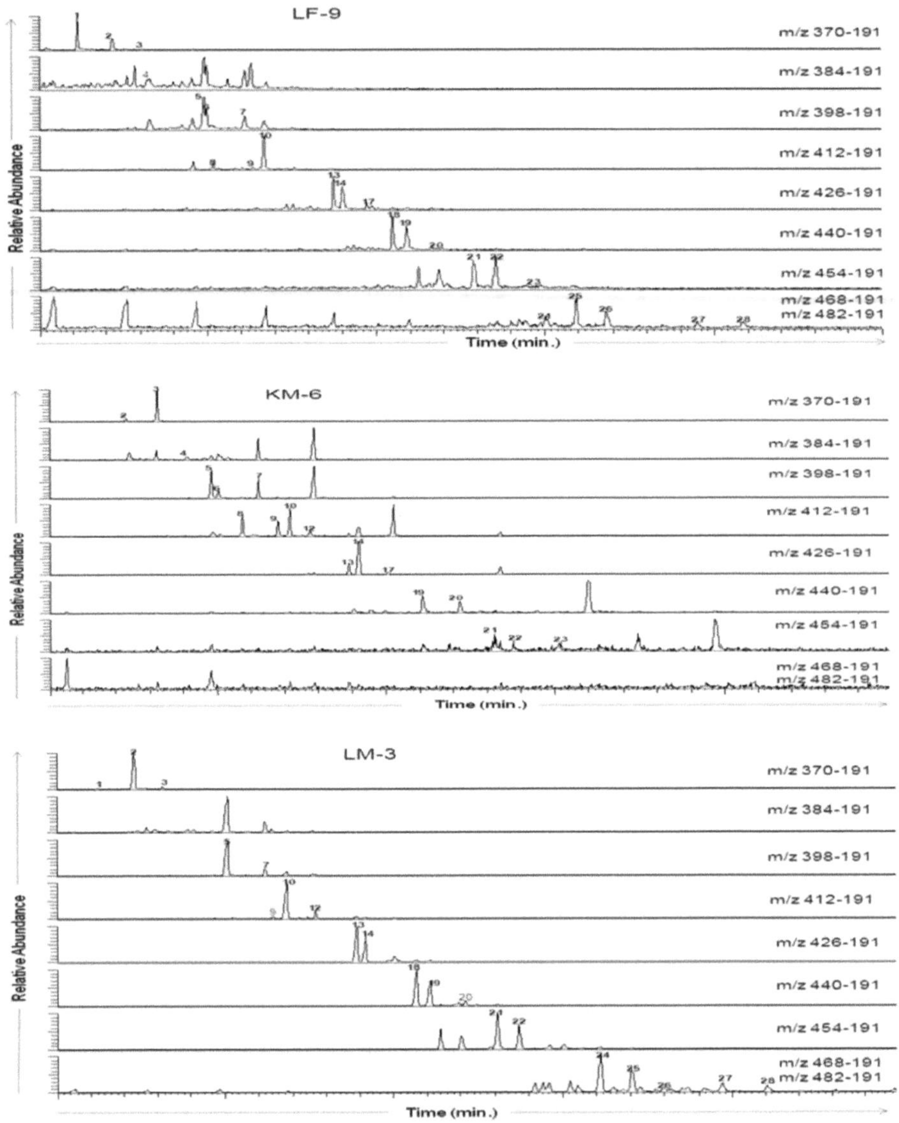

Fig. 5.3 GC-MS MRM mass chromatograms showing the distribution of C_{27}-C_{34} hopanes in Lafia-Obi (**LF-9**), Kumo (**KM-6**) and Lamza (**LM-3**) coal samples

Kumo samples recorded the highest ratio of 29.6 for 17β,21α(H)-moretane/17β,21α(H)-moretanes + 17α,21β(H)-hopane, i.e. βα/(βα+ αβ)−C_{30}, while the lowest ratios of 4.2–6.0 were observed in Lafia-Obi samples. Ratios in Lamza are in between 12.2 and 16.2. The ratio of 22*S* to 22*S* and 22*R* epimers of 17α(H),21β(H)-C_{31}-hopanes [22*S*/(22*S* + 22*R*)C_{31}αβ hopanes] of Lafia-Obi, and Lamza is ~60.0, while in Kumo samples they are in between 21.1 and 31.7. The ratios of Ts/(Ts + Tm)C_{27} hopanes in Lafia-Obi samples fall in between 51.4 and 85.4, while ratios in Kumo and Lamza samples are both less than 10.0.

Table 5.3 Identities of annotated hopanes peaks in Fig. 5.3

Peak	Compound
1	18α(H)-22,29,30-Trisnorneohopane (Ts)
2	17α(H)-22,29,30-Trisnorhopane (Tm)
3	17β(H)-22,29,30-Trisnorhopane
4	αβ C28
5	17α(H),21β(H)-30-Norhopane
6	18α(H)-Norneohopane (C29 Ts)
7	17β(H),21α(H)-30-Norhopane
8	17α(H)-Diahopane
9	Oleanane
10	17α(H),21β(H)-30-Hopane
11	17β(H),21β(H)-30-Norhopane
12	(22*S*)-17α(H),21β(H)-29-Homohopane
13	(22*R*)-17α(H),21β(H)-29-Homohopane
14	(22*S*)- and (22*R*)-17β(H),21α(H)-29-Homomoretane
15	(22*S*)-17α(H),21β(H)-29-Dihomohopane
16	(22*R*)-17α(H),21β(H)-29-Dihomohopane
17	(22*S*)- and (22*R*)-17β(H),21α(H)-29-Homohopane
18	(22*S*)-17α(H),21β(H)-29-Trihomohopane
19	(22*R*)-17α(H),21β(H)-29-Trihomohopane
20	(22*S*)- and (22*R*)-17β(H),21α(H)-29-Trihomohopane
21	(22*S*)-17α(H),21β(H)-29-Tetrahomohopane
22	(22*R*)-17α(H),21β(H)-29-Tetrahomohopane
23	(22*S*)- and (22*R*)-17β(H),21α(H)-29-Tetrahomohopane
24	(22*S*)-17α(H),21β(H)-29-Pentahomohopane
25	(22*R*)-17α(H),21β(H)-29-Pentahomohopane

5.3.4 Cholestanes

Representative GC-MS MRM mass chromatograms of C_{27} to C_{30} cholestanes in the aliphatic fraction of the coal extracts are shown in Fig. 5.4. The identities of the peaks annotated on the mass chromatograms are given on Table 5.5. The cholestanes are more abundant in samples from Lafia-Obi and Lamza than in Kumo. Kumo samples recorded the least ratio of 20*S*/(20*S* + 20*R*) 5α(H),14α (H),17α(H)-C_{29} of 21.2–28.5 %. Ratios in Lafia-Obi and Lamza samples were 40.3–49.3 % and 50.4–53.7 %, respectively. There is no significant difference in the ratios of 20*S*/(20*S* + 20*R*)5α(H),14β(H),17β(H)-C_{29} (i.e. 20*S*/(20*S* + 20*R*)αββ) of the coal samples. Ratios in all the samples fall in between 23.8 % and 33.3 % with the exception of a sample from Kumo which recorded the lowest value of 17.0 %. The (20*S* + 20*R*) αββ/(αββ + ααα) of C_{29} ratios start from 41.5 % in Kumo coal sample to a maximum value of 66.6 % in Lafia-Obi sample. Ratios in Kumo and Lamza samples are within the ranges 41.5–61.6 % and 54.4–61.4 %, respectively.

Table 5.4 Hopanoids and cholestanes biomarker ratios from the Benue trough coal samples

Locality	Identity	ααα 20*S*/(20*S* + 20*R*) C_{29} cholestanes	αββ 20*S*/(20*S* + 20*R*) C_{29} cholestanes	αββ/(αββ + ααα) 20*S* + 20*R* C_{29} cholestanes	βα/(βα + αβ) C_{30} hopanes	22*S*/(22*S* + 22*R*) C_{31}αβ hopanes	Ts/(Ts + Tm) C_{27} hopanes
Lafia-Obi	**LF-2**	45.2	29.8	66.6	5.7	59.4	51.4
	LF-4	40.3	29.4	61.8	6.0	61.8	85.4
	LF-6	49.3	23.8	62.6	4.2	59.7	66.2
	LF-9	45.5	33.3	60.5	5.0	63.2	74.6
Kumo	**KM-2**	21.2	31.2	44.1	29.3	31.7	6.2
	KM-3	28.5	26.3	43.2	28.7	21.1	8.8
	KM-5	21.4	29.6	41.5	27.7	22.3	5.6
	KM-6	27.5	17.0	61.6	29.6	25.4	4.4
Lamza	**LM-1**	53.7	30.0	54.7	15.2	59.8	2.9
	LM-3	50.9	24.0	54.4	15.7	60.6	2.3
	LM-4	50.4	23.9	57.2	16.2	59.4	2.5
	LM-5	50.7	28.9	61.4	12.2	59.4	3.4

ααα 20*S*/(20*S* + 20*R*) = epimers of 5α(H),14α(H),17α(H); αββ = 5α(H),14β(H),17β(H); βα/(βα + αβ)–C_{30} = 17β,21α(H)-moretane/(17β,21α(H) moretane + 17α,21β(H) hopane; 22*S*/(22*S* + 22*R*) = ratio for C_{31}-homohopane; *Ts* 18α-22,29,30-trisnorneohopane, *Tm* 17α-22,29,30-trisnorhopane

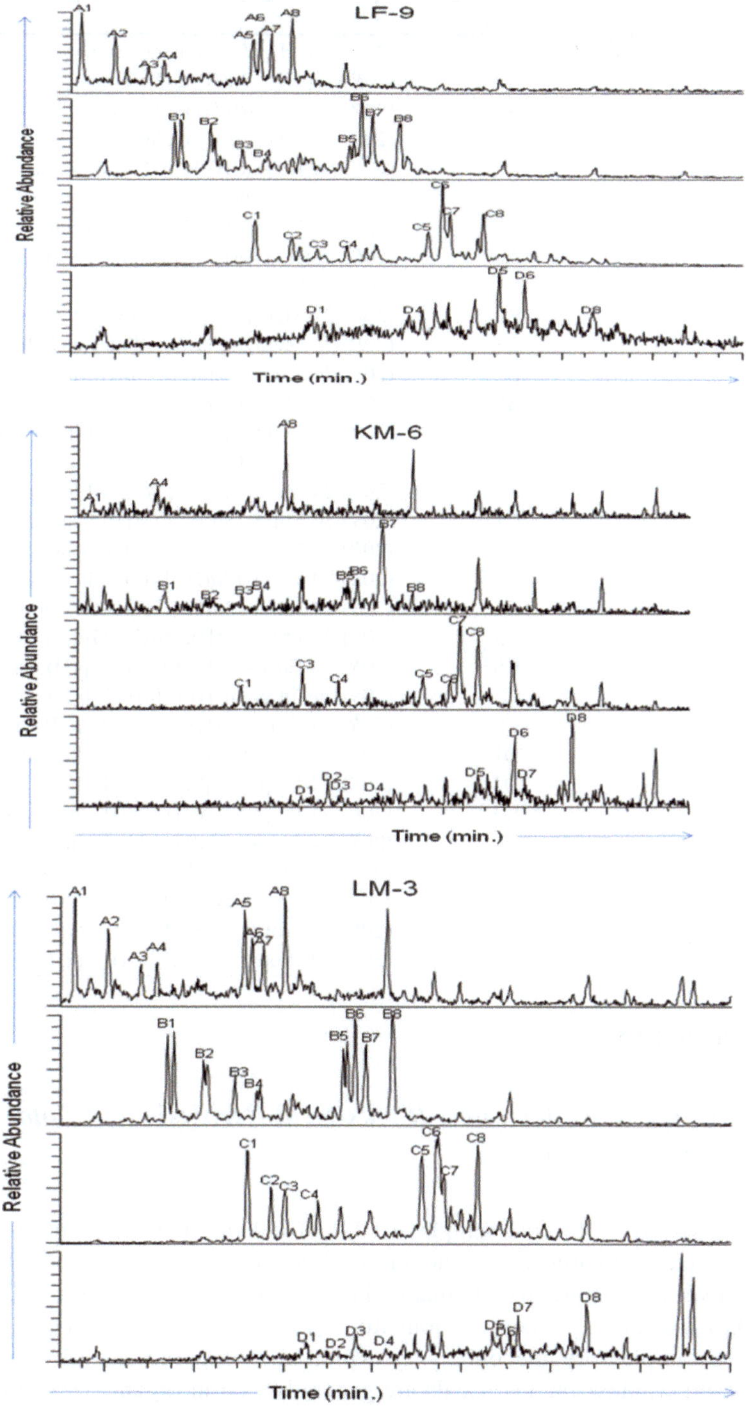

Fig. 5.4 GC-MS MRM partial mass chromatograms showing the distribution of C_{27}-C_{30} cholestanes in Lafia-Obi (**LF-9**), Kumo (**KM-6**) and Lamza (**LM-3**) coal samples

Table 5.5 Identities of annotated compounds of cholestanes in Fig. 5.4

Peak	Compound
A1	(20*S*)-13β(H),17α(H)-dicholestane
A2	(20*R*)-13β(H),17α(H)-dicholestane
A3	(20*R*)-13α(H),17β(H)-dicholestane
A4	(20*S*)-13α(H),17β(H)-dicholestane
A5	(20*S*)-5α(H),14α(H),17α(H)-cholestane
A6	(20*R*)-5α(H),14β(H),17β(H)-cholestane
A7	(20*S*)-5α(H),14β(H),17β(H)-cholestane
A8	(20*R*)-5α(H),14α(H),17α(H)-cholestane
B1	(20*S*)-24-Methyl-13β(H),17α(H)-dicholestane
B2	(20*R*)-24-Methyl-13β(H),17α(H)-dicholestane
B3	(20*R*)-24-Methyl-13α(H),17β(H)-dicholestane
B4	(20*S*)-24-Methyl-13α(H),17β(H)-dicholestane
B5	(20*S*)-24-Methyl-5α(H),14α(H),17α(H)-cholestane
B6	(20*R*)-24-Methyl-5α(H),14β(H),17β(H)-cholestane
B7	(20*S*)-24-Methyl-5α(H),14β(H),17β(H)-cholestane
B8	(20*R*)-24-Methyl-5α(H),14α(H),17α(H)-cholestane
C1	(20*S*)-24-Ethyl-13β(H),17α(H)-dicholestane
C2	(20*R*)-24-Ethyl-13β(H),17α(H)-dicholestane
C3	(20*R*)-24-Ethyl-13α(H),17β(H)-dicholestane
C4	(20*S*)-24-Ethyl-13α(H),17β(H)-dicholestane
C5	(20*S*)-24-Ethyl-5α(H),14α(H),17α(H)-cholestane
C6	(20*R*)-24-Ethyl-5α(H),14β(H),17β(H)-cholestane
C7	(20*S*)-24-Ethyl-5α(H),14β(H),17β(H)-cholestane
C8	(20*R*)-24-Ethyl-5α(H),14α(H),17α(H)-cholestane
D1	(20*S*)-24-Propyl-13β(H),17α(H)-dicholestane
D2	(20*R*)-24-Propyl-13β(H),17α(H)-dicholestane
D3	(20*R*)-24-Propyl-13α(H),17β(H)-dicholestane
D4	(20*S*)-24-Propyl-13α(H),17β(H)-dicholestane
D5	(20*S*)-24-Propyl-5α(H),14α(H)-cholestane
D6	(20*R*)-24-Propyl-5α(H),14β(H),17β(H)-cholestane
D7	(20*S*)-24-Propyl-5α(H),14β(H),17β(H)-cholestane
D8	(20*R*)-24-Propyl-5α(H),14αH),17α(H)-cholestane

5.4 Discussion

5.4.1 Vitrinite Reflectance, Rocks-Eval Pyrolysis and Total Organic Carbon

The %R_o of Lafia-Obi coal samples (0.9–1.1) indicate that they are at the late maturity stage for hydrocarbon generation, while most of Lamza samples (0.5–0.7) are in between immature to early mature [12]. The reflectance values for Kumo (% $R_o < 0.6$) indicate that they are immature (Table 5.1). Amongst coal samples, Lafia-Obi samples are the only ones within the proposed maturity range for effective oil window [4, 13]. Coals are generally rich in organic carbon content, thus the total organic carbon of coal samples are generally more than the 0.5 wt% threshold required for hydrocarbon generation [14]. The T_{max} of Lafia-Obi coals (449–455 °C) indicates that the samples are at their peak maturity for hydrocarbon generation while those for Lamza (434–438 °C) suggest that they are at the late immaturity or early maturity stage. T_{max} values for Kumo samples clearly indicate immaturity with respect to hydrocarbon generation. The "effective oil window" for humic coal is widely accepted to be at a T_{max} of 440–490 °C [15], thus only Lafia-Obi coal is matured enough to have reached the "effective oil window", which is in agreement with the vitrinite reflectance values. Hydrogen indices of Lafia-Obi and Lamza samples that fall within 136–246 mg HC/g TOC and 184–269 mg HC/g TOC (Table 5.1), suggest that samples from both sites are of type III kerogen, which may generate gas at peak maturity [16]. The samples from Kumo show different types of kerogen, ranging from type III, type II/III, and type II with probability of forming gas, oil/gas and oil, respectively. Hunt [15] stated that coals and terrestrial organic matter dispersed in shale capable of generating and expelling oil have HI values greater than 200 mg HC/g. The HI's of most samples from Lamza are greater than 200 mg HC/g, while most samples from Lafia-Obi have HI less than 200 mg HC/g. The lower HI values in Lafia-Obi could be due to maturity influence because at late maturity, there is a decrease in HI [9].

5.4.2 n-Alkanes Distribution

The unimodal *n*-alkane distribution and the dominance of n-C_{27}, n-C_{29}, and n-C_{31} in Lamza and Kumo coal samples (Table 5.2) indicate a terrestrial-derived organic matter input [17]. The lack of n-C_{27}, n-C_{29}, and n-C_{31} dominance in Lafia-Obi coal samples (Table 5.2) may be due to maturity influence [16, 18, 19]. The high pristane and low phytane which generally gave a Pr/Ph > 3 in all the samples (Table 5.2) indicate organic matter input deposited in an oxic to suboxic environment [18–20]. Even though terrigenous other interpretations of Pr/Ph ratios take other sources into account [21], they are probably less relevant in the case of coal [22]. The observed OEP value of ~1 in Lafia-Obi and Lamza samples (Table 5.2) suggest that they are higher in maturity than the Kumo samples with OEP of 2.1–5.9 [20, 23].

5.4.3 Terpanes

The presence of C_{27} to C_{35} hopanes in Lamza and Lafia-Obi coal samples as against C_{27} to C_{33} hopanes in Kumo coal samples (Fig. 5.3, Table 5.3) indicates that they are higher in maturity [24]. Oleanane is derived from β-amyrin which is a significant constituent of leaf waxes, wood, roots, and bark of angiosperms [25, 26]. Therefore, its presence indicates that the samples are cretaceous or younger in age [27, 28]. The lower concentration of the oleanane in Lamza and Lafia-Obi coals could be due to maturity influence because with increase in maturity, pentacyclic triterpenoids, such as oleanane, are converted to aromatic compounds [29].

The ratio of 17β,21α(H)-moretanes to their corresponding 17α,21β(H)-hopanes decreases with thermal maturity from ~80 % in immature bitumen to <15 % in a mature one [30, 31]. Therefore the percentage of βα/(βα + αβ) C_{30}-hopanes, which is high in all the samples and varies in the order Kumo > Lamza > Lafia-Obi (Table 5.4), suggest maturity trend in the reverse order. Lafia-Obi and Lamza samples have reached an equilibrium value of ~60 % for the 22*S*/(22*S* + 22*R*) 17α(H),21β(H) C_{31} hopanes (Table 5.4), thus indicating that Lafia-Obi and Lamza are in the oil window, while Kumo is immature [32, 33].

The presence of 17α(H),22,29,30-trisnorhopane (Tm) in higher concentration than 18α(H),22,29,30-trisnorneohopane (Ts) in Kumo and Lamza coal samples and the reverse case in Lafia-Obi samples (Fig. 5.3; Table 5.3) show that they are lower in maturity than Lafia-Obi coal samples [34]. This is supported by the Ts/(Ts + Tm) ratios shown in Table 5.4. Ratios in Lafia-Obi coal samples are above 50 %, indicating a wide maturity difference compared with Kumo and Lamza samples with ratios of less than 10 %.

5.4.4 Cholestanes

The higher concentration of cholestanes in Lafia-Obi and Lamza coal samples compared with samples from Kumo (Fig. 5.4, Table 5.5) could be due to maturity influence because such compounds are usually evolved from kerogen with increase in maturity [35, 36]. The ratios of (20*S*)/(20*S* + 20*R*) of 5α(H),14α(H),17α(H)-C_{29} cholestanes that are nearly the same in Lafia-Obi and Lamza samples but higher than those of Kumo samples (Table 5.4) support that the samples from Lafia-Obi and Lamza are more mature than Kumo [30, 32, 35]. Similarly, the αββ 20*S*/(20*S* + 20*R*) ratios also support that the Lafia-Obi coal is higher in maturity that both Kumo and Lamza (Table 5.4). The (20*S* + 20*R*) αββ/(αββ + ααα)−C_{29} cholestanes in the coal samples indicate that Kumo samples with ratios in the range of 41.5–61.5 are lower in maturity than both Lamza and Lafia-Obi with ratios of 54.4–61.4 and 60.5–66.6 respectively which are about to reach the equilibrium values of 67–71 % [32] (Table 5.4). The ratio (αββ/(αββ + ααα) 20*S* + 20*R* C_{29} cholestanes for Lafia-Obi samples, which is greater than 60 %, also supports that the coals are in the peak oil window [28]. Ratios for Lamza samples place them in the oil window, while Kumo samples are least in maturity.

5.5 Conclusions

The vitrinite reflectance and T_{max} values indicate that the Lafia-Obi coal samples are at their late maturity for hydrocarbon generation, Lamza is in between immature to early mature, while Kumo is immature. The dominance of high molecular weight odd *n*-alkanes (nC_{27}, nC_{29} and nC_{31}) (OEP > 1) in Kumo and its absence in Lafia-Obi coal samples both indicate that the organic matters are of terrestrial nature. Lack of high molecular weight odd *n*-alkanes dominance in Lafia-Obi (OEP ~ 1) is influenced by maturity. Ratio of Pr/Ph higher than 3 in all the samples indicate oxic to suboxic environment of deposition. Both Lafia-Obi and Lamza coals have attained an equilibrium value (60 %) for $22S/(22S + 22R)C_{31}\alpha\beta$ hopanes and are close to that of $20S/(20S + 20R)\alpha\beta\beta/(\alpha\beta\beta + \alpha\alpha\alpha)C_{29}$. The combined ratios of $20S/(20S + 20R)\alpha\alpha C_{29}$ cholestanes, $20S/(20S + 20R)\beta\beta C_{29}$ cholestanes, $\beta\alpha/(\beta\alpha + \alpha\beta)$ C_{30} hopanes, Ts/Ts + Tm C_{27} hopanes suggest that Lafia-Obi is higher in maturity (peak oil window), followed next by Lamza (early oil window) and then Kumo (immature).

Acknowledgements We thank the German Academic Exchange (DAAD) and European Association of Organic Geochemists (EAOG) for a study fellowship and travel grant to the first author.

Appendix I: Some Chemical Structures Cited

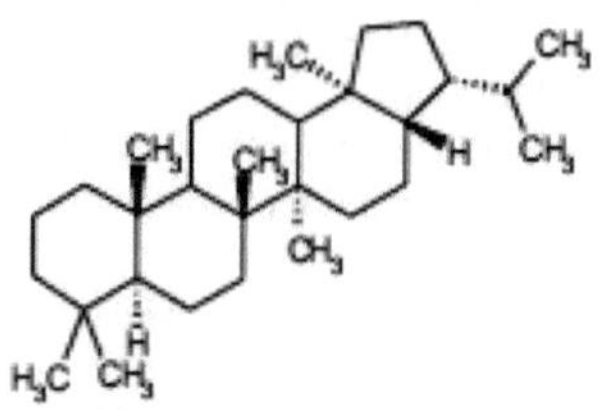

(a) 17β(H),21β(H)-Hopane

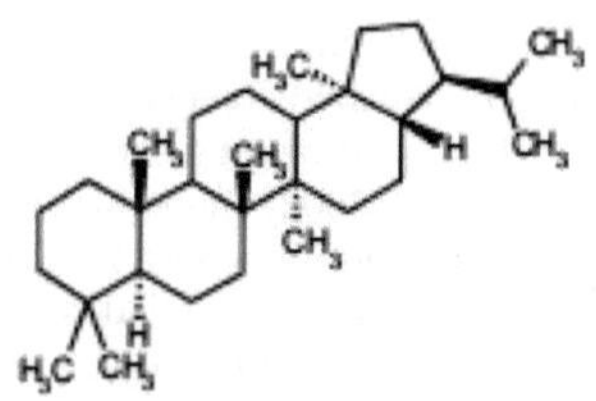

(b) 17β(H),21α(H)-Hopane

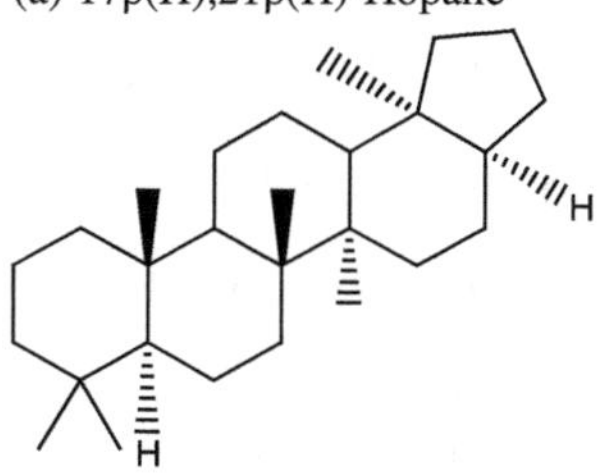

(c) 17α(H)-22,29,30-Trisnorhopane (Tm)

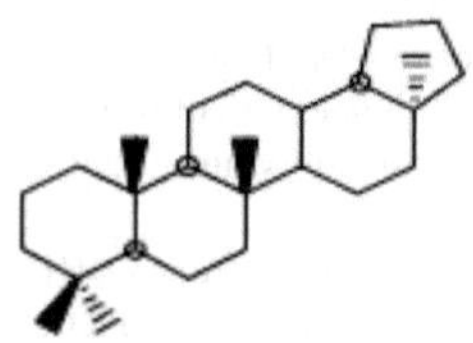

(d) 18α(H)-22,29,30-Trisnorneohopane (Ts)

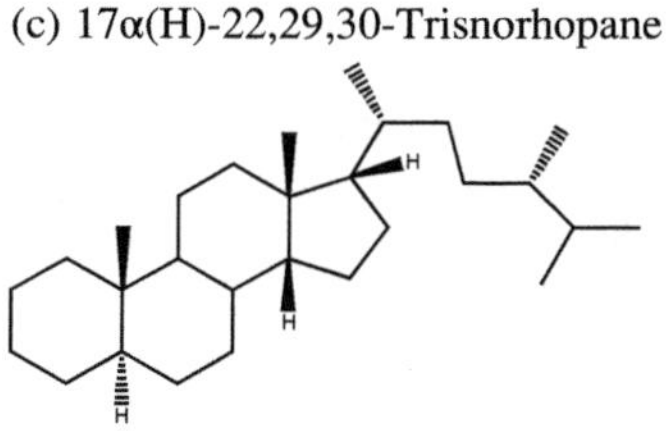

(e) 20*R*, 24*S* 5α(H),14β(H),17β(H)-Methylcholestane

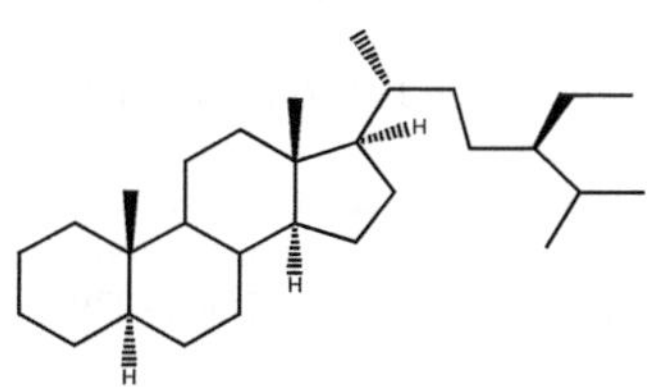

(f) 20*R*, 24*R* 5α(H),14α(H),17α(H)-Ethylcholestane

References

1. Durand B, Paratte M (1983) Oil potential of coals: a geochemical approach. In: Brooks J (ed) Petroleum geochemistry and exploration of Europe. Blackwell Scientific Publishing, Boston, pp 255–265
2. Horsfield B, Yordy KL, Crelling JC (1988) Determining the petroleum generating potential of coal using organic geochemistry and organic petrology. Org Geochem 13:121–129
3. Hendrix MS, Brassel SC, Carrol AR, Graham SA (1995) Sedimentology, organic geochemistry, and petroleum potential of Jurassic coal measures: Tarim, Junggar, and Turpan basins, northwest China. AAPG Bull 79:929–959
4. Petersen HI (2006) The petroleum generation potential and effective oil window of humic coal related to coal composition and age. Coal Geol 67:221–248
5. Obaje NG, Wehner H, Abubakar MB, Isah MT (2004) Nasara-I well, Gongola basin (Upper Benue trough, Nigeria): source-rock evaluation. Petrol Geol 27:191–206
6. Abubakar MB, Dike EFC, Obaje NG, Wehner H, Scheeder G, Jauro A (2008) Petroleum prospectivity of cretaceous formations in the Gongola basin, upper Benue trough, Nigeria. An organic geochemical perspective on a migrated oil controversy. Petrol Geol 31:387–408
7. Teerman SC, Hwang RJ (1991) Evaluation of the liquid hydrocarbon potential of coal by artificial maturation techniques. Org Geochem 17:749–764
8. Obaje NG, Wehner H, Scheeder G, Abubakar MB, Jauro A (2004) Hydrocarbon prospectivity of Nigeria's inland basins: from the viewpoint of organic geochemistry and organic petrology. AAPG Bull 88:325–353
9. Jauro A, Obaje NG, Agho MO, Abubakar MB, Tukur A (2007) Organic geochemistry of cretaceous Lamza and Chikila coals, upper Benue trough, Nigeria. Fuel 86:520–532
10. Benkhelil J (1989) The origin and evolution of cretaceous Benue trough, Nigeria. Afr Earth Sci 8:251–282
11. Weiss HM, Wilhems A, Mills N, Scotchmer J, Hall PB, Lind K, Brekke T (2000) The Norwegian industry guide to organic geochemical analyses [online]. Edition 4.0 Norsk Hydro Statoil, Geolab Nor. SINTEF Petroleum Research and the Norwegian Petroleum Directorate. http://www.npd.no/engelsk/nigoga/. Accessed 17 Mar 2006
12. Diessel CFK (1992) Coal-bearing depositional systems. Springer, Berlin/Heidelberg, p 721
13. Sykes R, Snowdon LR (2002) Guidelines for assessing the petroleum potential of coaly source rocks using rock-eval pyrolysis. Org Geochem 33:1441–1455
14. Peters KE, Cassa MR (1994) Applied source rock geochemistry. In: Magoon LB, Dow WG (eds) The petroleum systems from source to trap, AAPG Memoir 60. AAPG, Tulsa, pp 93–120
15. Hunt JM (1991) Generation of gas and oil from coal and other terrestrial organic matter. Org Geochem 17:673–680
16. Killops SD, Carlson RMK, Peters KE (2000) High temperature GC evidence for the early formation of C_{40+} *n*-alkanes in coals. Org Geochem 31:589–597
17. Illich HA (1983) Pristane, phytane and lower molecular weight isoprenoid distributions in oils. AAPG Bull 67:385–393
18. Tegelaar EW, Mathezing RM, Jansen JBH, Horsfield B, De Leeuw JW (1989) Possible origin of n-alkanes in high wax crude oils. Nature 342:529–531
19. Horsfield B (1997) The bulk composition of first-formed petroleum in source rocks. In: Welte DH, Horsfield B, Baker DR (eds) Petroleum and basin evolution. Springer, Berlin, pp 335–402
20. Didyk BM, Simoneit BRT, Brassel SC, Eglinton G (1978) Organic geochemical indicators of paleoenvironmental conditions of sedimentation. Nature 272:216–222
21. Haven HL, de Leeuw JW, Rullkötter J, Sinninghe DJS (1987) Restricted utility of the pristane/phytane ratio as a paleoenvironmental indicator. Nature 330:641–643
22. Amajiya H, Schwarzbauer J, Littke R (2006) Organic geochemistry of the Lower Suban coal seam, south Sumatra basin, Indonesia: paleoecological and thermal metamorphism implications. Org Geochem 37:261–279

23. Bray EE, Evans ED (1961) Distribution of *n*-paraffins as a clue to recognition of source bed. Geochim Cosmochim Acta 22:2–15
24. Charrié-Duhaut A, Lemoine S, Adam P, Connan J, Albrecht P (2000) Abiotic oxidation of petroleum bitumens under natural conditions. Org Geochem 31:977–1003
25. Murray AP, Sosrowidjojo IB, Alexander R, Kagi RI, Norgate CM, Summons RE (1997) Oleananes in oils and sediments: evidence of marine influence during early diagenesis? Geochim Cosmochim Acta 61:1261–1276
26. Karrer W, Cherbuliez E, Eugster CH (1977) Konstitution und Vorkommen der Organischen Pflanzenstoffe. Ergänzungsband I., Birkhaüser, Basel, Stuttgart
27. Ekweozor CM, Udo OT (1988) The Oleananes origin, maturation and limits of occurrence in southern Nigeria sedimentary basins. Org Geochem 13:131–140
28. Peters KE, Clifford CE, Moldowan JM (2005) The biomarker guide, 2nd edn. Prentice Hall, New Jersey
29. Borrego AG, Blanco CG, Püttmann W (1997) Geochemical significance of the aromatic hydrocarbons distribution in the bitumens of Puertollano oil shales, Spain. Org Geochem 26:219–228
30. MacKenzie AS, Patience RL, Maxwell JR, Vandenbroucke M, Durand B (1980) Molecular parameters of maturation in the Toarcian shales, Paris basin, France – I. Changes in the configuration of acyclic isoprenoid alkanes, steranes and triterpanes. Geochim Cosmochim Acta 44:1709–1721
31. Seifert WK, Moldowan JM (1980) The effects of thermal stress on source-rock quality as measured by hopane stereochemistry. Phys Chem Earth 12:229–237
32. Seifert WK, Moldowan JM (1986) Use of biological markers in petroleum exploration. In: Johns RB (ed) Methods in geochemistry and geophysics, Elsevier, Amsterdam, pp 261–290
33. Kolaczkowska E, Shougui NE, Watt DS, Maruca RE, Moldowan JM (1990) Thermodynamic stability of various alkylated, dealkylated and rearranged 17α- and 17β- hopane isomers using molecular mechanics calculations. Org Geochem 16:1033–1038
34. Mario GG, Ronald CS, Milton LL (1997) Generation and expulsion of petroleum and gas from almond formation coal, Greater Green River Basin, Wyoming. AAPG Bull 81:62–81
35. Farrimond P, Taylor A, Talnaes N (1998) Biomarker maturity parameters. The role of generation and thermal degradation. Org Geochem 29:1181–1197
36. Littke R, Lückge A, Wilkes H (1998) Organic matter in Neogene sediments of the Southern Canary Channel, Canary Island (Sites 955 and 956). In: Weaver PPE, Schminke HU, Firth JV, Duffield W (eds) Proceedings of the ODP scientific results, College Station, Texas, pp 157–372

Chapter 6
Risk Assessment and Toxic Effects of Exposure to Nanoparticles Associated with Natural and Anthropogenic Sources

Atar S. Pipal, Ajay Taneja, and Gautam Jaiswar

Abstract Humans have been exposed to airborne particles, especially to nanoscale particles, during their developing period, but this exposure has increased enormously over the last century due to anthropogenic sources. This increase is due mainly to combustion processes where these particles are released to the environment unintentionally but the rapidly growing field of nanotechnology is likely to become yet another source for such very small particles. Nanoparticles are natural products but their tremendous commercial use has encouraged the synthesis of these fine and ultrafine particles. Accelerated production of these particles may lead to increased risks to humans and the environment. Therefore, a detailed understanding of their sources, release interaction with environment and possible risk would provide a basis for safer use of engineered nanoparticles with minimal or hazardous impact on environment. Each of these particles carries different blends of chemicals and therefore poses a different health risk. Keeping all these points in mind the present work provides up to date information on sources, different types of interaction with environment and possible strategies for risk management of nanoparticles. An understanding makes a major contribution to the risk management that is needed to ensure that utilization of nanoparticles is made safely, is exploited to their full potential and disposed off safely.

6.1 Introduction

Airborne nano and microparticles are important for atmospheric chemistry and physics. It is also important for the biosphere, climate and public health. They influence the Earth's energy balance, the hydrological cycle and atmospheric circulation and the abundance of greenhouse and reactive trace gases. Moreover, they

A.S. Pipal • A. Taneja • G. Jaiswar (✉)
Department of Chemistry, Dr. B. R. Ambedkar University,
NH-2, Civil lines, 282002 Agra, India
e-mail: aspippal@gmail.com; ataneja5@hotmail.com; gjaiswar@gmail.com

M. Gupta Bhowon et al. (eds.), *Chemistry: The Key to our Sustainable Future*,
DOI 10.1007/978-94-007-7389-9_6,

play important roles in the reproduction of biological organisms and can cause or enhance diseases such as oxidative stress, hypertension, allergy and asthma. The primary parameters that determine the environmental and health effects of aerosol particles are their size, composition and concentration. These parameters, however, are spatially and temporally highly variable.

Particulate matter (PM) is a complex mixture of natural or geological particulates, natural soot or carbonaceous matter, and a host of anthropogenic or man-made particulate matter which includes a plethora of combustion particulate matter (including from power generation, agriculture burning, transportation, erosion and wear debris from paints and coatings, fire and pavement wear, other secondary reaction products and a wide range of biogenic PM). These PM include nanoparticulates ranging from one to several hundred nanometers in diameter to primary particulates. Aggregates are categorised and they are defined as ultrafine, fine and coarse PM as $< PM_{0.1} < PM_1 < PM_{10}$ [1, 2], whereas very small particles with size ranging 1–100 nm are called nanoparticles. Nanomaterial means an insoluble or biopersistent and intentionally manufactured material with one or more external dimension or an internal structure, on the scale in the range 1–100 nm [3]. According to EU the nanoparticles are defined as a natural, incidental or manufactured material containing particles, in an unbound state or as an aggregate or as an agglomerate and where, for 50 % or more of the particles in the number size distribution, one or more external dimensions is in the size range 1–100 nm [4]. It has the ability to enter translocation within and damage living organisms. This ability results primarily from their small size, which allows them to percolate psychological barriers and travel within the circulatory systems of a host. Nanoparticles in the atmosphere may be present due to release directly from a source (primary emissions) as nanoparticles or formation as a result of reactions in the atmosphere (secondary emissions). Global pollution is estimated to cause roughly 40 % of deaths worldwide [5]. While ultrafine or nanoparticles account for <1 % of general outdoor mass abundance of PM, they account for a significant fraction (>90 %) of the PM abundance or number concentration [6].

Nanoparticles generally have been demonstrated to be associated with adverse respiratory health issues in contrast to the larger size PM, where the number concentrations are smaller while the corresponding mass concentrations are much greater [7, 8]. The air in different micro-environments already contains varying concentrations of different kinds of small airborne particles that are regarded as pollutants. Large scale epidemiological studies confirm the relationship between exposure to airborne particles and mortality in cardiopulmonary diseases. Increased attention is directed at airborne nanoparticles in our indoor environments. The reason for this is the increased application of nanotechnology for new materials possessing attractive mechanical, electrical, magnetic and optical properties. A large increase in such materials could, when they are handled, give rise to an increase in the concentration of nanoparticles in air in different environments. There is concern that there is a risk of health effects when humans inhale these particles [9–11] and also causing cardiovascular disease in humans, such as chronic obstructive pulmonary disease (COPD) and asthma in children and compromised adults

[12–14]. To identify nanoparticles, exposure scenario that may occur during the whole life cycle from production to use of manufactured nanomaterials is an important component of the overall risk characterization and management programme [15]. The characterizations of nanoparticles in terms of mass and number concentration, chemical composition, physical and chemical properties and their sources are necessary to identifying their effects on human health and environment. Therefore the present article provides information on sources, different types, interactions with environment, chemical composition and possible strategies for risk management of nanoparticles.

6.2 Types of Nanoparticles

Figure 6.1 shows the size and scale of nanoparticles and based on dimensionality, morphology, composition, uniformity and agglomeration, the nanoparticles are classified into the following categories:

- 1D nanoparticles (nano fibers, rods and carbon nanotubes (CNT))
- 2D nanoparticles (nanocoats and nanofilms)
- 3D nanoparticles such as polycrystals

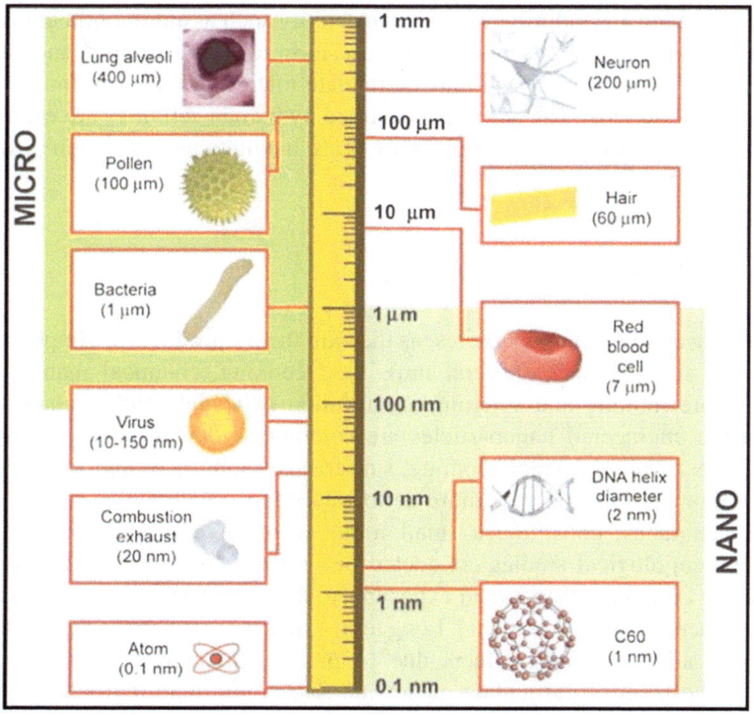

Fig. 6.1 Length *scale* showing size of nanomaterials compared to biological component [16]

- Nanotubes
- Nanowires
- Magnetic nanoparticles (dispersed aerosols, suspension/colloid)
- Engineering nanoparticles

6.3 Sources of Nanoparticles

The composition and mass concentrations of nanoparticles in clean background areas are strongly affected by long-range transport. During transport and aging, particles of different origin may change their properties due to coagulation and cloud processes as well as due to reactions with gases via various heterogeneous pathways [17]. The origins of these particles can be natural as well as anthropogenic.

6.3.1 Natural Sources

Nanoparticles are abundant in nature, as they are produced in many natural processes, including photochemical reactions, volcanic eruptions, forest fires, simple erosion and by plants and animals, like shedding of skin and hair. Though we usually associate air pollution with human activity such as automobile industry and charcoal burning, natural events such as dust storm, volcanic eruption and forest fire can produce such a quantity of nanoparticulate matter that they profoundly affect air quality worldwide. The aerosols generated by human activities are estimated to be about 10 % of the total, the remaining 90 % having a natural origin [18].

6.3.2 Anthropogenic Sources

Humans have created nanoparticles, as they are the by-products of simple combustion (with sizes down to several nm), food cooking, chemical manufacturing, welding, ore refining and smelting, combustion in vehicle and airplane engines [19], while engineered nanoparticles are commonly used in cosmetics, sporting goods, tyres, stain-resistance clothing, sunscreens, toothpastes and food additives. These nanomaterials and new more deliberately fabricated nanoparticles, such as carbon nanotubes, constitute a small minority of environmental nanomaterials [20]. Epidemiological studies conducted on diesel locomotive drivers showed a correlation between occupational exposures to diesel engine exhaust and incidence of lung cancer in the workers [21]. Long term exposure to indoor cooking emissions may pose adverse health effects due to particulate matter inhalation. During cooking, the level of particulate matter increases more than tenfold compared to non-cooking hours [22]. In developing countries, approximately half of the world population and up to 90 % of rural households still used the unprocessed biomass fuels such as wood, dung and crop residues for cooking. In the energy ladder, these biomass fuels which are known as dirtiest fuels lie at the bottom of the energy ladder. It is also found that about 78 % of Indian population relied upon the biomass fuels and 3 % on coal for cooking. Moreover, the stoves or chullas which are used for cooking are not energy efficient and the fuels are not burned completely. The incomplete combustion of biomass releases complex mixtures of organic compounds such as particulate matter, carbon monoxide, polyorganic material, polyaromatic hydrocarbons and formaldehyde. It is estimated that about half a million women and children die each year from indoor air pollution which are emitted from biomass fuels used for cooking in India. On comparing these data to other countries, India has among the largest burden of diseases due to the use of dirty household fuels and 28 % of all deaths due to indoor air pollution in developing countries occur in India [23–25].

In many regions of the world, death caused from indoor smoke from solid fuels is considerable, especially in Asia and Africa. World Health Organization estimates that more than 50 % of the world population uses solid fuels for cooking and heating, including biomass fuels [26].

6.4 Nanotoxicology

Human skin, lungs and the gastro-intestinal tract are in constant contact with the environment. The lungs and the gastro-intestinal tract are more vulnerable while the skin is generally an effective barrier to foreign substances. These three body parts are the most likely points of entry for natural or anthropogenic nanoparticles. Injections and implants are other possible routes of exposure, primarily limited to engineered materials. The possible adverse health effects associated with inhalation, ingestion, and contact with nanoparticles depend on various factors including size, aggregation, crystalline and surface fictionalization. Figure 6.2a, b show the publications in the field of nanomaterials or ultrafine particles and their toxicity.

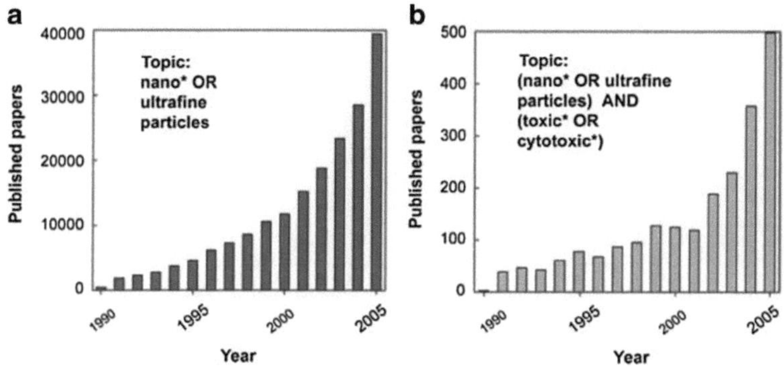

Fig. 6.2 (**a**) and (**b**) Statistics on scientific articles published on (**a**) nanomaterials and (**b**) their toxicity (ISI Web of Science) [16]

6.5 Characteristics of Particles

6.5.1 *Size and Agglomeration*

Nanoparticles such as engineered nanoparticles, unintentionally created nanoparticles, and ambient ultrafine particles are smaller than 100 nm in diameter (Fig. 6.3). Their physical and chemical characteristics are as varied as the processes that create the particles [22]. As the particle size becomes smaller, a greater fraction of atoms are at the surface and quantum effects tend to increase the surface reactivity. At the same time, nanoparticles have a tendency to agglomerate and form larger structures. Thus, agglomeration can lead to a reduction in the number of atoms at the surface with a reduction in surface energy.

6.5.2 *Particle Shape*

Prior experience with asbestos and other fibrous aerosols indicates that the shape of the particles (i.e. their length and diameter) has a profound effect on toxicity. Smaller diameter fibers penetrate deeper into the respiratory tract, while longer fibers are cleared more slowly [27, 28]. Engineered nanoparticles come in various

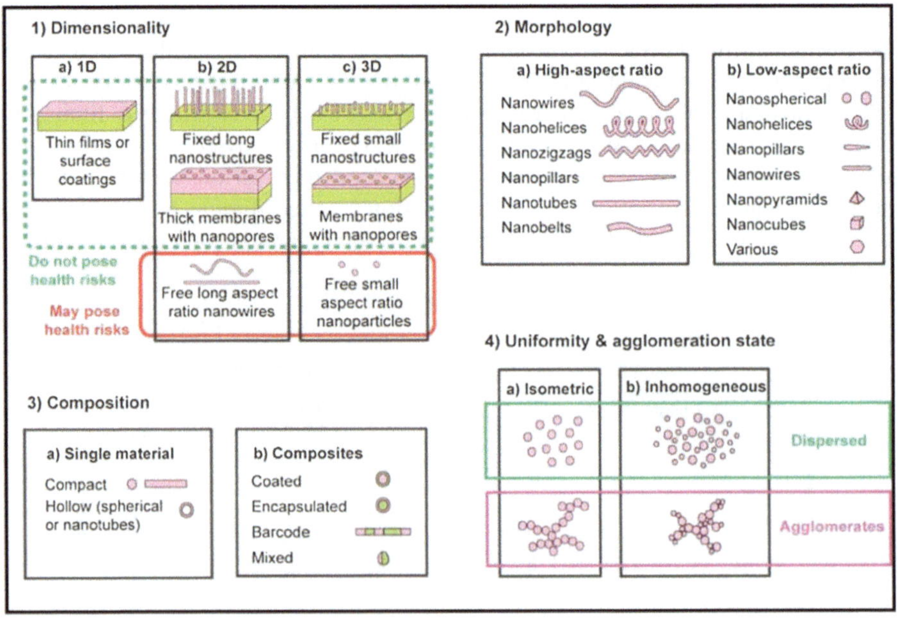

Fig. 6.3 Classification of nanostructure materials based on dimensions, morphology, composition, uniformity and agglomeration state [16]

shapes such as spheres (e.g. dendrimers), tubes (e.g. single wall carbon nanotubes (SWCNT) and multi wall carbon nanotubes (MWCNT)), plates (e.g. nanoclay flakes), fullerenes, and needles. In the human body it seems that the deposition, fate and toxicity of the particles will be affected by the shape of the particles [29].

6.5.3 Chemical Composition

The chemical composition of particulate matter is an essential information for assessment of its sources and health effects [26]. The chemical composition of the surface and the bulk of engineered nanoparticles will affect toxicity (Fig. 6.3). Surface coatings that modify the agglomeration properties of nanoparticles can also have biological effects [30, 31]. Experiments using fullerene soot with different impurities (e.g. metallic endohedral fullerene) indicate that the pulmonary toxicity response depends on the type of nanomaterials and their impurities [32].

6.6 Importance and Interaction of Nanoparticles in Environments

The risk associated with exposure to engineered nanomaterials (ENM) will be determined in part by the environmental processes that control fate, transport and transformation. The toxicity of environmental nanoparticles (ENP) depends upon the exposure levels of the nanomaterials during transformation from one ecosystem to another ecosystem. Atmospheric nanoparticles are of growing interest to many investigators for two main reasons. First, nanoparticles are important precursors for the formation of larger particles, which are known to strongly influence global climate, atmospheric chemistry, visibility, and the regional and global transport of pollutants and biological nutrients. Second, atmospheric nanoparticles may play critical roles in the deleterious human health effects associated with air pollution. In addition to these two well-recognized roles, nanoparticles may also significantly influence the chemistry of the atmosphere. Because their composition and reactivity can be quite different from larger particles, the presence of nanoparticles may open novel chemical transformation pathways in the atmosphere. There are possibilities that an important role for nanostructures can be visualized within larger particles, which often contain nano-scale features such as mineral grain agglomerates, soot spherules, or layer coatings of sulfates and nitrates. These complex morphological features likely influence a number of properties. For example, nanostructures probably affect water uptake via capillary condensation and nanoscale aqueous surface films may provide a medium for heterogeneous chemistry. In addition, nanoscale active sites on surfaces may influence particle phase transitions through heterogeneous nucleation.

However, with a few notable exceptions, the potential roles and implications in atmospheric chemistry for nanoparticles and nanostructures have not been quantitatively examined.

6.7 Risk Assessment of Nanoparticles

Research on the effects of particulate matter on human health was initiated in the 1970s [33] and has been strongly developed since then. It indicates that small particles cause serious problems to human health, contributing to increased mortality and sickness [34]. The mechanisms underlying these adverse effects are not well understood and major questions concerning the species size fraction, chemical composition, and causative mechanisms leading to the observed health effects remain [35]. This is partly due to results from epidemiological surveys showing that there is a correlation between the mass concentration of particles and mortality and hospitalization [36]. From a health perspective, the size of the particles is important as it affects their ability to penetrate into the lungs and cause adverse health effects. The larger particles (e.g. those greater than 10 μ) tend to settle in the nose and mouth and are unlikely to pose a health risk [37]. However, ultrafine particles (i.e. diameters less than 0.1 μm) are considered especially detrimental to human health, since these particles can be inhaled and deposited deep in the alveoli of human lungs [38]. It has been increasingly recognized through epidemiological investigations that particulate matter in agricultural air contributes to the progression and exacerbation of respiratory diseases (such as asthma and other ailments), and in urban air leads to an increase in morbidity and mortality from respiratory and cardiac conditions [39]. Children's susceptibility to health problems as a result of exposure to air pollution is of concern. Children may receive an increased dose of particulate matter to their lungs compared to adults [40]. Moreover, PM has been linked to cancer and premature death [41].

Nanoparticles have been reported to potentially induce adverse health effects [6, 42, 43] and the need for thorough risk assessment has been identified [44]. The risk of nanoparticles is not only a function of their potential hazard, but also of exposure thereto. Assessing the exposure to the nanoparticles at workplaces during nanoparticles production is therefore one important step towards a sustainable nanotechnology [45].

A health risk assessment has to consider data from various lines of evidence (e.g. human epidemiological and clinical studies, experimental animal and in vitro studies, *in silico* studies) and integrate these into a cohesive evaluation. It is furthermore essential to have relevant information on exposure. A risk can then be deduced from exposure data together with the hazard assessment. Needless to say, the assessment becomes more reliable when more relevant information is available. Data on assessment of human exposure to ENPs are very sparse. However, there is at present very little reason to expect that the general public is exposed to any significant amounts of air-borne ENPs, although ENPs are present in certain consumer products. It is more likely that occupational exposures can be a factor in at least some settings.

6.8 Conclusions

The aim of the present article is to provide updated information on physico-chemical properties, sources and risk assessment of atmospheric nanoparticles. Apart from this information it also provides a survey of what is currently known about nanoparticles in the atmosphere, especially in terms of their formation and growth, number concentrations and chemical composition, and chemical, physical, and mechanical properties. Although combustion process in microenvironments is an important source of nanoparticles, the production of nanoparticles may cause hazardous biological effects. Therefore, detailed understanding of their sources, release interaction with environment and possible risk assessment would provide a basis for safer use of engineered nanoparticles with minimal or hazardous impact on environment. Uncertainties in conventional quantitative risk assessment typically relate to values of parameters in risk models. For many environmental contaminants, there is a lack of sufficient information about multiple components of the risk assessment framework. In such cases, the use of default assumptions and extrapolations to fill in the data gaps is a common practice. Nanoparticles risks, however, pose a new form of risk assessment challenge. Besides a lack of data, there is deep scientific uncertainty regarding every aspect of the risk assessment framework: (a) particle characteristics that may affect toxicity; (b) their fate and transport through the environment; (c) the routes of exposure and the metrics by which exposure ought to be measured; (d) the mechanisms of translocation to different parts of the body; and (e) the mechanisms of toxicity and disease. In each of these areas, there are multiple and competing models and hypotheses. These are not merely parametric uncertainties but uncertainties about the choice of the causal mechanisms themselves and the proper model variables to be used, i.e. structural uncertainties.

The present state of knowledge is unsatisfactory for a proper risk assessment of nanoparticles and improvements of the study qualities as well as an increased number of relevant studies are strongly recommended.

References

1. Murr LE, Garza KM (2009) Natural and anthropogenic environmental nanoparticulates: their microstructural characterization and respiratory health implications. Atmos Environ 43:2683–2692
2. Massey DD, Masih J, Taneja A (2010) Real room measurements of fine particles in the range of 250–1000 nanometers in photocopier and printer centers in India. In: Proceedings of the international conference on nanoscience and technology in chemistry, health, environment, and energy, Tata McGraw Hill, Agra, pp 296–300
3. Jensen KA, Koponen LK, Clausen PA, Schneider T (2009) Dustiness behaviour of loose and compacted Bentonite and organoclay powders: what is the difference in exposure risk. J Nanopart Res 11:133–146

4. Becher S, Soukup JM, Sioutas C, Cassee FR (2003) Response of human alveolar macrophages to ultrafine, fine and coarse urban air pollution particles. Exp Lung Res 29:29–44
5. Pimentel D, Cooperstein S, Randell H, Filiberto D, Sorrentino S, Kaye B (2007) Ecology of increasing diseases: population growth and environmental degradation. Hum Ecol 35:653–668
6. Oberdorster G (2001) Pulmonary effects of inhaled ultrafine particles. Int Arch Occup Environ Health 74:1–8
7. Englert N (2004) Fine particles and human health. Toxicol Lett 149:235–242
8. Pope CA, Dockery DW (2006) Health effects of fine particulate air pollution: lines that connect. J Air Waste Manag Assoc 56:709–742
9. Frampton MW (2001) Systematic and cardiovascular effects of airway injury and inflammation: ultrafine particle exposure in humans. Environ Health Perspect 109:S529–S532
10. Donaldson K, Tran CL (2002) Inflammation caused by particles and fibers. Inhal Toxicol 14:5–27
11. Nel A (2005) Atmosphere. Air-pollution-related illness: effects of particles. Science 308:804–806
12. D'Amato G, Liccordi G (1998) Outdoor environmental injury of the airways and development of allergic respiration diseases. Pulm Pharmacol 11:369–374
13. Johnson RL Jr (2004) Relative effects of air pollution on lungs & heart. Circulation 109:5–7
14. Koenig JQ, Mar TF, Allen RW (2005) Pulmonary effects of indoor and outdoor generated particles in children with asthma. Environ Health Perspect 113:499–503
15. Schulte P, Geraci C, Zumwalde R, Hoover M, Kuempel E (2008) Occupational risk management of engineered nanoparticles. J Occup Environ Hyg 5:239–249
16. Buzea C, Blandino IIP, Robbie K (2007) Nanomaterials and nanoparticles: sources and toxicity. Biointerphases 2:MR17–MR172
17. Poschl U (2005) Atmospheric aerosols: composition, transformation, climate and health effects. Angew Chem Int Ed 44:7520–7540
18. Niemi JV, Saarikoski S, Tervahattu H, Makela T, Hillamo R, Vehkamaki H, Sogacheva L, Kulmala M (2006) Changes in background aerosol composition in Finland during polluted and clean periods studied by TEM/EDX individual particle analysis. Atmos Chem Phys 6:5049–5066
19. Taylor DA (2002) Dust in the wind. Environ Health Perspect 110:A80–A87
20. Rogers F, Arnott P, Zielinska B, Sagebiel J, Kelly KE, Wagner D, Lighty JS, Sarofim AF (2005) Real time measurements of jet aircraft engine exhausts. J Air Waste Manag Assoc 55:583–593
21. Bigert C, Gustavsson P, Hallqvist J, Hogstedt C, Lewne M, Plato N, Reuterwall C, Scheele P (2003) Myocardial infarction among professional drivers. Epidemiology 14:333–339
22. See SW, Balasubramanian R (2006) Risk assessment of exposure to indoor aerosols associated with Chinese cooking. Environ Res 102:197–204
23. World Health Organisation (1999) Air Management Information System (AMIS). World Health Organisation, Geneva
24. Smith KR (2002) Indoor air pollution in developing countries: recommendations for research. Indoor Air 2:198–207
25. National Family Health Survey (1995) National health family survey (MCH) and family planning: India. International Institute for Population Sciences, Bombay
26. Garshick E, Schenker MB, Mumoz A, Segal T, Smith TJ, Woskiw S, Hammond SK, Speizer FE (1988) A retrospective cohort study of lung cancer and diesel enhaust exposure in railroad workers. Am Rev Respir Dis 137:820–825
27. Balakrishnan K, Mehta S, Kumar P, Ramaswamy PSS, Kumar SK (2004) Indoor air pollution associated with household fuel use in India, an exposure assessment and modeling exercise in rural districts of Andhra Pradesh, India. The World Bank Report, Washington, DC
28. Pipal AS, Kulshrestha A, Taneja A (2011) Characterization and morphological analysis of airborne $PM_{2.5}$ and PM_{10} in Agra located in north central India. Atmos Environ 45:3621–3630

29. Mossman BT, Bignon J, Corn M, Seaton A, Gee JBL (1990) Asbestos: scientific developments and implications for public policy. Science 247:294–301
30. Fubuni B, Aust AE, Bolton RE, Borm PJ, Bruch J, Ciapett G (2008) Non-animal tests for evaluating the toxicity of solid xenobiotics. Altern Lab Anim 26:579–615
31. Jia G, Wang H, Yan L, Wang X, Pel R, Yan T, Zhao Y, Guo X (2005) Cytotoxicity of carbon nanomaterials: single wall nanotube, multiwall nanotube, and fullerene. Environ Sci Technol 39:1378–1383
32. Oberdorster G, Oberdorster E, Oberdorster T (2005) Nanotoxicology: an emerging discipline evolving from studies of ultrafine particles. Environ Health Perspect 113:823–839
33. Warheit DB, Brock WJ, Lee KP, Webb TR, Reed KL (2005) Comparative pulmonary toxicity inhalation and instillation studies with different TiO2 particle formulations: impact of surface treatments on particle toxicity. Toxicol Sci 88:514–524
34. Derbyshire E (2005) Natural aerosolic mineral dusts and human health. In: Selinus B, Allowway B, Centebo JA, Finkelman RB, Ruge R, Lindh U, Smedley P (eds) Essentials of medical geology. Elsevier Academic Press, London, pp 459–480
35. Ferris BG, Speizer FE, Spengler JD, Dockery D, Bishop YMM, Wolfson M, Humble C (1979) Effects of sulfur oxides and respirable particles on human health. Methodology and demography of populations in study. Am Rev Respir Dis 120:767–779
36. Shi ZB, Shao LY, Jones TP, Whittaker AG, Lu SL, Berube KA, He T, Richards RJ (2003) Characterization of airborne individual particles collected in an urban area, a satellite city and a clean air area in Beijing, 2001. Atmos Environ 37:4097–4108
37. Sondreal EA, Benson SA, Pavlish JH, Ralston NVC (2004) An overview of air quality III: mercury, trace elements, and particulate matter. Fuel Process Technol 85:425–440
38. Ketzel M, Wahlin P, Kristensson A, Swietlicki E, Berkowicz R, Nielsen OJ, Palmgren F (2004) Particle size distribution and particle mass measurements at urban, near-city and rural level in the Copenhagen area and southern Sweden. Atmos Chem Phys 4:281–292
39. Carbon B (2004) Health effects of PM10 in New Zealand, Air quality technical report no. 39–2003. Ministry for the Environment, Wellington
40. Mueller-Anneling L, Avol E, Peters JM, Thorne PS (2004) Ambient endotoxin concentrations in PM10 from southern California. Environ Health Perspect 112:583–588
41. Wheeler AJ, Williams I, Beaumont RA, Hamilton RS (2000) Characterisation of particulate matter sampled during a study of children's personal exposure to airborne particulate matter in a UK urban environment. Environ Monit Assess 65:69–77
42. Shah AP, Pietropaoli AP, Frasier LM, Speers DM, Chalupa DC, Delehanty JM, Huang L-S, Utell MJ, Frampton MW (2008) Effect of inhaled carbon ultrafine particles on reactive hyperemia in healthy human subjects. Environ Health Perspect 116:375–380
43. Donaldson K, Li XY, MacNee W (1998) Ultrafine (nanometre) particle mediated lung injury. J Aerosol Sci 29:553–560
44. Maynard AD, Kuempel ED (2005) Airborne nanostructured particles and occupational health. J Nanopart Res 7:587–614
45. Fissan H (2008) Nachhaltige Nanotechnologie, Nordrhein-Westfalische Akademie der Wissenschaften-Vortrage I 21. Frediand Schoningh, Paderborn. ISBN 978-3-506-76565-9

Chapter 7
Immunomodulatory Activity of Phenolic Fraction from *Piper Borbonense* and *Cassytha Filiformis* Growing in Comoros Islands

Said H. Soidrou, Dalila Bousta, Mohammed Lachkar, Said O.S. Hassane, Amal El Youbi-Hamsas, Latifa El Mansouri, Jamal Benjilali, Hanane El-Hajaji, and Abdellah Farah

Abstract *Cassytha filiformis* L. and *Piper borbonense* C. DC. are used as ethnical medicine for diverse diseases in Comoros islands. This study aims to evaluate the potential immunomodulator properties ex-vivo of the phenolic fraction provided of the methanol extract of these plants by flow cytometer. The results highlighted an immunosuppressive effect on lymphocyte subpopulations for all doses administered intraperitoneally (i.p.) compared to the control group. The most immunosuppressive activity was observed at doses of 25 and 100 mg/kg b.w. respectively for *Cassytha filiformis* and *Piper borbonense*. However, an interesting immunostimulant activity was noted on monocytes for all doses and on granulocytes for doses of 50 and 100 mg/kg b.w. in treated group with *Cassytha filiformis*. The same result was observed for *P. borbonense* on granulocytes for all doses. This immunostimulant effect observed on granulocytes can suggest a possible anti-inflammatory potency of these plants and an eventual use of these plants in the treatment of inflammatory symptoms diseases.

S.H. Soidrou • D. Bousta (✉) • A. El Youbi-Hamsas • L. El Mansouri • J. Benjilali • A. Farah
National Institute of Medicinal and Aromatic Plants, Taounate, PAMSN, University Sidi Mohamed Ben Abdellah, Fez, Morocco
e-mail: mizidjou.ssh@hotmail.com; boustadalila@gmail.com; amal.elyoubi@gmail.com; lati64mn@hotmail.com; mc_org@hotmail.fr; hybride_farah@yahoo.fr

M. Lachkar • H. El-Hajaji
Faculty of Sciences Dhar el Mehraz, LIMOM, University Sidi Mohamed Ben Abdellah, Fez, Morocco
e-mail: lachkar.mohammed@gmail.com; elhajaji.hanane@gmail.com

S.O.S. Hassane
Faculty of Sciences and Technology, University of Comoros, BP 2585 Moroni, Comoros
e-mail: said_omar2000@yahoo.fr

M. Gupta Bhowon et al. (eds.), *Chemistry: The Key to our Sustainable Future*,
DOI 10.1007/978-94-007-7389-9_7,

7.1 Introduction

Utilization of natural products for their therapeutic properties is as old as human civilization. Approximately 80 % of the world's population uses plants for their primary care [1]. Many of these natural products have been shown to present interesting biological and pharmacological activities and are used as chemotherapeutic agents or serve to the development of modern medicines [2, 3]. Plants are recognized for their ability to produce a wealth of secondary metabolites and mankind has used many species for centuries to treat a variety of diseases [4]. Many natural plant products and synthetic compounds are known for their capacity to modulate immune responses [5–9]. Immune system plays a vital role as the main line of defense. Further, its efficiency and integrity are important in chemotherapeutic interventions for treatment of many diseases [10]. However, various diseases which we are confronted daily affect and weaken our immune system. In order to reduce inflammatory response and damage induced to the human body, many efforts have been made in looking for natural immunomodulatory products [11].

In Comoros, traditional medicine is considered as an integral part of ancient traditions and is passed down orally from generation to generation. There is very little written information on the traditional plants used in this folk medicine [12]. As part of our work on upgrading and diversifying the use of medicinal plants used in Comoros, we are interested in *Cassytha filiformis* and *Piper borbonense* plants known locally as "Irundakanga" and "Gwe la rambuwu msiru" respectively.

C. filiformis is used for several diseases. In the Indian Ocean, it is considered as one of the 215 plants most used in folkloric medicine [13]. In Comoros, it is largely used for dermal problems especially for treatment of scabies [14]. Our ethnobotanical investigation showed the use of this plant to treat inflammation and diarrhea. In Nigeria, this plant is used to treat diabetes, ulcers, venereal discharge, hemorrhoids and cough [15–18]. It is also used in Polynesia against cancers [19].

Twigs of *P. borbonense* are largely used to treat the tooth diseases in Comoros. The leaves are used against diseases like diabetes, malaria, intestinal worms, and diarrhea. *P. borbonense* was less studied. In Reunion island, it is used as antidiarrheic and diuretic [20].

Supported by the INCO-FP7 MAP2ERA project, this work aims to investigate the phytochemical screening and dosage of phenolic and flavonoids compounds. In the present work, we have evaluated the immunomodulatory effects of phenolic fraction from *Cassytha filiformis* and *Piper borbonense* in rats.

7.2 Methodology

7.2.1 Plant Material

Aerial part of *C. filiformis* from Mbeni Hamahamet (N/E of Ngazidja) and twigs of *P. borbonense* were collected from Hantsongoma forest (950 m altitude, Oichili province, NE of Ngazidja, Comoros) in September 2009 and identified by Professor Jean-Noël LABAT from Musée National d'Histoire Naturelle (Paris, France). Voucher specimens were deposited in the Herbarium of Botany Department of the Faculty of Science and Technology, University of Comoros.

7.2.2 Extract Preparation

Dried and ground leaves of plants materials were successively extracted with hexane, ethyl acetate, dichloromethane and methanol in soxhlet. Phenolic fraction was obtained using method described by Maiza-Benabdesselam et al. [21] with minor modifications. Methanol extract was concentrated and recovered by an aqueous solution of 5 % HCl, and then filtered. The residue was dried and stored for different tests.

7.2.3 Phytochemical Screening

Flavonoids, tannins, alkaloids, saponosids and total phenolic constituents were determined according to the work of Hajaji et al. [22]. Total flavonoids contents were determined by spectrophotometry using method described in the work of Salem with some modifications [23].

7.2.4 Animals

Male Wistar rats (100–150 g) were obtained from ECWP (Emirates Center for Wildlife Propagation, Missour, Morocco). They were kept in well-ventilated environment and had a free access to water and food ad libitum and were housed in a quiet room under a "12-h light: 12-h dark" cycle for 2 weeks before experimentation.

7.2.5 Flow Cytometer

The animals were anesthetized by "i.p." with sodium pentobarbital at the dose of 30 mg/kg, 24 h after treatment with the phenolic fraction of *C. filiformis* and *P. borbonense*. The blood was collected in heparinized tubes, and then subjected to analysis by FCM (flow cytometer Epics-XL MCL type). FCM was used to evaluate the proportion of leukocyte sub-populations. This is based on cell morphology, size and structure.

7.2.6 *Statistical Analysis*

Data were expressed as the mean ± SEM. Comparisons of means were performed by using the student *t*-test. The level of statistical significance was set at $p < 0.05$.

7.3 Results and Discussion

7.3.1 *Phytochemical Screening*

As shown in Table 7.1, phytochemical screening of residue of *C. filiformis* and *P. borbonense* revealed the presence of flavonoids and tannins in both plants. However, alkaloids and saponosids were not detected. The total phenolic content (mg/l) was determined from regression calibration curve ($y = 0.0005x + 0.131$; R^2: 0.9392) and expressed in gallic acid equivalent (GAE). The result was calculated and found to be 142 and 34 mg/l GAE respectively for *C. filiformis* and *P. borbonense*.

The total flavonoids content was also determined from regression calibration curve ($y = 0.0224x + 0.6183$; $R^2 = 0.86$) and expressed in quercetin equivalent (QE). The results calculated were 59 and 80 mg/l QE respectively for *C. filiformis* and *P. borbonense*.

Presence of flavonoids in *C. filiformis* was confirmed by Tsai et al. [24], but the major chemical family characterized in this plant was alkaloids aporphins [16, 24, 25].

However no phytochemical description was found in the literature about *P. borbonense*.

7.3.2 *Immunomodulatory Effect*

The immunomodulatory effect of extract was studied by evaluation of relative percent of leukocyte subpopulations in flow cytometer. We noted an immunosuppressive effect of lymphocytes subpopulation at all doses tested of *C. filiformis* and *P. borbonense*.

For *C. filiformis*, the dose of 25 mg/kg b.w. (9.12 %) was the most active compared to the other doses ($p < 0.001$). Further a significant difference was observed when we compare this dose and the group treated by prednisolone

Table 7.1 Phytochemical screening of phenolic fraction of *C. filiformis* and *P. borbonense*

	Alkaloids	Flavonoids	Saponosids	Tannins
C. filiformis	–	+	–	++
P. borbonense	–	++	–	+

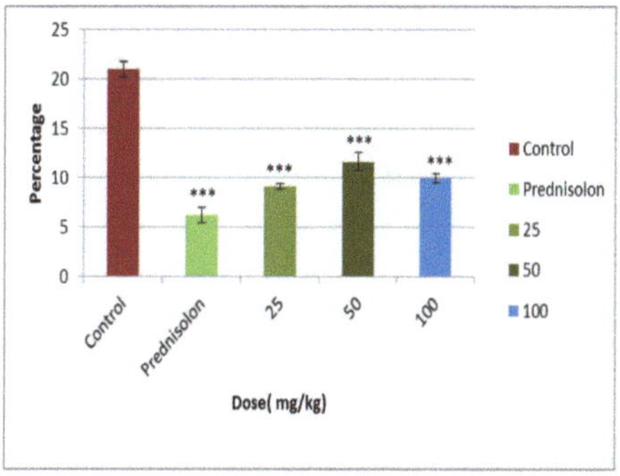

a. Lymphocytes

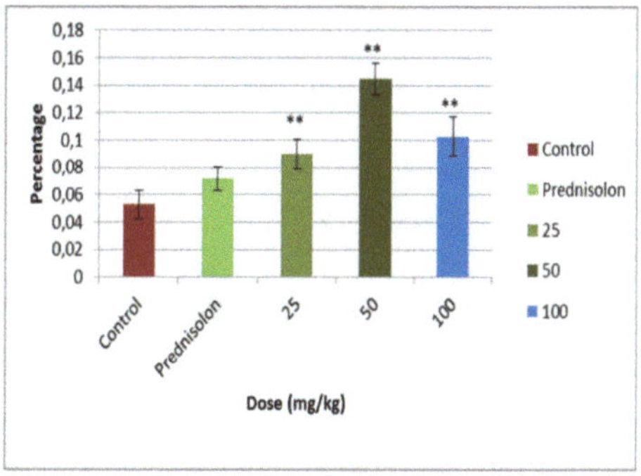

b. Monocytes

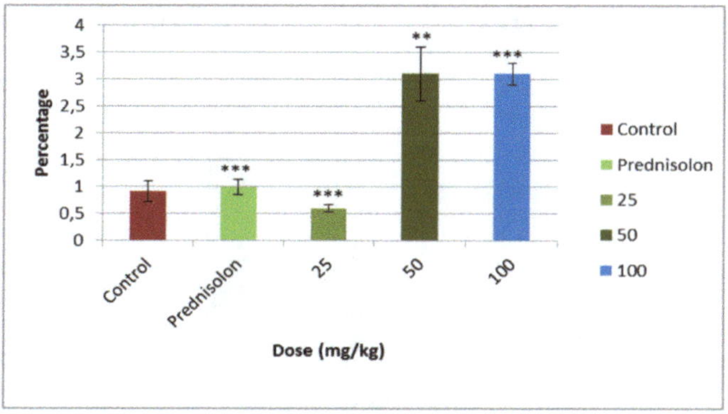

c. Granulocytes

Fig. 7.1 Effect of phenolic fraction of *C. filiformis* in leukocytes subpopulations of rats (**a**) Lymphocytes; (**b**) Monocytes; (**c**) Granulocytes. Values are expressed as mean ± SEM; n = 10 for each group (*$p < 0.05$; **$p < 0.01$; ***$p < 0.001$)

(6.23 %; $p < 0.001$). However we noted a significant enhancement of monocytes and granulocytes sub-populations percent at doses of 50 and 100 mg/kg b.w. compared to control and to the prednisolone groups (Fig. 7.1).

P. borbonense showed also an immunosuppressive activity on lymphocytes sub-population. Two doses were tested and 100 mg/kg was more immunosuppressive on lymphocytes and monocytes subpopulations. However, the results showed an immunostimulant effect on granulocytes percent for 50 and 100 mg/kg b.w. (Fig. 7.2). Thus, a probable anti-inflammatory effect of phenolic fraction of *C. filiformis* and *P. borbonense* could be suggested.

The immunosuppressive activity shown in this study could be related to the effect of flavonoids that was earlier demonstrated. In this context, genistein, an isoflavon isolated from *Genista tinctoria* showed a potential inhibition of T lymphocytes proliferation, IL-2 synthesis and IL-2 receptor expression [26]. Baicalein, another flavonoid, inhibited the growth of T-lymphoid leukemia cells and reduced the Protein Kinase C activity [27]. The work of Li et al. [28] demonstrated a significant suppressive effect on human neutrophil chemotaxis, mitogen-induced lymphocyte transformation, mixed lymphocyte culture, NK cell

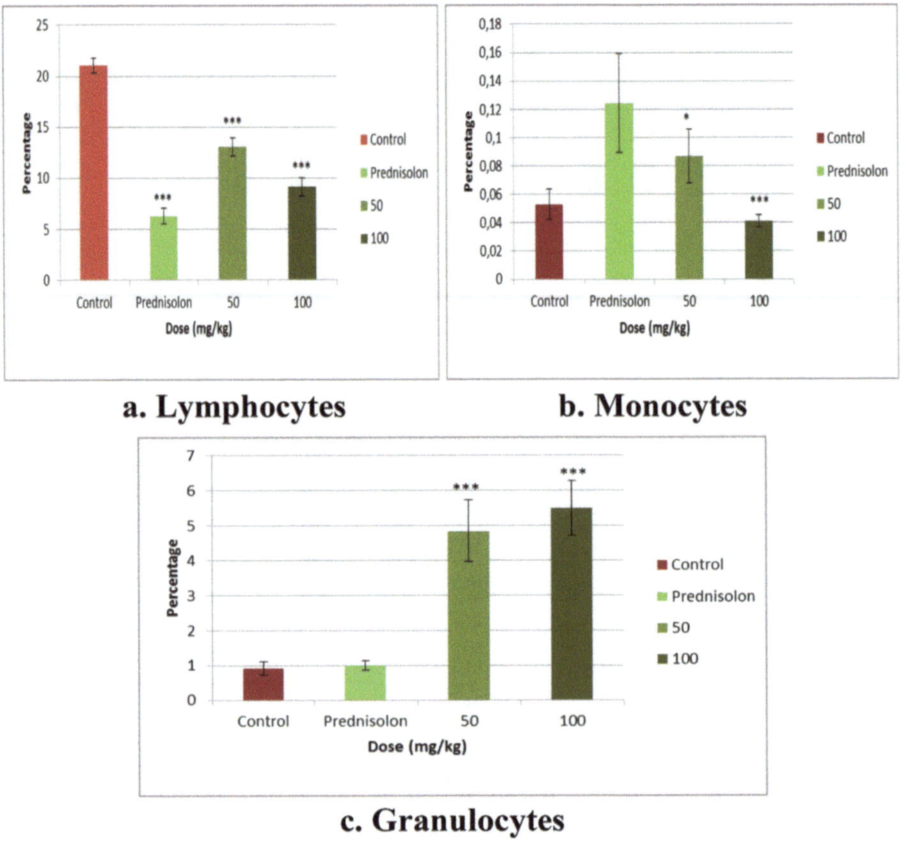

Fig. 7.2 Effect of phenolic fraction of *P. borbonense* in leukocytes subpopulations of rats (**a**) Lymphocytes; (**b**) Monocytes; (**c**) Granulocytes. Values are expressed as Mean ± SEM; n = 10 for each group (*$p < 0.05$; **$p < 0.01$; ***$p < 0.001$)

cytotoxic activity, and IL-2 synthesis by Baohuoside-1 (3,5,7-trihydroxy-4′-methoxy-8-prenylflavone-3-*O*-α-L-rhamnopyranoside). Other works noted that eupalitin-3-O-β-D-galactopyranoside, flavonoid extracted from *Boerhaavia diffusa*, blocked the activation of DNA binding of nuclear factor-κB and AP-1, two major transcription factors centrally involved in expression of the IL-2 and IL-2R gene, which are necessary for T cell activation and proliferation [29]. It is described also by Chang et al. [30] that centaurein and centaureidin, both flavonoids isolated from *Bidens pilosa*, induced cell death on T cells. Implication of tannins in immunosuppressive activity was discussed. In this sense, methyl gallate, tannin compounds isolated from *Acacia nilotica*, showed an immunosuppressive activity [31].

However reaction between various compounds of our extracts would also explain this immunomodulatory dose-dependent effect. In this study, the most immunosuppressive activity was manifested by the highest dose for *P. borbonense*. The increase in the dose of *P. borbonense* enhances probably the quantity of tannins in extract and also the immunosuppressive activity on lymphocytes subpopulation. In this regard, the quantity of flavonoids was also increased and their ability to stimulate other immune cellular subpopulations is noted (granulocytes case). However for *C. filiformis*, the highest immunosuppressive activity on lymphocytes is noted after administration of low doses. This same dose of *C. filiformis* increased also the quantity of flavonoids and thus induced an immunostimulating effect on granulocytes subpopulation.

7.4 Conclusions

Phytochemical screening of *C. filiformis* and *P. borbonense* extracts revealed the presence of flavonoids and tannins. The results demonstrated a potential immunosuppressive effect on percent of lymphocytes sub-population. They also revealed an immunostimulant effect of high doses especially on granulocytes. This work suggests that these plants could be used as a source of diverse bioactive compounds and that it is necessary to identify and isolate the compounds that are responsible for the immunomodulatory properties.

Acknowledgments We are grateful to Professor Jean-Noël LABAT for botanical identification of this plant and the Emirates Center for Wildlife Propagation (ECWP) in Missour, Morocco for the donation of animals. This work was supported by the MAP2ERA project (7PCRD).

References

1. World Health Organization (2002) WHO traditional medicine strategy 2002–2005. World Health Organization, Geneva
2. Verpoorte R (1998) Exploration of nature's chemodiversity: the role of secondary metabolites as leads in drug development. Drug Dev Trends 3:232–238
3. Verpoorte R (2000) Pharmacognosy in the new millennium: lead finding and biotechnology. J Pharm Pharmacol 52:253–262
4. Cragg GM, Boyd MR, Khanna R, Kneller R, Mays TD, Mazan KD, Newman DJ, Sausville EA (1999) International collaboration in drug discovery and development: the NCI experience. Pure Appl Chem 71:1619–1633
5. Davis L, Kuttan G (2000) Immunomodulatory activity of *Withania somnifera*. J Ethnopharmacol 71:193–200
6. Bin-Hafeez B, Haque R, Parvez S, Pandey S, Sayeed I, Raisuddin S (2003) Immunomodulatory effects of fenugreek (*Trigonella foenumgraecum* L.) extract in mice. Int Immunopharmacol 3:257–265
7. Barnes J, Anderson LA, Gibbons S, Phillipson JD (2005) Echinacea species (*Echinacea angustifolia* (DC.) Hell., *Echinacea pallida* (Nutt.) Nutt., *Echinacea purpurea* (L.) Moench): a review of their chemistry, pharmacology and clinical properties. J Pharm Pharmacol 57:929–954
8. Cheng W, Li J, You T, Hu C (2005) Anti-inflammatory and immunomodulatory activities of the extracts from the inflorescence of *Chrysanthemum indicum* Linné. J Ethnopharmacol 101:334–337
9. Patwardhan B, Manish G (2005) Botanical immunodrugs: scope and opportunities. Drug Discov Today 10:495–502
10. Ishizuka M, Kawatsu M, Yamashita T, Ueno M, Takeuchi T (1995) Low molecular weight immunomodulators produced by microorganisms. Int J Immunopharmacol 17:133–139

11. Hackett CJ (2003) Innate immune activation as a broad-spectrum biodefense strategy: prospects and research challenges. J Allergy Clin Immunol 112:686–694
12. Kaou AM, Mahiou-Leddet V, Hutter S, Aïnouddine S, Hassani S, Yahaya I, Azas N, Ollivier E (2008) Antimalarial activity of crude extracts from nine African medicinal plants. J Ethnopharmacol 116:74–83
13. PLARM (Plantes Aromatiques et Médicinales) (1990–1997) Rapport sur le projet: inventaire et Etude des plantes médicinales et aromatiques des Etats de l'Océan Indien. Plantes médicinales des Comores, de Madagascar, des Mascareignes (Maurice et Rodrigues), et des Seychelles. Ethnobotaniques et phytochimiques, COI/UE, p 94
14. Tramil in Faujour A (2002) Contribution à l'amélioration des soins de santé primaires par une investigation scientifique de la pharmacopée traditionnelle populaire des Comores – Bilan de la phase pilote du projet. CNDRS (Centre National de Documentation et de la Recherche Scientifique), p 248
15. Abdullahi M, Mohammed G, Abdulkadir NU (2003) Medicinal and economic plants of Nupeland. Jube-Evans Books and Publications, Bida, p 140
16. Quetin-Leclercq J, Hoet S, Block S, Wautier MC, Stévigny C (2004) Studies on *Cassytha filiformis* from Benin: isolation, biological activities and quantification of aporphines. Proceed Biores Towards Drug Discov Dev 81–107
17. Neuwinger HD (2000) African traditional medicine. A dictionary of plants' use and applications. Med Pharm 99:1–12
18. Babayi HM, Udeme JJ, Abalaka JA, Okogun JI, Salawu OA, Akumka DD, Adamu, Zarma SS, Adzu BB, Abdulmumuni SS, Ibrahime K, Elisha BB, Zakariys SS, Inyang US (2007) Effect of oral administration of aqueous whole extract of *Cassytha filiformis* on haematograms and plasma biochemical parameters in rats. J Med Toxicol 3:146–151
19. Nelson SC (2008) *Cassytha filiformis*, Plant disease PD-42. College of Tropical Agriculture and Human Resources, Honolulu
20. Fortin H, Robin V, Vigor C, Le Bosse B, Lohezic-Le Devehat F, Boustie J, Amoros M (2002) In vitro antiviral activity of thirty-six plants from La Reunion Island. Fitoterapia 73:346–350
21. Maiza-Benabdesselam F, Khentache S, Bougoffa K, Chibane M, Adach S, Chapleur Y, Max H, Laurain-Mattar D (2007) Antioxidant activities of alkaloid extracts of two Algerian species of Fumaria: *Fumaria capreolata* and *Fumaria bastardii*. Rec Nat Prod 1:28–35
22. Hajaji HE, Lachkar N, Alaoui K, Cherrah Y, Farah A, Ennabili A, Bali BE, Lachkar M (2011) Antioxidant activity, phytochemical screening, and total phenolic content of extracts from three genders of carob tree barks growing in Morocco. Arab J Chem 4:321–324
23. Salem JH (2009) Extraction, identification, caractérisation des activités biologiques de flavonoïdes de *Nitraria retusa* et synthèse de derivés acyles de ces molecules par voie enzymatique. Thèse de doctorat, Institut National Polytechnique de Lorraine, Université de Nancy, France
24. Tsai TH, Wang GJ, Lin LC (2008) Vasorelaxing alkaloids and flavonoids from *Cassytha filiformis*. J Nat Prod 71:289–291
25. Chang FR, Chao YC, Teng CM, Wu YC (1998) Chemical constituents from *Cassytha filiformis* II. J Nat Prod 61:863–866
26. Atluru S, Atluru D (1991) Evidence that genistein, a protein-tyrosine kinase inhibitor, inhibits CD28 monoclonal-antibody-stimulated human T cell proliferation. Transplantation 51:448–450
27. Huang HC, Hsieh LM, Chen HW, Lin YS, Chen JS (1994) Effects of baicalein and esculetinon transduction signals and growth factors expression in T-lymphoid leukemia cells. Eur J Pharmacol 268:73–78
28. Li SY, Teh BS, Seow WK, Liu YL, Thong YH (1991) In vitro immunopharmacological profile of the plant flavonoid baohuoside-1. Int J Immunopharmacol 13:129–134
29. Pandey R, Maurya R, Singh G, Sathiamoorthy B, Naika S (2005) Immunosuppressive properties of flavonoids isolated from *Boerhaavia diffusa Linn.* Int J Immunopharmacol 5:541–553
30. Chang SL, Chiang YM, Chang CLT, Yeh HH, Shyur LF, Kuo YH, Wu TK, Yan WC (2007) Flavonoids, centaurein and centaureidin, from *Bidens pilosa*, stimulate IFN-γ expression. J Ethnopharmacol 112:232–236
31. Chaubal R, Deshpande VH, Deshpande NR (2005) Methyl gallate, the medicinally important compound: a review. Electron J Environ Agric Food Chem 4:956–962

Chapter 8
Need for Smoking Cessation Support for Better Health of Employees

Marie Chan Sun, Jevisha Erriah, and Deerajen Ramasawmy

Abstract Tobacco is a silent killer, with more than 400 poisonous chemical substances. The use of tobacco is driven by dependence on nicotine, the primary addictive substance in tobacco. Smoking in the workplace has adverse effects on both the employee health and productivity. Hence, it is important to raise awareness on the need to quit smoking. The objectives of the study were to determine the intention to quit smoking and cessation attempts among employees of an educational institution, to assess the association between level of dependence and the Trans-Theoretical Model of Change and to identify the smoking cessation facilities available in Mauritius. This study had a mixed approach. In the quantitative part, smokers were recruited using a convenience sampling. The inclusion criteria were employees who were current smokers. A validated questionnaire adapted from the Global Adult Tobacco Survey questionnaire was used. The qualitative part evaluated the facilities available at the Smoking Cessation Clinics set up in Mauritius. The findings showed that 81 % of the participants intended to quit smoking and most of them planned to quit within the next month. Three major obstacles to relapse were identified. Awareness of smoking cessation techniques was low but willingness to use nicotine replacement therapy and other support therapy was high. This study highlighted the need to implement free smoking cessation programs on work premises to cater for the needs of employees so that they stop inhaling poisonous chemical substances which jeopardize the health and future of employees.

M. Chan Sun (✉) • J. Erriah
Department of Medicine, Faculty of Science, University of Mauritius, Réduit, Mauritius
e-mail: lan.sun@uom.ac.mu; jevi290990@gmail.com

D. Ramasawmy
Department of Management, Faculty of Law and Management, University of Mauritius, Réduit, Mauritius
e-mail: deerajen@gmail.com

M. Gupta Bhowon et al. (eds.), *Chemistry: The Key to our Sustainable Future*,
DOI 10.1007/978-94-007-7389-9_8, © Springer Science+Business Media Dordrecht 2014

8.1 Introduction

Tobacco, scientifically known as "*Nicotiana tobacum*", is the only legal consumer product that can harm everyone exposed to it mainly in the form of cigarettes [1, 2]. Tobacco smoking has been described as "a global problem with serious consequences for public health" [3] consisting of over 4,000 chemicals among which 300 are poisonous products and more than 50 are cancer-causing products [4, 5]. Despite the fact that many countries have imposed restrictions on its use, the tobacco epidemic is still one of the biggest public health threats the world has ever faced [6]. There are practically six million deaths each year due to tobacco with a high prevalence of mortality mostly in low- and middle-income countries [6]. Global consumption of cigarettes is expanding with over 150 billion of cigarettes smoked worldwide every day [7]. Trends in smoking prevalence vary among countries but the prevalence of tobacco use among males is higher than among females [8]. The biggest rise in tobacco-related diseases is expected in China, where an estimated two to three million deaths will occur by 2020 [9] while in the African Region (AFRO) the highest increase in the rate of tobacco-related diseases is being experienced with a 4.3 % increase per year [10]. According to a study done in 14 countries, the highest rate in the prevalence of smokers was found to be in Bangladesh with 43 %, followed by Russian Federation at 39 % and India at 35 % in the period 2008–2010 [6].

According to the Non-Communicable Diseases survey carried out by the Ministry of Health and Quality of Life (MoHQL) of Mauritius, the prevalence of smoking was estimated to be 21.7 % in 2009 with 3.7 % in women and 40.3 % in men; this prevalence is among the highest in Africa. Moreover, the prevalence of smoking was highest in the younger age groups with over 50 % of men aged between 19 and 24 years; smoking decreased with age in both men and women. Considering the general trend over the 22-year period 1987–2009, the prevalence of smoking decreased from 30.7 % in 1987 to 18.0 % in 2004, but increased to 21.9 % in 2009. Paying particular attention to the data with respect to men, it is observed that the smoking prevalence among men experienced a fall from 1987 (57.9 %) to 2004 (35.9 %), but increased to 40.3 % in 2009 [11]. The wide range of serious tobacco-related health problems has been extensively reviewed [5, 12–16].

In view of its commitment towards controlling and reducing tobacco-related mortality and morbidity, Mauritius ratified in May 2004 the World Health Organization (WHO) Framework Convention on Tobacco Control (FCTC) which is the world's first public health treaty [3]. Since then, Mauritius has taken vital steps to fulfill its obligations under the FCTC. In 2007, the Mauritian government, in collaboration with the WHO and several other stakeholders, developed a National Action Plan on Tobacco Control 2008–2012 and revised the 1999 policies [17]. Regulations passed under the sections 193 and 194 of the Public Health Act 2008 have resulted in indoor public places, public transport, and workplaces being declared non-smoking areas. This represents a major step towards protection of non-smokers from second-hand smoke. Moreover, access to tobacco products is limited by restricting the sale of cigarettes to packs of 20, controlling packaging, imposing health warnings with at least 65 % of the display area on each side of cigarette packs, increasing tobacco taxes, enforcing minor access laws and reducing exposure to tobacco advertising and promotion [17]. In fact, Mauritius became the first African country to enforce mandatory picture warnings on cigarette packs in 2008 [6] and has become a world leader for tobacco control in Africa [18]. Penalties have also been implemented to those who contravene these regulations. These range from a fine not exceeding Rs 10,000 to an imprisonment up to 12 months. In spite of these measures, the control of tobacco consumption is still a major public health challenge due to growing burden of non-communicable diseases accounting for about 1,000 deaths (out of a total population of 1.2 million) from tobacco-related diseases every year [17]. Only cessation has been shown to lead to a considerable reduction in deaths due to tobacco in the coming three and four decades [6, 19]. In this respect, a smoking cessation clinic comprising of psycho-therapy and counseling services was set up on a pilot basis in December 2008 at the Odette Leal Community Health centre at Beau-Bassin.

In order to promote smoking cessation and achieve an overall decline in smoking prevalence, workplaces should be one of the many targeted environments [20]. Indeed, the workplace is one important venue for tobacco control because smoking harms employees' health [21] and these costs are borne by employers [22]. Basically, smoking contributes to absenteeism and a decline in economic productivity; smokers, on average, miss 6.16 days of work per year due to sickness including smoking-related acute and chronic conditions, compared to non-smokers, who miss only 3.86 days of work per year [23]. In addition to harming the smoker's health, it is a long-known fact that passive smoking affects non-smokers [24]. Second-hand smoke has been classified as a potential cancer-causing agent [25] and thus the prevention to its exposure at work is needed in health and safety legislation. Nevertheless, smoke-free workplace policies lead to less smoking among workers and the elimination of second-hand smoke exposure, thus creating a healthier workforce, whereas smoking restrictions, which permit smoking areas or provide separately ventilated sections, are not effective at all in preventing or eliminating exposure to second-hand smoke [16]. Yet, implementing smoking legislations are not adequate and it has been revealed that providing cessation assistance greatly increases quit rates [26].

Moreover, the readiness of the smoker to stop smoking plays an important role as is demonstrated by Prochaska and Velicer [27] in the Trans-Theoretical Model of Behaviour Change which assesses an individual's readiness to act on a healthy behaviour. This model puts forward that health behaviour change involves progress through six stages of change: (1) Precontemplation (Not Ready) – "People are not intending to take action in the foreseeable future, and can be unaware that their behaviour is problematic"; (2) Contemplation (Getting Ready) – "People are beginning to recognize that their behaviour is problematic, and start to look at the pros and cons of their continued actions"; (3) Preparation (Ready) – "People are intending to take action in the immediate future, and may begin taking small steps toward behaviour change"; (4) Action – "People have made specific overt modifications in modifying their problem behaviour or in acquiring new healthy behaviours"; (5) Maintenance – "People have been able to sustain action for a while and are working to prevent relapse"; and (6) Termination – "Individuals have zero temptation and they are sure they will not return to their old unhealthy habit as a way of coping". In addition, the researchers conceptualized "relapse" (recycling) which is not a stage in itself but rather the "return from Action or Maintenance to an earlier stage".

Taking into consideration the Trans-Theoretical Model of Behaviour Change, this study was designed with the following objectives: To determine the intention to quit smoking and cessation attempts among employees of an educational institution, to assess the association between Nicotine Dependence (ND) and the Trans-Theoretical Model of Behaviour Change (TTM) and to identify the smoking cessation facilities available in Mauritius.

8.2 Methodology

It is a mixed approach project involving both qualitative and quantitative methods.

8.2.1 Quantitative

Sample Population

This questionnaire-based study was carried out in a workplace setting which comprised of 1,009 employees. Participants were recruited by convenience sampling method based on their willingness to participate in the study, as participation was entirely voluntary. Employees who were current smokers, either daily or occasional smokers, either male or female, were included in the study. The minimum sample size targeted was 50.

Survey Instrument

A validated questionnaire adapted from the Global Adult Tobacco Survey questionnaire was used. The questionnaire which required about 15 min for completion comprised of six sections with mainly close-ended questions. In the first section, background information pertaining to the gender, age, current occupation and current monthly household income of the participant were collected. The second section inquired about the tobacco smoking behaviour whereas the third and fourth sections inquired about exposure to second-hand smoke and intention in quitting tobacco smoking respectively. Finally, the fifth and sixth sections dealt with tobacco cessation and the awareness of tobacco cessation techniques.

Measures

Stage of change with respect to the TTM
Response of participants was classified as follows:

1. "Plan to quit within the next month" – preparation stage
2. "Intend to quit in the next 6 months" – contemplation stage
3. "Intend to quit some day but not in the next 6 months" – precontemplation stage

Fagerstrom Test for Nicotine Dependence (FTND)
FTND has six items which provide a total score ranging between 0 and 10, with higher scores reflecting greater dependence. There are three yes/no items, each item is given a score of 0 (No) or 1 (Yes). There are two multiple-choice items, each item can have a score between 0 and 3, and one having two options (score of 0 or 1). FTND score was classified in three stages, low dependence (0–4), medium dependence (5) and high dependence (6–10).

Ethical Considerations

Ethical clearance from the University of Mauritius Research Ethics Committee and authorization to carry out the survey were obtained in November 2011. All participants were offered a brochure provided by the MoHQL and were counselled about the Smoking Cessation Clinic situated in Beau-Bassin. The brochure outlined the prevalence of smoking in Mauritius, the harmful and poisonous chemicals and substances contained in tobacco smoke and sensitized about the health effects of smoking, passive smoking, benefits of smoking cessation, and tips on how to quit smoking. Anonymity was respected by allocating a code number to each questionnaire, whereby the list of identification numbers was known by only one investigator.

Data Collection and Analysis

The survey was carried out from January 2012 to February 2012. Before filling the questionnaire, the participants were provided a participant information sheet, explaining the aim and nature of the study and ensuring that all personal data would be kept strictly confidential. After the participants have read the information sheet and understood all the implications involved, they were required along with the investigator to sign the participation consent form prior to the filling of the questionnaire.

The data collected from the questionnaires were recorded and analysed using the software IBM SPSS (Statistical Package for the Social Sciences) Statistics 20.0. Results were exported on Microsoft Excel (Office 2007) to generate tables. The Chi-square test at 5 % level of significance was used to evaluate relationships between categorical variables.

8.2.2 Qualitative

Study Design

This is a descriptive study to identify the facilities available in the seven smoking cessation clinics in Mauritius.

Data Collection and Processing

The qualitative phase of the study was conducted in March 2012. Data were obtained from each smoking cessation clinic which comprised of the operational date, working hours, smoking cessation strategies and the number of new cases and follow-up.

8.3 Results

8.3.1 Quantitative

Sample Characteristics

Sixty-three volunteers participated in the survey. The profile of the participants is summarised in Table 8.1.

Intention to Quit Smoking and Cessation Attempts

Among the 63 smokers who participated in the study, 81 % representing 51 smokers wanted to quit smoking at the time of the survey. Data obtained are summarized in Table 8.2.

Most of the respondents had the intention to quit and 71.4 % of the participants have previously attempted to quit smoking with the highest frequency of cessation attempts being one or two. The highest proportion (23.9 %) relapsed after an abstinence period of less than 1 day and had recourse to no special therapy in their attempts to stop smoking. The main reason for cessation was to improve health and physical fitness while “other people smoking around” was the main ground for relapse (Table 8.2).

Craving to smoke was reported by 41.3 % of participants as the reason for relapse. While withdrawal symptoms were reported by many participants, 8.7 % reported the absence of support for smoking cessation. These findings demonstrate the need to provide smoking cessation support.

Table 8.1 Characteristics of the sample population (n = 63)

Participant characteristics	Percentage (%)
Gender	
Male	98.4
Female	1.6
Age group (years)	
20–30	11.0
31–40	30.2
41–50	30.2
51–60	28.6
Current occupation	
Administrative and secretarial staff	49.2
Professionals	7.9
Service workers	15.9
Skilled agricultural, forestry & fishery workers	27.0
Current monthly household income	
<Rs 10,000	25.4
Rs 10,000–Rs 25,000	52.4
>Rs 25,000	22.2
Smoking status	
Daily smoker	79.4
Occasional smoker	6.4
Occasional smoker, formerly daily	7.9
Occasional smoker, never daily	6.3
Number of cigarettes per day	
≤10	52.3
11–20	25.4
21–30	17.5
≥31	4.8

Tobacco Cessation Techniques

Awareness and willingness to use NRT (Nicotine Replacement Therapy) was greater than medications such as bupropion. However, many respondents would be agreeable to take the help of a telephone quit line, counselling and an internet quit line if ever they decide to quit smoking one day as detailed in Table 8.3.

Association Between ND and TTM

The FTND score of the participants varied from 0 to 9. Participants who had low ND were mostly in the preparation stage, as shown in Table 8.4. There was no statistically significant relationship (p-value being 0.860) between level of ND and TTM.

Table 8.2 Summary of data obtained for intention to quit smoking and cessation attempts

Variable	Percentage (%)
Want to quit smoking cigarettes now	81.0
Tried to stop cigarettes during the past year	71.4
Number of cessation attempts (times)	
1–2	65.2
3–5	23.9
6–9	2.2
≥10	8.7
Duration of cessation attempts	
Less than a day	23.9
1–2 days	15.2
3–7 days	21.7
More than 7 days but less than 30 days	13.0
More than 30 days but less than 6 months	13.0
More than 6 months but less than a year	13.0
Do you think that was a successful attempt?	
Yes	67.4
No	32.6
Quitting method used[a]	
No special therapy	71.7
Use self-help smoking cessation booklets	19.6
Used nicotine replacement therapy (nicotine patch/gum)	13.0
Seek help from healthcare professionals	6.5
Joined smoking cessation program	2.2
Used support group therapy	–
Process used to quit smoking during longest quit attempt	
Abrupt cessation	60.9
Gradually reduced the amount of cigarettes smoked & tried to stop smoking	39.1
Reasons for cessation[a]	
To improve my health and physical fitness	54.3
To save money	50.0
The fear of a future smoking-related disease	45.7
As a self-achievement	41.3
People around me did not approve of it	21.7
To avoid causing troubles at workplace	21.7
The new anti-smoking laws	13.0
I was tired of smoking	8.7
I have been encouraged by a doctor	8.7
Reasons for relapse[a]	
Other people smoking around	54.3
Habitual smoking	43.5
Craving to smoke	41.3
Under pressure/stress	26.1
Depressed mood	19.6

(continued)

Table 8.2 (continued)

Variable	Percentage (%)
Restlessness	13.0
Difficulty in concentrating	13.0
Anxiety	13.0
Insomnia	10.9
Increased appetite	8.7
No support/encouragement	8.7
Irritability/Frustration/Anger	7.9

[a]Multiple answers apply

Table 8.3 Awareness and willingness to use tobacco cessation techniques

Tobacco cessation technique	Percentage (%)
Awareness	
Nicotine Replacement Therapy (NRT)	33.3
Medication such as bupropion	14.3
Willing to use if ever you decide to stop smoking[a]	
Counselling	61.9
NRT	20.6
Medications	4.8
Telephone quit lines	60.3
Internet quit lines	60.3

[a]Multiple answers apply

8.3.2 Smoking Cessation Clinics in Mauritius: A Qualitative Insight

Background

In Mauritius, seven smoking cessation clinics have been set up to provide free and high-quality assistance to smokers willing to quit smoking cigarettes. Initially, only one smoking cessation clinic was implemented on a pilot basis in 2008 at the Odette Leal Community Health Centre at Beau-Bassin. The aim was to reduce the prevalence of tobacco use and tobacco-related illnesses especially NCDs. As part of the WHO initiative, there has been extension of smoking cessation clinics in the five regional hospitals and recently, in February 2012, one was launched at Souillac Hospital. Patients at the smoking cessation clinics include smokers who either come on their own, are referred by doctors in the public sectors usually in the outpatient department (OPD) or are referred by private doctors. From 2008 to 2012, a total of 377 new cases were recorded in all the seven smoking cessation clinics, among which 328 were males and 49 were females (Table 8.5).

Table 8.4 Level of nicotine dependence and stage of Trans-Theoretical Model of change of smokers

			TTM			
			Preparation	Contemplation	Precontemplation	Total
Level of dependence	Low dependence	Count	14	11	9	34
		Row %	41.2 %	32.4 %	26.5 %	100 %
		Column %	63.6 %	64.7 %	75.0 %	66.7 %
	Medium dependence	Count	6	5	3	14
		Row %	42.9 %	35.7 %	21.4 %	100 %
		Column %	27.3 %	29.4 %	25.0 %	27.5 %
	High dependence	Count	2	1	0	3
		Row %	66.7 %	33.3 %	0 %	100 %
		Column %	9.1 %	5.9 %	0 %	5.9 %
Total		Count	22	17	12	51
Row %			43.1 %	33.3 %	23.5 %	100 %
Column %			100 %	100 %	100 %	100 %

Table 8.5 Facilities available and the distribution of the new cases in the seven smoking cessation clinics in Mauritius

Smoking cessation clinic	Open since	Working hours	Smoking cessation services	New cases up to March 2012		
				Male	Female	Total
Odette Leal community health centre	Dec 2008	Wed 13.00–16.00	Counselling Pharmacotherapy	174	37	211
SSRN hospital	Oct 2011	Sat 09.00–12.00	Counselling Pharmacotherapy	18	–	18
Victoria hospital	Nov 2011	Sat 09.00–12.00	Counselling Pharmacotherapy	34	8	42
Jawaharlal Nehru hospital	Nov 2011	Wed 13.00–15.30	Counselling Pharmacotherapy	17	1	18
Dr A.G Jeetoo hospital	Dec 2011	Thur 13.00–16.00	Counselling Pharmacotherapy	24	3	27
Flacq hospital	Dec 2011	Sat 09.00–12.00	Counselling Pharmacotherapy	57	–	57
Souillac hospital	Feb 2012	Thur 13.00–16.00	Only counselling	4	–	4
Total				**328**	**49**	**377**

Infrastructure and Human Resources

Each smoking cessation clinic consists of an information desk where general information regarding the services can be obtained and where appointments are scheduled. It also comprises of a spacious consultation room especially designed for comfort and convenience. Moreover, a pharmacy is available to cater patients on pharmacotherapy with medications and a record office where patients' files are kept confidentially. Hence, each smoking cessation clinic comprises of a physician, a record clerk, an attendant and a pharmacist.

Smoking Cessation Intervention Strategies

The services provided by the public Smoking Cessation Clinics include preliminary assessment, counselling on quitting smoking, pharmacotherapy and follow-up. In the preliminary assessment also known as brief intervention, the socio-demographic data, the smoking status and the past medical history are recorded. Then, the nicotine dependence of smokers is calculated using FTND score. The vital signs including the blood pressure, pulse, height, weight and body mass index (BMI) are also measured. Counselling includes guidance on physiological and psychological adaptation, behavioural and lifestyle modification. The motivation and compliance of quitting smoking and the need of pharmacotherapy are assessed by counsellors. They also provide information about the treatment that will be offered. Counselling usually lasts for around 20–30 min. Previously counselling was done by psychologists but since last year, counsellors include nursing staff and doctors trained by experts of the WHO in a 3 days' workshop. Pharmacotherapy includes medications such as bupropion hydrochloride tablets (Zyban) and nicotine replacement products which include nicotine patches (Nicorette) and nicotine gums. Nicotine patches of 5, 10 and 15 mg doses, nicotine gums of 2 mg and Zyban of 150 mg are all available. Nonetheless, a revised version of the protocol for the smoking cessation intervention strategies for a better management will soon be available. A follow-up is done every 2 weeks so as to prevent relapses.

8.4 Discussion

8.4.1 Quantitative

Intention to Quit Smoking, Cessation Attempts and Tobacco Cessation Techniques

This study revealed that smokers had a high degree of readiness to quit smoking. This is higher than the findings of Griffiths et al. [28] which showed that 70 % (vs. 81 %) of smokers in a UK employee population expressed a desire to quit. Findings on motivational variables for cessation were consistent with other studies [29–31], which showed that the major reason for cessation was to improve health and physical fitness. This was closely followed by "to save money" and "fear of future smoking-related diseases". The observed results could be attributed to the implementation of pictorial warnings on packs of cigarettes, increasing awareness on the ill-effects of smoking. In fact, it has been revealed that for the majority of Mauritian smokers (55 %), the introduction of pictorial warnings has raised smokers' awareness of the health risks of smoking, and has made an increased contribution to leading smokers to think about quitting [18].

One key finding of this study is that more than half of the participants (52.4 %) had recourse to no special therapy in their attempts to stop smoking. This is slightly lower than Shiffman et al. findings which showed that 64.2 % tried to quit without the use of any cessation treatments [32]. This might be an explanation for the failure of the smokers to successfully quit smoking. International studies of adult smokers showed that a greater number of quit attempts was associated with an intention to quit smoking [33–36] and smokers may need to have recourse to as many as 12–14 quit attempts on average before being successful [37]. However, in this study, the highest frequency was one to two attempts. Therefore, it is imperative to motivate these smokers to quit and encourage them to initiate attempts.

This study, which revealed that the main reason for relapse was "other people smoking around", was in line with several other studies [38–43]. However, it was in contrast with the findings of the Wisconsin Tobacco Survey [44] and those of Norregaard et al. [45], whereby "craving for cigarette" was the number one barrier to relapse. Nonetheless, in this study, 43.5 % acknowledged that "habitual smoking" influences relapse and 41.3 % claimed that "craving for a cigarette" was the reason of their relapse. Undoubtedly, stress, weight gain, depression during attempts to quit smoking increased the likelihood of relapse [46–48]. This was concordant with results from this study where significant number of respondents reported these reasons as well as other obstacles to relapse. Hence, eliminating barriers to quitting will help more smokers achieve their goal of becoming tobacco-free.

Although pharmacotherapy interventions have been proven to provide therapeutic effects in assisting smokers in smoking cessation [49–51], most participants were unaware of such smoking cessation techniques. While participants were agreeable to have recourse to counselling as a smoking cessation technique, they reported their unwillingness to use NRT and medications such as bupropion as cessation aids if ever they decide to stop smoking. These findings were in conformity with a number of related studies which reported that knowledge about smoking cessation aids was low among adult smokers and they were less likely to use them [35, 52–55]. On the other hand, the findings of the current study are different from those of the ITC Mauritius Survey 2011 which showed that smokers in Mauritius were most interested in accessing stop smoking medications [18]. Both psychotherapy and pharmacotherapy are required on the workplace so as to improve the likelihood of quitting among smokers who seek help as found in the literature [56, 57].

Association Between ND and TTM

Employees of the educational institution surveyed presented a minimal level of ND as most of the participants had a FTND score between 0 and 4. Similar levels of ND were found in Spain [58, 59] and other countries [60, 61], whereas Bang [62] found high ND among employees of a university. With regard to the correlation between physiological factors and the intention to quit, this study which indicated low ND and high intention to quit smoking supports the findings of Girma et al. [63] which demonstrated that smokers who had no intention to quit smoking had high level of ND.

The application of the TTM to the sample under study was not consistent with the results of Duaso et al. [60] which showed that 59.6 % (vs. 19.0 %) of the participants in their study were in precontemplation stage, 32.3 % (vs. 27.0 %) were in contemplation stage and 8.1 % (vs. 34.9 %) were in preparation stage. However, this difference may arise from the limitation of this study, whereby the sample size was relatively small. Although convenience sampling was used, the use of a validated questionnaire adapted from the Global Adult Tobacco Survey questionnaire constitutes the main strength of this study. Considering the TTM, stage of change was found to be unrelated to ND in accordance with the study of DiClemente et al. [64] but in contradiction with other studies conducted by Kleinjan et al. [65, 66]. Follow-up studies are thus needed to shed light on behaviour change of employees.

8.4.2 Qualitative

Facilities Available in Mauritius to Support Smokers

The main focus of this qualitative part was to identify the facilities offered in Mauritius to support the smokers who were willing to make quitting-related actions. The findings of this research provide an array of information about smoking cessation clinics available in Mauritius, on their working hours, smoking cessation strategies, infrastructure and human resources. Hence, employees of the educational institution surveyed can find relevant information to gain control over their life.

This study revealed that smoking cessation clinics have been extended to regional areas making the facilities accessible to a larger proportion of the Mauritian population. Moreover, each smoking cessation clinic, having the same infrastructure and human resources, offers similar smoking cessation intervention strategies. On the other hand, the different working hours in each smoking cessation clinic allow smokers to benefit from the smoking cessation facilities at their convenience. Yet, more smoking cessation sessions per week in each smoking cessation clinic and extension of these services in Area and Community Health Centres are advocated for. It has been noticed that all the smoking cessation clinics offer both counselling and pharmacotherapy except that of Souillac which provides only counselling as part of the smoking cessation intervention. However, pharmacotherapy will soon be offered at Souillac Hospital too. It is important to note that the counsellors were trained by the WHO which constitutes the greatest asset of the smoking cessation clinics. Nonetheless, there is need to address the training of more health professionals for improved practice with respect to smoking cessation interventions. Results from this research indubitably demonstrated that a high number of new cases have been registered in each smoking cessation clinic.

Despite the various facilities offered by the smoking cessation clinics, there are numerous limitations which need to be addressed. Paper records are used as a medium to collect and record patients' data and these impede the continuity and quality of care. Illegible handwriting, ambiguous and incomplete data, data fragmentation, poor availability and lack of space for storage files constitute barriers to a first-class record service. Hence, using an automated health record will be more beneficial. Indeed, electronic health records will represent a huge potential for cost savings and decreasing staff inefficiencies. As per the records at Odette Leal Community Health Centre, 194 patients have dropped out due to unknown reasons and seven have relapsed and only eight patients have reported to have successfully responded to treatment and have given up smoking. However, in the other smoking cessation clinics, these data are not available as they have recently been operational and the relapse rate requires a minimum of 1 year to be calculated. Thus, there is need to monitor the various shortcomings in the monitoring of patients' response to treatment and to identify the reasons for drop-outs and relapses. Another limitation is that telephone quit lines are not included in the services provided.

8.5 Conclusions

This study has contributed to shed light on the smoking burden in an employee population of an educational institution as they try to stop inhaling poisonous chemical substances in cigarettes. Indeed, this study clearly showed that the overall intention to quit among the participants was high. There is need for health promotion on work premises to cater for the needs of employees who want to quit. Through successful smoking cessation techniques, they will stop inhaling poisonous chemical substances which jeopardize their health and those of other employees. Counselling with relevant therapies and sustained follow-up by health professionals on the workplace is recommended to reduce smoking relapse, especially for those who are nicotine-dependent. Eliminating barriers to quitting will help more smokers achieve their goal of becoming tobacco-free, which will contribute to a tobacco-free workplace.

References

1. World Health Organization (2008) Report on the global tobacco epidemic, The MPOWER package, Geneva. http://www.who.int/tobacco/mpower/mpower_report_full_2008.pdf. Accessed 11 Sep 2011
2. Dixon T (2009) Should there be further restrictions on the advertising, selling, and smoking of tobacco? Should it be banned? http://www.idebate.org/debatabase/topic_details.php?topicID=16. Accessed 20 Sep 2011
3. World Health Organization (2003) WHO framework convention on tobacco control. http://www.who.int/tobacco/framework/WHO_FCTC_english.pdf. Accessed 22 Oct 2011
4. Dunnington J (1993) Tobacco almanac: a collection of tobacco facts. University of Texas, Houston
5. Cancer Research UK (2009) Smoking and cancer. http://info.cancerresearchuk.org/healthyliving/smokingandtobacco/whatsinacigarette/#Tar. Accessed 14 Sep 2011
6. World Health Organization (2011) Report on the global tobacco epidemic. Warn about the dangers of tobacco. http://whqlibdoc.who.int/publications/2011/9789240687813_eng.pdf. Accessed 22 Oct 2011
7. The Tobacco Atlas (2009). Third edition. http://www.tobaccoatlas.org/downloads/TobaccoAtlas_sm.pdf. Accessed 11 Sep 2011
8. World Health Organization (2010) WHO framework convention on tobacco control. http://apps.who.int/gb/fctc/PDF/cop4/FCTC_COP4_12-en.pdf. Accessed 22 Oct 2011
9. WHO/EMRO (2004) Collaboration on a national level: working together to bring about anti-tobacco legislation: a success story, World Health Organization, Geneva
10. WHO/AFRO (2004) Tobacco-free initiative: WHO/AFRO overview, World Health Organization, Geneva
11. Ministry of Health & Quality of Life (2009) Non-communicable diseases survey, republic of Mauritius. http://www.gov.mu/portal/goc/moh/file/ncd/ncd-2009.pdf. Accessed 12 Sep 2011
12. United States Environmental Protection Agency (1992) Respiratory health effects of passive smoking: lung cancer and other disorders, Office of Research and Development, Office of Air and Radiation. http://www.epa.gov/ncea/ets/pdfs/acknowl.pdf. Accessed 24 Nov 2011
13. Gonzales AC (2003) Passive smoking. Br Med J 326:1094

14. Secretan B, Straif K, Baan R, Grosse Y, El Ghissassi F, Bouvard V, Benbrahim-Tallaa L, Guha N, Freeman C, Galichet L, Cogliano V (2009) A review of human carcinogens – part E: tobacco, areca nut, alcohol, coal smoke, and salted fish. Lancet Oncol 10:1033–1034
15. Carlo WA, Goudar SS, Jehan I, Chomba E, Tshefu A, Garces A, Parida S, Althabe F, McClure EM, Derman RJ, Goldenberg RL, Bose C, Krebs NF, Panigrahi P, Buekens P, Chakraborty H, Hartwell TD, Wright LL (2010) Newborn-care training and perinatal mortality in developing countries. N Engl J Med 362:614–623
16. Centers for Disease Control and Prevention (2011) Current cigarette smoking prevalence among working adults, United States, 2004–2010. Morb Mortal Wkly Rep 60:1305–1309
17. Ministry of Health & Quality of Life (2011) Report on strengthening the health information system regarding tobacco use among NCD patients attending public health institutions. http://www.gov.mu/portal/goc/mih/file/atsa4.pdf. Accessed 11 Apr 2012
18. International Tobacco Control Policy Evaluation Project (2011) ITC Mauritius wave 2 national report. www.itcproject.org/download/.../itc20mauritius20reportprintdraftpdf. Accessed 20 Feb 2012
19. Jha P, Chaloupka FJ, Moore J, Gajakshmi V, Gupta PC, Peck R, Asma S, Zatonski W (2006) Tobacco addiction (chap 46). In: Jamison DT, Breman JG, Measham AR, Alleyne G, Claeson M, Evans DB, Jha P, Mills A, Musgrove P (eds) Disease control priorities in developing countries, 2nd edn. World Bank, Washington, pp 869–895
20. Håkansta (2004) Workplace smoking. Working paper: a review of national and local practical and regulatory measures. http://www.ilo.org/wcmsp5/groups/public/@ed_protect/@protrav/@safework/documents/publication/wcms_108424.pdf. Accessed 5 Apr 2012
21. Bunn WB, Stave GM, Downs KE, Alvir JM, Dirani R (2006) Effect of smoking status on productivity loss. J Occup Environ Med 48:1099–1108
22. Max W (2001) The financial impact of smoking on health related costs: a review of the literature. Am J Health Promot 15:321–331
23. Halpern MT, Shikiar R, Rentz AM, Khan ZM (2001) Impact of smoking status on workplace absenteeism and productivity. Tob Control 10:233–238
24. US National Research Council (1986) Environmental tobacco smoke: measuring exposures and assessing health effects, National Academy Press, Washington, DC
25. American Cancer Society (2012) Guide to quitting smoking. http://www.cancer.org/Healthy/StayAwayfromTobacco/GuidetoQuittingSmoking/guide-to-quitting-smoking-types-of-nrt. Accessed 28 Feb 2012
26. Fiore MC (2008) Treating tobacco use and dependence. Clinical practice guideline. http://www.surgeongeneral.gov/tobacco/treating_tobacco_use08.pdf. Accessed 17 Feb 2012
27. Prochaska JO, Velicer WF (1997) The transtheoretical model of health behavior change. Am J Health Promot 12:38–48
28. Griffiths J, Grieves K (2002) Why smoking in the workplace matters: an employer's guide, World Health Organization, Copenhagen. http://www.euro.who.int/document/e74820.pdf. Accessed 5 Apr 2012
29. International Tobacco Control (2009) International Tobacco Control Policy Evaluation Project. China Summary
30. Abdullah AS, Lam TH, Chan SS, Hedley AJ (2006) Smoking cessation among Chinese young smokers: does gender and age difference matters and what are the predictors? Addict Behav 31:913–921
31. Wong DC, Chan SS, Ho SY, Fong DY, Lam TH (2010) Predictors of intention to quit smoking in Hong Kong secondary school children. J Public Health 32:360–371
32. Shiffman S, Brockwell SE, Pillitteri JL, Gitchell JG (2008) Use of smoking cessation treatments in the United States. Am J Prev Med 34:102–111
33. Fagan P, Augustson E, Backinger CL, O'Connell ME, Vollinger RE, Kaufman A, Gibson JT (2007) Quit attempts and intention to quit cigarette smoking among young adults in the United States. Am J Public Health 97:1412–1420

34. Joseph A, Lexau B, Willenbring M, Nugent S, Nelson D (2004) Factors associated with readiness to stop smoking among patients in treatment for alcohol use disorder. http://www.ncbi.nlm.nih.gov/pubmed/15370939. Accessed 5 Apr 2012
35. Leatherdale ST, Shields M (2009) Smoking cessation: intentions, attempts and techniques. Health Rep 20:31–39
36. Burns DM, Shanks TG, Major JM, Gower KB, Shopland DR (2000) Restrictions on smoking in the workplace. In: USDHHS and National Cancer Institute (ed) Population based smoking cessation. National Institutes of Health, Bethesda, p 99
37. Zhu SH (2007) Differential cessation rates across populations: what explains it and how to reduce it? Paper presented at the meeting of the Oceania tobacco control conference, Auckland. http://otcc.confex.com/otcc/otcc07/techprogram/S1069.HTM. Accessed 14 Oct 2011
38. Chen PH, White HR, Pandina RJ (2001) Predictors of smoking cessation from adolescence into young adulthood. Addict Behav 26:517–529
39. Ellickson PL, McGuigan KA, Klein DJ (2001) Predictors of late-onset smoking and cessation over 10 years. J Adolesc Health 29:101–108
40. Macy JT, Seo DC, Chassin L, Presson CC, Sherman SJ (2007) Prospective predictors of long-term abstinence versus relapse among smokers who quit as young adults. Am J Public Health 97:1470–1475
41. Paavola M, Vartiainen E, Puska P (2001) Smoking cessation between teenage years and adulthood. Health Educ Res 16:49–57
42. Rose JS, Chassin L, Presson CC, Sherman SJ (1996) Prospective predictors of quit attempts and smoking cessation in young adults. Health Psychol 15:261–268
43. Tucker JS, Ellickson PL, Orlando M, Klein DJ (2005) Predictors of attempted quitting and cessation among young adult smokers. Prev Med 41:554–561
44. Wisconsin Tobacco Survey (2003) www.ctri.wisc.edu/Publications/.../BarrierstoQuitting.2.28.pdf. Accessed 5 Apr 2012
45. Norregaard J, Tonnesen P, Petersen L (1993) Predictors and reasons for relapse in smoking cessation with nicotine and placebo patches. http://www.ncbi.nlm.nih.gov/pubmed/8483863. Accessed 19 Oct 2011
46. Catley D, Harris KJ, Okuyemi KS, Mayo MS, Pankey E, Ahluwalia JS (2005) The influence of depressive symptoms on smoking cessation among African Americans in a randomized trial of bupropion. Nicotine Tob Res 7:859–870. http://www.ncbi.nlm.nih.gov/pubmed/16298721. Accessed 3 Oct 2011
47. Perkins KA (1993) Weight gain following smoking cessation. J Consult Clin Psychol 61:768–777. http://psycnet.apa.org/index.cfm?fa=buy.optionToBuy&id=1994-07255-001. Accessed 18 Sep 2011
48. Hughes JR (1992) Tobacco withdrawal in self-quitters. J Consult Clin Psychol 60:689–697
49. Doggrell SA (2007) Which is the best primary medication for long-term smoking cessation–nicotine replacement therapy, bupropion or varenicline? http://www.ncbi.nlm.nih.gov/pubmed/18001252. Accessed 5 Apr 2012
50. Wu P, Wilson K, Dimoulas P, Mills JE (2006) Effectiveness of smoking cessation therapies: a systematic review and meta-analysis. http://www.biomedcentral.com/1471-2458/6/300. Accessed 5 Apr 2012
51. Fiore MC, Bailey WC, Cohen SJ, Dorfman SF, Goldstein MG, Gritz EF, Heyman RB, Jae'n CR, Kottke TE, Lando JA, Meckleburg RE, Mullen PD, Nett LM, Robinson R, Stitzer ML, Tommasello AC, Villejo L, Wewers ME (2000) Treating tobacco use and dependence: clinical practice guidelines. U.S. Department of Health and Human Services, Public Health Service, Rockville
52. Curry SJ, Sporer AK, Pugach O, Campbell RT, Emery S (2007) Use of tobacco cessation treatments among young adult smokers: 2005 National Health Interview Survey. Am J Public Health 97:1464–1469

53. Ellis JA, Perl SB, Davis K, Vichinsky L (2008) Gender differences in smoking and cessation behaviors among young adults after implementation of local comprehensive tobacco control. Am J Public Health 98:310–316
54. Hughes JR, Cohen B, Callas PW (2009) Treatment seeking for smoking cessation among young adults. J Subst Abuse Treat 32:211–213
55. Messer K, Trinidad DR, Al-Delaimy WK, Pierce JP (2008) Smoking cessation rates in the United States: a comparison of young adult and older smokers. Am J Public Health 98:317–322
56. McDonald P (2003) Consideration and rationale for a national action plan to help Canadian tobacco users. Ontario Tobacco Research Unit, Canada
57. Pierce JP, Gilpin EA (2002) Impact of over-the-counter sales on effectiveness of pharmaceutical aids for smoking cessation. JAMA 288:1260–1264
58. Becona E, Gomez-Duran B, Alvarez-Soto E, Garcia MP (1992) Scores of Spanish smokers on Fagerstrom's Tolerance Questionnaire. Psychol Rep 71:1227–1233
59. Jimenez-Ruiz CA, Masa F, Miravitlles M, Gabriel R, Viejo JL, Villasante C, Sobradillo V (2001) Smoking characteristics: differences in attitudes and dependence between healthy smokers and smokers with COPD. Chest 119:1365–1370
60. Duaso MJ, De Irala J, Canga N (2005) Employee's perceived exposure to environmental tobacco smoke, passive smoking risk beliefs and attitudes towards smoking: a case study in a university setting. Health Educ Res 21:26–33
61. Sanglam L, Bayraktar R, Kadioglu EE, Acemoglu H (2010) Smoking prevalence and the degree of nicotine dependence among healthcare workers at the Ataturk University Medical Facility. http://www.eajm.org/pdf/pdf_EAJM_346.pdf. Accessed 5 Apr 2012
62. Bang H (1996) Smoking among employees at University of Pittsburgh/Upmc. http://d-scholarship.pitt.edu/8061/1/bangh_etd.pdf. Accessed 5 Apr 2012
63. Girma E, Assefa T, Deribew A (2010) Cigarette smokers' intention to quit smoking in Dire Dawa town Ethiopia: an assessment using the Transtheoretical Model. BMC Public Health 10:320
64. DiClemente CC, Prochaska JO, Fairhurst SK, Velicer WF, Velasquez MM, Rossi JS (1991) The process of smoking cessation: an analysis of precontemplation, contemplation, and preparation stages of change. J Consult Clin Psychol 59:295–304
65. Kleinjan M, Brug J, Eijnden RJJM, Vermulst AA, Zundert RMP, Engels RCME (2007) Associations between the transtheoretical processes of change, nicotine dependence and adolescent smokers' transition through the stages of change. http://fataids.org/assets/pdf/00008514-200802000-00026.pdf. Accessed 5 Apr 2012
66. Kleinjan M, Eijnden RJJM, Leeuwe JV, Brug J, van de Ven MOM, Engels RCME (2008) Adolescent's movement towards cessation of smoking: role and relative value of the processes of change and nicotine dependence. http://www.tandfonline.com/doi/abs/10.1080/08870440701757344#preview. Accessed 5 Apr 2012

Chapter 9
Preparation and Characterization of Some Imidazoles and Formimidoyl-1*H*-Imidazoles from Formamidines

Asieh Yahyazadeh

Abstract Imidazole and formimidoyl-1*H*-imidazole derivatives were prepared in good yields by the reaction between formamidines and a weak or strong base depending on reaction conditions. Starting from readily available ethyl (*Z*)-*N*-(2-amino-1,2-dicyanovinyl)formimidate **1**, *N*-aryl-*N′*-[2-amino-1,2-dicyanovinyl] formamidines **2** can be prepared in good yields by the reaction with aromatic amines at room temperature in the presence of an acid catalyst. Treatment of the amidines with the base, DBU (1,8-diazabicyclo[5.4.0]undec-7-ene) at room temperature gave the corresponding 5-amino-1-aryl-4-(cyanoformimidoyl)-1*H*-imidazoles **3** in high yields, whereas the reaction with potassium hydroxide solution afforded the respective 5-amino-4-cyano-1-arylimidazoles **4**. Compounds **4** can also be prepared from compounds **3** by reaction with potassium hydroxide solution. All the derivatives were fully characterized by spectroscopic data.

9.1 Introduction

5-Amino-4-cyanoimidazoles have long been recognized as useful synthetic precursors to purines, but there is no simple, general synthesis available for 1-aryl derivatives of these compounds. In 1984, Sen and Mukhopadhyay [1] reported the preparation of 5-amino-4-cyano-1-(*p*-aminosulfonylphenyl)imidazole *via* a multi-step synthesis from the corresponding 1-methyl derivative. Frank and Zeller [2] described the synthesis of a number of 1-aryl- and 1-heteroaryl-5-amino-4-cyanoimidazoles (aryl=2- and 4-ClC_6H_4, 2,4-$Cl_2C_6H_3$, 2,4-$Me_2C_6H_3$; heteroaryl= 2- and 3-pyridyl, 5-Cl-2-pyridyl, 3,5-Cl_2-2-pyridyl, 2-Cl-3-pyridyl, and 2-pyrimidinyl) in low to moderate yields by reaction of the corresponding ethyl *N*-substituted formimidate with 2-aminomalodinitrile tosylate in acetic acid.

A. Yahyazadeh (✉)
Department of Chemistry, University of Guilan, P.O. Box 41335-1914, Rasht, Iran
e-mail: yahyazadehphd@yahoo.com; yahyazadeh@guilan.ac.ir

M. Gupta Bhowon et al. (eds.), *Chemistry: The Key to our Sustainable Future*, 131
DOI 10.1007/978-94-007-7389-9_9, © Springer Science+Business Media Dordrecht 2014

We have been interested in the chemistry of diaminomaleonitrile (DAMN) and its derivatives, in particular, ethyl-2-(2-amino-1,2-dicyanovinyl)formimidate **1**. The latter can be prepared in good yield from the reaction between DAMN and triethyl orthoformate in 1,4-dioxane [3–6]. From our previous work in this area, it appeared to us that **1** would be a useful starting material for the preparation of new *N*-aryl-*N′*–[2-amino-1,2-dicyanovinyl]formamidines **2**. Using procedures developed in our laboratories it was envisaged that these could be readily converted into 5-amino-1-aryl-4-(cyanoformimidoyl)-1*H*-imidazoles **3** [7–13], which are expected to be useful precursors to new 6-carbamoyl-1, 2-dihydropurines and 6-substituted purines [14–20]. In addition, reactions of compound **2** could provide a simple route to the desired 5-amino-1-aryl-4-cyanoimidazoles **4**. The results of this investigation are now reported.

9.2 Experimental

9.2.1 Materials and Methods

All solvents were purified and dried using established procedures. The ^{1}H NMR spectra were recorded on a Bruker XL 500 (500 MHz) instrument, ^{13}C NMR spectra on a DRX-500 AVANCE spectrometer, and IR spectra on a Shimadzu IR-470 spectrophotometer. Mass spectra were recorded on a Kratos Concept instrument. The melting points were measured on an Electrothermal digital melting point apparatus and are uncorrected.

9.2.2 General Procedure for the Preparation of the N-aryl-N′-[2-amino-1,2-dicyanovinyl] formamidines 2a-c

The aromatic amines (5.64 mmol) were added to a suspension of formimidate **1** (5.62 mmol) in dry ethanol or ethyl acetate, which contained anilinium chloride (0.01 g). The mixture was stirred at room temperature until TLC showed that all the formimidate had disappeared (usually 3–4 h) and the amidine was isolated by filtration. In a few cases precipitation had to be assisted by concentrating the solution and addition of a 1:1 mixture of light petroleum (b.p. 40–60 °C) and chloroform. In most cases the products obtained were yellow to pale green. The precipitates were washed with diethyl ether or a light petroleum-chloroform mixture and were dried under vacuum to give the analytically pure products in the yields 82–92 %.

***N*-Phenyl-(*Z*)-*N'*-[2-amino-1,2-dicyanovinyl]formamidine 2a:** Recrystallization of the product from dry chloroform/petroleum ether (1:1) gave yellow crystals (1.05 g, 4.97 mmol, 82 %). M.p. 132–134 °C (decomp.). [Found: C, 62.7; H, 4.2; N, 33.0. $C_{11}H_9N_5$ requires: C, 62.6; H, 4.3; N, 33.2 %]; *m*/*z* (EI): 212 $(M + 1)^+$ 100 %, 211 $(M)^+$ 12.5 %, 185 $(M\text{-}CN)^+$ 18.4 %, 106 $[(M + 1)\text{-}C_4H_2N_4]^+$ 2 %, 94 $[(M + 1)\text{-}C_5H_2N_4]^+$ 21 %, 93 $[(M + 1)\text{-}C_5H_3N_4]^+$ 7.5 %; δ_H (500 MHz, d_6-DMSO): 6.60 (br. s, NH_2), 7.10–7.15 (m, 1H, H9), 7.38–7.42 (m, 2H, H8 & H10), 7.55–7.70 (br. s, 2H, H7 & H11), 7.74 (s, 1H, H5), 8.16 (br. s, 1H, NH); δ_C (75 MHz, d_6-DMSO): 151.7 (C5), 143.3 (C6), 133.1 (C8 & C10), 126.8 (C9), 122.9 (C7 & C11), 122.3 (C2), 119.7 and 118.8 (C3 & C4), 109.2 (C1) ppm; ν_{max} (Nujol mull): 3,450 s, 3,420 w, 3,410 w, 3,345 s, 3,300 m, 3,240 w (N-H str.), 2,210 s (CN str.), 2,195 s (CN str.), 1,655 s (C=N str.), 1,600 s (N-H bend), 1,590 s, 1,500 w, 1,320 m, 960 s, 870 s, 730 m cm^{-1}; λ_{max} (EtOH): 202.1 (ε 16631), 205.2 (ε 13869), 262.5 (ε 10989), 262.8 (ε 10894) and 353.0 (ε 32456) nm.

***N*-Benzyl-(*Z*)-*N'*-[2-amino-1,2-dicyanovinyl]formamidine 2b:** Recrystallization of the product from dry chloroform/petroleum ether (1:1) gave white crystals (1.22 g, 5.4 mmol, 89 %). M.p. 91–92 °C (decomp.). [Found: C, 64.2; H, 4.9; N, 31.1. $C_{12}H_{11}N_5$ requires: C, 64.0; H, 4.9; N, 31.1 %]; *m*/*z* (EI): 226 $(M + 1)^+$ 100 %, 225 $(M)^+$ 62 %, 199 $(M\text{-}CN)^+$ 11.4 %, 108 $[(M + 1)\text{-}C_5H_2N_4]^+$ 66.0 %; δ_H (500 MHz, d_6-DMSO): 4.52 (d, 2H, $^3J_{6,\,NH}$ 6 Hz, H6), 6.10 (s, 2H, NH_2), 7.20–7.38 (m, 5H, H8, H9, H10, H11 & H12), 7.74 (d, 1H, $^3J_{5,NH}$ 6 Hz, H5), 8.18 (br. d, 1H, $^3J_{NH,5}$ 6 Hz, NH) ppm; δ_C (75 MHz, d_6-DMSO): 154.4 (C5), 148.8 (C7), 132.4 (C10), 131.8 (C9 & C11), 131.0 (C8 & C12), 121.1 and 120.2 (C3 & C4), 119.2 (C2), 110.1 (C1), 47.9 (C6) ppm; ν_{max} (Nujol mull): 3,460 s, 3,355 s (N-H str.), 2,225 s (CN str.), 2,210 s (CN str.), 1,640 s (C=N str.), 1,600 s (N-H bend), 1,580 m, 1,530 m, 1,350 m, 1,290 m, 1,200 m, 1,170 s, 1,070 m, 960 s, 790 s, 700 s cm^{-1}; λ_{max} (EtOH): 205.3 (ε 22688), 206.6 (ε 1907), 228.0 (ε 20813), 228.1 (ε 1931), 330.1 (ε 39938), and 330.2 (ε 3878) nm.

***N*-(4-Methylphenyl)-(*Z*)-*N'*-[2-amino-1,2-dicyanovinyl]formamidine 2c:** Recrystallization of the product from dry chloroform/petroleum ether (1:1) gave pale green crystals (1.35 g, 6.0 mmol, 92 %). M.p. 123–124 °C (decomp.). [Found: C, 63.9; H, 4.8; N, 31.6. $C_{12}H_{11}N_5$ requires: C, 64.0; H, 4.9; N, 31.1 %]; *m*/*z* (EI): 226 $(M + 1)^+$ 100 %, 225 $(M)^+$ 3.0 %, 195 $(M\text{-}CN)^+$ 40.9 %, 120 $[(M + 1)\text{-}C_4H_2N_4]^+$ 6.0 %, 91 $[(M + 1)\text{-}C_5H_4N_5]^+$ 6.0 %; δ_H (500 MHz, d_6-DMSO): 2.50 (s, 3H, CH_3), 6.40 (br. s, NH_2), 6.90 (d, 2H, $^3J_{8,7}$ 8 Hz, H8 & H10), 7.30−7.42 (d, 1H, 6 Hz, H5), 7.50−7.90 (br. complex m, 2H, H7 & H11), 9.90 (br. s, NH) ppm; δ_C (75 MHz, d_6-DMSO): 159.1 (C9), 150.9 (C5), 136.6 (C6), 124.5 (C7 & C11), 122.3 (C2), 119.9 and 118.8 (C3 & C4), 118.1 (C8 & C10), 109.6 (C1), 39.1 (C12) ppm; ν_{max} (Nujol mull): 3,460 w, 3,450 s, 3,340 s, 3,300 m, 3,250 s, 3,120 m (N-H str.), 2,210 s (CN str.), 1,650 s (C=N str.), 1,590 s (N-H bend), 1,575 w, 1,515 m, 1,310 s, 1,295 m, 1,245 s, 1,220 s, 1,175 m, 1,040 s, 960 m, 820 s, 780 m cm^{-1}; λ_{max} (EtOH): 201.1 (ε 14971), 202.2 (ε 15641), 269.0 (ε 12659) and 356.6 (ε 29334) nm.

9.2.3 General Procedure for the Preparation of the 5-Amino-1-aryl-4-(cyanoformimidoyl)-1H-imidazoles 3a-c

To a stirred suspension of the formamidine (1.0 g) in either dry ethyl acetate, ethanol or a 1:1 mixture of ethyl acetate and isopropanol, was added DBU (10 drops) and the reaction was monitored by TLC. After 1–3 h the product precipitated as an off-white to pale yellow solid. This was filtered off, washed with diethyl ether or light petroleum and dried under vacuum to give the title compounds in yields of 85–91 %.

5-Amino-1-phenyl-4-(cyanoformimidoyl)-1*H*-imidazole 3a: Recrystallization of the product from dry chloroform/petroleum ether (1:1) gave pale yellow crystals (0.9 g, 4.2 mmol, 90 %). M.p. 114–116 °C (decomp.). [Found: C, 64.8; H, 4.2; N, 31.0. $C_{11}H_9N_5$ requires: C, 64.9; H, 4.3; N, 30.4 %]; *m/z* (EI): 212 $(M + 1)^+$ 3.9 %, 211 $(M)^+$ 14.6 %, 159 $[(M + 1)\text{-CN}]^+$ 14.8 %; δ_H (500 MHz, d_6-DMSO): 6.68 (br. s, 2H, NH_2), 7.42–7.60 (m, 5H, H9, H10, H11, H12 & H13), 7.72 (s, 1H, H2), 11.20 (br. s, 1H, NH) ppm; δ_C (75 MHz, d_6-DMSO): 147.9 (C6), 147.1 (C5), 137.9 (C8), 136.1 (C2), 134.0 (C11), 132.6 (C9 & C13), 128.8 (C10 & C12), 120.5 (C7), 117.6 (C4) ppm; ν_{max} (Nujol mull): 3,340 s, 3,300 m, 3,250 s, 3,180 m, 3,120 m (N-H str.), 2,200 s (CN str.), 1,650 s (C=N str.), 1,600 s (N-H bend), 1,580 s, 1,290 m, 1,230 s, 1,170 m, 960 m, 820 s, 780 m cm^{-1}; λ_{max} (EtOH): 202.0 (ε 20057), 205.4 (ε 13194), 206.2 (ε 13218), 219.9 (ε 13531), 220.2 (ε 14046), 347.2 (ε 9207) and 347.3 (ε 9017) nm.

5-Amino-1-benzyl-4-(cyanoformimidoyl)-1*H*-imidazole 3b: Recrystallization of the product from dry chloroform/petroleum ether gave white crystals (0.85 g, 3.7 mmol, 85 %). M.p. 136–137 °C (decomp.). [Found: C, 64.2; H, 4.8; N, 31.0. $C_{12}H_{11}N_5$ requires: C, 64.0; H, 4.9; N, 31.1 %]; *m/z* (EI): 226 $(M + 1)^+$ 100 %, 225 $(M)^+$ 12.6 %, 199 $(M\text{-CN})^+$ 22 %, 91 $(M\text{-}C_5H_4N_5)^+$ 20.8 %; δ_H (500 MHz, d_6-DMSO): 5.10 (s, 2H, H8), 6.76 (br. s, 2H, NH_2), 7.20–7.38 (m, 6H, H10, H11, H12, H13 & H14), 7.42 (s, 1H, H2), 10.88 (br. s, 1H, NH) ppm; δ_C (75 MHz, d_6-DMSO): 148.3 (C6), 147.1 (C5), 140.4 (C9), 136.4 (C2), 132.4 (C12), 131.7 (C11 & C13), 131.2 (C10 & C14), 120.2 (C7), 117.5 (C4), 49.7 (C8); ν_{max} (Nujol mull): 3,350 s, 4,260 s, 3,180 m, 3,120 m (N-H str.), 2,225 s (CN str.), 1,659 s (C=N str.), 1,595 s (N-H bend), 1,550 s, 1,520 m, 1,350 s, 1,300 s, 1,220 s, 1,195 s, 970 s, 940 s, 820 cm^{-1}; λ_{max} (EtOH): 207.5 (ε 15935), 208.5 (ε 16015), 225.0 (ε 12222) and 348.4 (ε 10945) nm.

5-Amino-1-(4-methylphenyl)-4-(cyanoformimidoyl)-1*H*-imidazole 3c: Recrystallization of the product from dry diethyl ether (1:1) gave pale green crystals (0.97 g, 4.02 mmol, 91 %). M.p. 124–125 °C (decomp.). [Found: C, 59.7; H, 4.9; N, 28.9. $C_{12}H_{11}N_5$ requires: C, 59.8; H, 4.6; N, 29.0 %]; *m/z* (EI): 242 $(M + 1)^+$ 22 %, 241 $(M)^+$ 55.5 %, 240 $(M\text{-}1)^+$ 27.3 %, 106 $[(M + 1)\text{-}C_5H_4N_5]^+$ 17 %; δ_H (500 MHz, d_6-DMSO): 2.54 (s, 3H, CH_3), 6.72 (br. s, 2H, NH_2), 7.22 (d, 2H, $^3J_{10,9}$ 8 Hz, H10 & H12), 7.46 (s, 1H, H2), 7.54 (d, 2H, $^3J_{13,12}$ 8 Hz, H9 & H13), 11.12 (br. s, 1H, NH) ppm; δ_C (75 MHz, d_6-DMSO): 163.5 (C6), 148.3 (C11), 147.2 (C5), 136.5 (C2),

130.6 (C9 & C13), 130.6 (C8), 120.3 (C7), 119.2 (C10 & C12), 117.5 (C4), 34.4 (C14) ppm; ν_{max} (Nujol mull): 3,420 m, 3,260 s, 3,120 m (N-H str.), 2,230 s (CN str.), 1,635 m (C=N str.), 1,600 s (N-H bend), 1,580 s, 1,550 s, 1,525 s, 1,250 s, 1,180 m, 1,170 m, 1,015 s, 930 s, 800 m cm^{-1}; λ_{max} (EtOH): 201.8 (ε 22405), 205.0 (ε 13933), 205.5 (ε 14021), 206.2 (ε 13977), 211.9 (ε 11736), 227.3 (ε 16098), 228.1 (ε 15539), 348.1 (ε 9722) and 348.5 (ε 9571) nm.

9.2.4 General Procedure for the Preparation of the 5-Amino-1-aryl-4-cyanoimidazoles 4a-c

An aqueous solution of 1 M potassium hydroxide (1 ml) was added to a suspension of the formamidine (0.5 g) in ethanol (1 ml) and the mixture was stirred at room temperature for approximately 1 h. The white solid which precipitated was washed with water, a few drops of ethanol and finally diethyl ether before drying under vacuum. The yields of these reactions were 75–90 %.

5-Amino-1-phenyl-4-cyanoimidazole 4a: Recrystallization of the product from a mixture of dry diethyl ether/ethanol (10:1) gave an off-white solid (0.38 g, 2.07 mmol, 90 %). M.p. 194–196 °C (decomp.). [Found: C, 65.1; H, 4.6; N, 30.4. $C_{10}H_8N_4$ requires: C, 65.2; H, 4.3; N, 30.4 %]: *m/z* (EI): 185 $(M + 1)^+$ 5.8 %, 184 $(M)^+$ 24 %, 77 $(M\text{-}C_4H_3N_4)^+$ 12 %; δ_H (500 MHz, d_6-DMSO): 6.20 (s, 2H, NH_2), 7.42 (s, 1H, H2), 7.48–7.68 (m, 5H, H8, H9, H10, H11 & H12) ppm; δ_C (75 MHz, d_6-DMSO): 151.2 (C5), 137.9 (C2), 136.6 (C7), 133.8 (C10), 132.6 (C9 & C11), 129.3 (C8 & C12), 121.1 (C6), 95.0 (C4) ppm; ν_{max} (Nujol mull): 3,370 m, 3,300 s, 3,180 s, 3,110 m (N-H str.), 2,210 s (CN str.), 1,680 s (C=N str.), 1,600 s (N-H bend), 1,530 s, 1,265 s, 1,230 s, 1,100 s, 915 s, 780 s, 755 s cm^{-1}; λ_{max} (EtOH): 202.7 (ε 16643), 223.8 (ε 16407), 248.0 (ε 11075) and 346.2 (ε 177) nm.

5-Amino-1-benzyl-4-cyanoimidazole 4b: Recrystallization of the product from a mixture of dry diethyl ether/ethanol (10:1) gave an off-white solid (0.36 g, 1.8 mmol, 82 %). M.p. 197–198 °C (decomp.). [Found: C, 66.7; H, 5.3; N, 28.3. Calc. for $C_{11}H_{10}N_4$: C, 66.7; H, 5.1; N, 28.3 %] ; *m/z* (EI): 199 $(M + 1)^+$ 100 %, 198 $(M)^+$ 4.5 %, 108 $[(M + 1)\text{-}C_7H_7]^+$ 9.6 %; δ_H (500 MHz, d_6-DMSO): 5.12 (s, 2H, CH_2), 6.42 (s, 2H, NH_2), 7.22–7.28 (m, 2H, H10 & H12), 7.32 (s, 1H, H2), 7.34–7.48 (m, 3H, H9, H11 & H13) ppm; δ_C (75 MHz, d_6-DMSO): 152.6 (C5), 140.3 (C8), 136.7 (C2), 132.4 (C11), 131.6 (C10 & C12), 131.5 (C13 & C9), 121.4 (C6), 94.2 (C4), 49.9 (C7) ppm; ν_{max} (Nujol mull): 3,380 s, 3,340 w, 3,200 s, 3,100 w (N-H str.), 2,200 s (CN str.), 1,650 s (C=N str.), 1,590 s (N-H bend), 1,530 s, 1,180 s, 1,000 s, 850 m, 800 s cm^{-1}; λ_{max} (EtOH): 202.2 (ε 13538), 245.7 (ε 12771) and 340.0 (ε 250) nm.

5-Amino-1-(4-methylphenyl)-4-cyanoimidazole 4c: Recrystallization of the product from a mixture of dry diethyl ether/ethanol (10:1) gave a white solid (0.33 g, 1.68 mmol, 75 %). M.p. 167–169 °C (decomp.). [Found: C, 66.3; H, 5.2; N, 28.8.

$C_{11}H_{10}N_4$ requires: C, 66.7; H, 5.1; N, 28.3 %]; *m/z* (EI): 198 $(M)^+$ 100 %, 91 $(M\text{-}C_4H_3N_4)^+$ 4.0 %; δ_H (500 MHz, d_6-DMSO): 2.42 (s, 3H, CH_3), 6.12 (s, 2H, NH_2), 7.14 (d, 2H, $^3J_{8,9}$ 9 Hz, H9 & H11), 7.39 (d, 2H, $^3J_{12,11}$ 9 Hz, H8 & H12), 7.41 (s, 1H, H2) ppm; δ_C (75 MHz, d_6-DMSO): 163.4 (C10), 151.5 (C5), 136.8 (C2), 131.1 (C8 & C12), 130.5 (C7), 121.4 (C6), 119.4 (C9 & C11), 94.2 (C4), 35.6 (C13) ppm; ν_{max} (Nujol mull): 3,360 w, 3,340 s, 3,200 s, 3,170 m (N-H str.), 2,240 s (CN str.), 1,650 s (C=N str.), 1,580 s (N-H bend), 1,530 s, 1,260 s, 960 s, 810 s cm^{-1}; λ_{max} (EtOH): 202.5 (ε 12808) and 232.8 (ε 13632) nm.

9.3 Results and Discussion

Treatment of **1** with an equimolar amount of the appropriate aromatic amine at room temperature in ethanol in the presence of a catalytic amount of anilinium chloride afforded the corresponding formamidines **2a-c** in 82–92 % yields (Scheme 9.1). Isolation of product was achieved by simple filtration of the final reaction mixture and little or no further purification was required to give analytically pure products. The anilinium chloride is believed to act as a general acid catalyst and a significant acceleration of the rate of reaction was observed. When it is absent, appreciable decomposition occurs and the amidine product becomes difficult to isolate from tarry by-products with consequent low yields. The amine hydrochloride salts of the arylamine reactions also catalyse these reactions, but it is often more convenient to use the anilinium salt.

The spectroscopic data obtained on these compounds **2a-c** were satisfactory. The high resolution mass spectra gave molecular ions at 212, 226, 226 $(M + 1)^+$ consistent with the molecular weights of 211, 225, 225 for the formamidines **2a-c**. The ^{1}H NMR spectra of compounds **2a-c** in deuterated dimethyl sulfoxide had some interesting features. The HC=N proton appeared at δ 7.30–7.74 ppm and NH proton appeared as a broad singlet at δ 8.16–9.90 ppm and was confirmed by D_2O exchange. The ^{13}C NMR spectra were fully consistent with the assigned structures. The infrared spectra of amidines **2a-c** showed two strong absorptions in the region

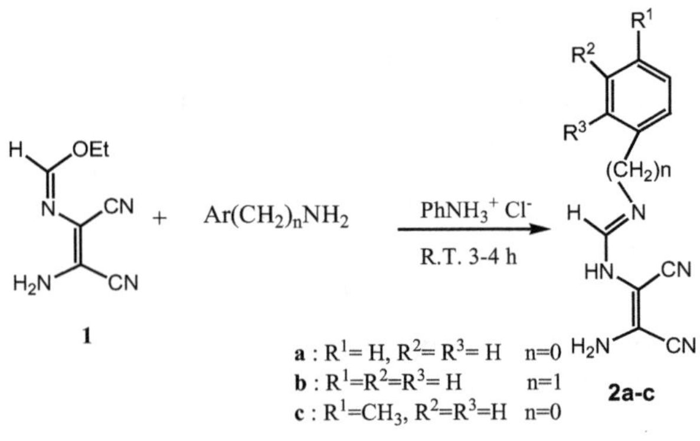

Scheme 9.1 Synthetic route toward preparation of formamidines **2a-c**

a : R^1= H, R^2= R^3= H n=0
b : R^1=R^2=R^3= H n=1
c : R^1=CH_3, R^2=R^3=H n=0

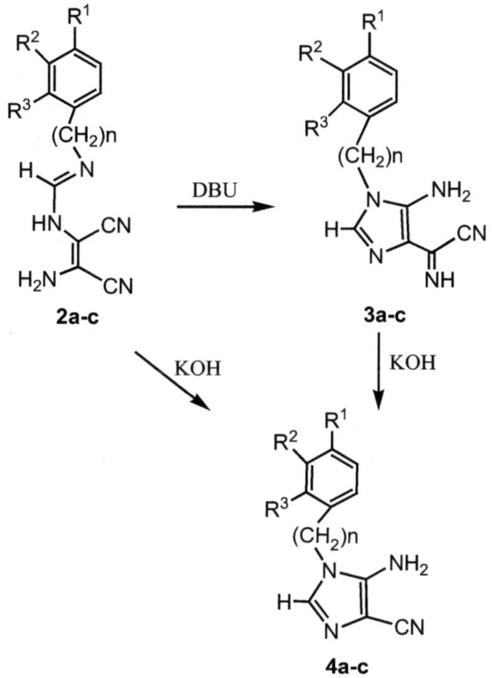

Scheme 9.2 Synthetic route toward preparation of 5-amino-1-aryl-4-(cyanoformimidoyl) imidazoles **3a-c** and 5-amino-1-aryl-4-cyanoimidazoles **4a-c**

2,210–2,225 cm^{-1} characteristic of C≡N stretching vibrations, together with an NH and a C=N stretching vibrations at 3,100–3,460 cm^{-1} and 1,610–1,650 cm^{-1} respectively.

When several drops of DBU were added to a suspension of the amidines **2a-c** in ethyl acetate or ethanol, cyclization occurred in 1–3 hours to give the corresponding 5-amino-1-aryl-4-(cyanoformimidoyl)imidazoles **3a-c** in good yields (85–91 %). All the compounds were obtained in an analytically pure state and IR spectra showed strong bands in the 2,200–2,230 cm^{-1} region for a C≡N bond and C=N stretching vibrations within the region of 1,660–1,635 cm^{-1}. In all cases, the signals were sharp and in the ^{1}H NMR spectra the CH proton of the imidazole ring appeared in the range δ 7.42–7.72, the NH_2 protons at δ 6.68–6.76 and the NH proton at δ 10.88–11.20 ppm. The ^{13}C NMR spectra were fully consistent with the assigned structures.

When a saturated solution of potassium hydroxide in ethanol was added to a suspension of the amidines **2a-c** in an alcohol at room temperature the corresponding 5-amino-1-aryl-4-cyanoimidazoles **4a-c** were formed in good yields (75–90 %). These compounds can also be made in comparable yields by the reaction of the compounds **3a-c** with a saturated solution of potassium hydroxide in ethanol under conditions similar to those described above (Scheme 9.2).

Compounds **4a-c** were recrystallised from a mixture of ethanol/methanol (1:1) and gave pale yellow to off-white crystals respectively. These were fully characterized by microanalysis, IR, ^{1}H, ^{13}C NMR and mass spectroscopy. The high resolution mass spectra gave molecular ions at 185, 199, 199 $(M + 1)^+$ consistent with the expected molecular weights of 184, 198 and 198 for the imidazoles **4a-c**. The infrared spectra confirmed the presence of the NH and C=N stretching vibrations within the region of 3,100–3,380 cm^{-1} and 1,650–1,680 cm^{-1} respectively. The infrared spectra also showed sharp absorption bands within the range of 2,210–2,240 cm^{-1} for the C≡N stretching vibration. In the ^{1}H NMR spectra of the isolated 5-amino-1-aryl-4-cyanoimidazoles, the primary amine protons were observed in the region of δ 6.12–6.42 ppm and in several cases the assignment was confirmed by D_2O exchange. The protons of the imidazole rings appeared as sharp singlets in the range of δ 7.34–7.42 ppm. The ^{13}C NMR spectra of the compounds **4a-c** had the expected number of peaks. The signal for the C-2 carbon of the imidazole ring appeared in the region of 136.7–137.9 ppm.

Acknowledgement We are thankful to the University of Guilan Research Council for partial support of this work.

References

1. Sen AK, Mukhopadhyay AK (1984) Preparation of 5-amino-4-cyano-1-(*p*-aminosulfonylphenyl)imidazole. Ind J Chem 238:870–877
2. Frank I, Zeller M (1990) Imidazole and benzimidazole synthesis. Synth Commun 20:2519–2526
3. Alves MJ, Booth BL, Proenca MFP (1987) Some reactions of 5-amino-4-(C-cyanoformimidoyl) imidazoles obtained from nitrilium trifluoromethane sulphonate salts and diaminomaleonitrile. J Chem Soc Perkin Trans 1:1521–1526
4. Booth BL, Dias AM, Proenca MF (1992) Synthesis of 9-hydroxyalkyl-substituted purines from the corresponding 4-(C-cyanoformimidoyl) imidazole-5-amines. J Chem Soc Perkin Trans 1:2119–2122
5. Yahyazadeh A, Booth BL (2001) Synthesis of aryl (or benzyl)-(Z)-N-[2-amino-1,2-dicyanovinyl]formimidines. Synth Commun 23:3617–3622
6. Yahyazadeh A, Booth BL (2001) Synthesis of 5-amino-1-aryl-4-cyanoimidazoles from N-aryl-N-(1,2-dicyanovinyl)-formamidines. Synth Commun 21:3225–3230
7. Woodward DW (1950) 2-Ethylamino-2-methyl-propanol-(1). U.S. Patent 2 534 331-340
8. Booth BL, Alves MJ, Carvalho A, Eastwood PR, Nezhat L (1994) Synthesis of 9-aryl-6-carbamoyl-1,2-dihydropurines and a study of their tautomerism. J Chem Soc Perkin Trans 2:1949–1956
9. Yahyazadeh A, Abbasi J (2006) Synthesis and characterization of some new 5-amino-1-aryl-1*H*-imidazole-4-carbonitrile. Int J Chem 16:257–259
10. Yahyazadeh A, Sharifi Z (2006) Synthesis of 5-amino-4-cyano formimidoylimidazoles and 5-amino-4-cyanoimidazoles from amidines. Phosphorus Sulfur 181:1339–1343
11. Yahyazadeh A (2007) Synthesis and characteristics of 2-bromophenyl-6-iminopurine from 2-bromophenyl-4-cyanoimidazole. E-J Chem 4:310–312
12. Montgomery JA (1986) Synthesis of 9-benzyl-6-aminopurines from 5-amino-1-benzyl-4-cyanoimidazole. Acc Chem Res 19:293–298

13. Birkett PR, King H, Chapleo CB, Ewing DF, Mackenzie G (1993) Synthesis of optically active-α-(imidazol-1-yl, purin-9-yl or uracil-1-yl)propanoic and succinic acid derivatives. Tetrahedron 49:11029–11036
14. Matsumoto H, Hara S, Nagata N, Ikeda K (1995) Synthetic studies toward the novel peptide nucleic acid. Heterocycles 41:47–52
15. Yahyazadeh A (2003) Synthesis of 9-aryl-6-aminopurines from 5-amino-1-aryl-1*H*-imidazole-4-carbonitriles. Russ J Org Chem 39:1649–1651
16. Yahyazadeh A, Pourrostam B (2003) Synthesis of 9-benzyl-6-aminopurines from 5-amino-1-benzyl-4-cyanoimidazoles. Bull Korean Chem Soc 24:1723–1724
17. Yahyazadeh A, Hossani F (2007) Synthesis and characterization of 6-carbamoyl-1,2-dihydropurines. E-J Chem 4:376–380
18. Carvalho MA, Alvares Y, Zaki ME, Proenca MF, Booth BL (2004) Synthesis of novel 6-enaminopurines. Org Biomol Chem 2:2340–2345
19. Viktor O, Iaroshenko AM, Dmytro O, Alexander V, Anke S, Peter L (2011) Efficient synthesis of purines by inverse electron-demand Diels-Alder reactions of 1-substituted-1*H*-imidazol-5-amines with 1,3,5-triazines. Tetrahedron 67:8321–8330
20. Yahyazadeh A, Hadisinea S, Daneshmandi MS (2012) Silica sulfuric acid as an efficient and reusable heterogeneous catalyst for the synthesis of formamidines from the diaminomaleonitrile. Iran J Cat 2:153–156

Chapter 10
Synthesis and Characterization of 6-Carbamoyl-2-Alkyl-9-(Phenyl or Benzyl)-9*H*-Purines

Asieh Yahyazadeh

Abstract The 2-(5-amino-1-phenyl or benzyl)-4-(cyanoformimidoyl)-1*H*-imidazoles proved to be important intermediates not only in the preparation of new 6-carbamoyl purines but also in the synthesis of 6-amino-purine, 6-cyano-purine and 1,2-dihydropurines. 6-Carbamoyl-2-alkyl-9-(phenyl or benzyl)-9*H*-purines have been synthesized in high yields by reactions between 2-(5-amino-1-phenyl or benzyl)-4-(cyanoformimidoyl)-1*H*-imidazoles and acetylacetone at room temperature. Intramolecular hydrogen bonding appears between H of NH_2 and N1 at purine ring. All the compounds have been fully characterized by spectroscopic data.

10.1 Introduction

Purines are extremely important compounds with a wide array of synthetic and industrial applications. They are integral parts of DNA and RNA, playing an essential role in several biological processes with considerable chemical and pharmacological importance. The biological importance of the purine structure is evident from the countless derivatives that have been prepared and are active, especially as antiviral and antitumor agents [1–4]. The substituent in the 6-position plays an important role in the potency and selectivity of the purine derivatives. As a result, great efforts have been directed to the synthesis and biological evaluation of a number of 6-substituted purines.

A. Yahyazadeh (✉)
Department of Chemistry, University of Guilan, P.O. Box 41335-1914, Rasht, Iran
e-mail: yahyazadehphd@yahoo.com; yahyazadeh@guilan.ac.ir

M. Gupta Bhowon et al. (eds.), *Chemistry: The Key to our Sustainable Future*,
DOI 10.1007/978-94-007-7389-9_10,

10.2 Experimental

10.2.1 General Techniques

All solvents were purified and dried using established procedures. The ^{1}H NMR spectra were recorded on Bruker XL 500 (500 MHz) instruments (with J-values given in Hz), and IR spectra on a Shimadzu IR-470 spectrophotometer. The melting points were measured on an Electro-thermal digital melting point apparatus with the thermometer uncorrected.

10.2.2 General Procedure for the Preparation of the 2-(5-Amino-1-phenyl or benzyl)-4-(cyanoformimidoyl)-1H-imidazoles 3a-d

To a suspension of the corresponding aryl-(*Z*)-*N*-[2-amino-1,2-dicyanovinyl] formamidine **2a-d** (1.00 g) in dry ethanol (8 ml) was added DBU (9 drops) [6–8, 13]. The mixture was stirred under an argon atmosphere at room temperature for 0.5–3 h until TLC showed that all the amidine had been consumed. The reaction mixture was then filtered off, washed with dry diethyl ether and dried under vacuum to give **3a-d**.

10.2.3 General Procedure for the Preparation of the 6-Carbamoyl-2-alkyl-9-(phenyl or benzyl)-9H-purines 4a-d

1,3-Dione (5 ml) was added to a solution of 2-(5-amino-1-phenyl or benzyl)-4-(cyanoformimidoyl)-1*H*-imidazoles **3a-d** (0.12 g) in acetonitrile (10 ml) and the mixture was stirred at room temperature for 20–24 h. The reaction mixture was then filtered off, washed with dry diethyl ether and dried under vacuum to give **4a-d**.

Synthesis of 6-carbamoyl-2-methyl-9-(4-ethoxyphenyl)-9*H*-purine 4a: Recrystallization of the product from dry diethyl ether and air-drying gave white crystals of 6-carbamoyl-2-methyl-9-(4-ethoxyphenyl)-9*H*-purine **4a** (0.87 g, 72 %); mp 287–288 °C; Analysis: [Found: C, 59.80; H, 5.24; N, 24.10. Calc. for $C_{15}H_{15}N_5O_2$: C, 60.60; H, 5.05; N, 23.56 %]; IR (KBr): 3,400 s, 3,180 s, 3,100 w, 2,980 w, 1,650 m, 1,610 m, 1,520 s, 1,480 w, 1,400 s, 1,345 w, 1,250 s, 1,220 s, 1,040 m cm^{-1}; ^{1}H NMR (DMSO) (δ ppm): 8.92 (s, 1H, H_8), 8.56 (s, br, NH), 8.06 (s, br, NH), 7.72 (d, J = 8.5 Hz, 2H, H_{10} & H_{14}), 7.14 (d, $^{3}J_{13,14}$ = 8.6 Hz, 2H, H_{11} & H_{13}), 4.10 (q, 2H, CH_2), 2.50 (s, 3H, CH_3), 1.35 (t, 3H, CH_3); MS (EI, 70 eV): *m*/*z* (%) 298 (45) $(M + 1)^+$, 297 (100) $(M)^+$.

Synthesis of 6-carbamoyl-2-methyl-9-(3,4-dimethoxyphenyl)-9*H*-purine 4b: Recrystallization of the product from dry diethyl ether and air-drying gave green crystals of 6-carbamoyl-2-methyl-9-(3,4-dimethoxy phenyl)-9*H*-purine **4b** (0.84 g, 70 %); mp 218 °C; Analysis: [Found: C, 57.80; H, 4.81; N, 22.40. Calc. for $C_{15}H_{15}N_5O_3$: C, 57.50; H, 4.79; N, 22.36 %]; IR (KBr): 3,450 s, 3,380 s, 3,100 w, 1,680 m, 1,600 m, 1,520 s, 1,450 w, 1,345 w, 1,250 s, 1,220 s, 1,020 m cm^{-1}. ^1H NMR ($CDCl_3$): 8.56 (s, br, NH), 8.35 (s, 1H, H_8), 7.23 (d, J = 3.6 Hz, 1H, H_{10}), 7.16 (dd, J = 3.5 Hz, 1H, H_{14}), 7.01 (d, J = 8.5 Hz, 1H, H_{13}), 6.29 (s, br, NH), 3.95 (s, 6H, OCH_3), 2.88 (s, 3H, CH_3) ppm; MS (EI, 70 eV): *m*/*z* (%) 314 (24) $(M + 1)^+$, 313 (100) $(M)^+$.

Synthesis of 6-carbamoyl-9-(3,4-dimethoxybenzyl)-2-(4-isopropylphenyl)-9*H*-purine 4c: Recrystallization of the product from dry diethyl ether and air-drying gave white crystals of 9-(3,4-dimethoxybenzyl)-9*H*-purine-6-carboxamide **4c** (0.12 g, 0.27 mmol, 64 %); mp 218–219 °C; Analysis: [Found: C, 65.80; H, 6.04; N, 16.40. Calc. for $C_{24}H_{26}N_5O_3$: C, 66.66; H, 6.01; N, 16.20 %]; IR (KBr): 3,330 s, 3,200 s (N-H), 3,100 w (C-H Ar), 2,900 m (C-H aliphatic), 1,690 s (C=O), 1,650 s (C=N), 1,475 m, 1,400 m, 1,350 s, 1,250 s, 1,100 m, 1,020 m cm^{-1}; ^1H NMR (DMSO): 9.09 (s, 1H, $H_{imidazole}$), 8.59 (br. s, 1H, NH_2), 8.44 (d, 2H, J = 7.73 Hz, H_{Ar}), 8.13 (br. s, 1H, NH_2), 7.64 (s, 1H, H_{Ar}), 7.53 (d, 1H, J = 8 Hz, H_{Ar}), 7.42 (d, 2H, J = 7.72 Hz, H_{Ar}), 7.25 (d, 1H, J = 8 Hz, H_{Ar}), 3.90, 3.88 (s, 6H, 2 × OCH_3), 2.98 (m, 1H, $H_{isoprop}$), 2.53 (s, 2H, CH_2), 1.29, 1.25 (s, 6H, $2CH_3$) ppm; MS (EI, 70 eV): *m*/*z* (%) 433 (42) $(M + 1)^+$, 432 (87) $(M)^+$.

Synthesis of 6-carbamoyl-9-(3,4-dimethoxybenzyl)-2-(3,4-dimethoxyphenyl)-9*H*-purine 4d: Recrystallization of the product from dry diethyl ether and air-drying gave white crystals of 6-carbamoyl-9-(3,4-dimethoxybenzyl)-2-(3,4-dimethoxyphenyl)-9*H*-purine **4d** (0.14 g, 0.31 mmol, 74 %); mp 140–142 °C; Analysis: [Found: C, 61.65; H, 5.48; N, 16.01. Calc. for $C_{23}H_{21}N_5O_5$: C, 61.33; H, 5.33; N, 15.55 %]; IR (KBr): 3,320 s (N-H), 3,010 m (C-H Ar), 2,930 m (C-H aliphatic), 1,615 s (C=O), 1,560 s (C=N), 1,510 s (C=C), 1,475 m, 1,420 m, 1,265 s, 1,020 s, 1,140 m, 818 s, 760 s cm^{-1}; ^1H NMR (DMSO): 8.18 (s, 1H, $H_{imidazole}$), 7.84 (br. s, 1H, NH_2), 7.70 (s, 2H, H_{Ar}), 7.47 (d, 2H, J = 7.5 Hz, H_{Ar}), 7.04 (d, 2H, J = 7.5 Hz, H_{Ar}), 3.84, 3.83 (s, 12H, 4 × OCH_3), 3.59 (s, 2H, CH_2) ppm; MS (EI, 70 eV): *m*/*z* (%) 451 (44) $(M + 1)^+$, 450 (88) $(M)^+$.

10.3 Results and Discussion

In our research, we have been studying the reactivity of 2-(5-amino-1-phenyl or benzyl)-4-(cyanoformimidoyl)-1*H*-imidazoles **3a-d**, as versatile precursors to 6-substituted purines [2, 5–10]. Amidines **2a-d** were prepared *via* a multistep synthesis from diaminomaleonitrile [11–18]. We thus attempted to cyclise the amidines **2a-d** to obtain compounds **3a-d**, in 72–80 % yield by treating with DBU (1,8-diazabicyclo[5.4.0]undec-7-ene) in ethanol [9–13] (Scheme 10.1 and Table 10.1).

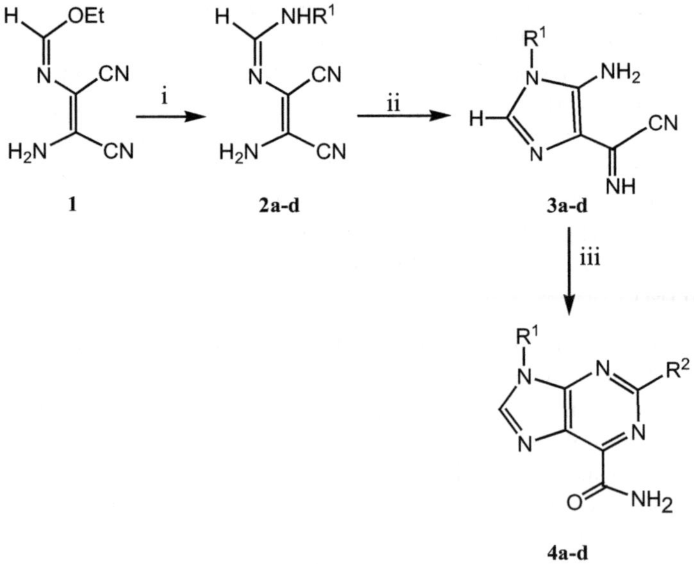

Scheme 10.1 Synthesis of 6-carbamoyl-2-alkyl-9-(phenyl or benzyl)-9*H*-purines; Reagents and conditions: i, RNH_2, $PhNH_3^+Cl^-$, r.t., 3–4 h; ii, DBU, EtOH, r.t.; iii, Dione, CH_3CN, r.t.

Table 10.1 Synthesis of purines **4a-d**

Entry	Product	R^1	R^2	Yield (%)
3a	**4a**	4-$(C_2H_5O)C_6H_4$-	CH_3-	72
3b	**4b**	3,4-$(CH_3O)_2C_6H_3CH_2$-	CH_3-	70
3c	**4c**	3,4-$(CH_3O)_2C_6H_3CH_2$-	4-$(CH_3)_2CH$-C_6H_4-	64
3d	**4d**	3,4-$(CH_3O)_2C_6H_3CH_2$-	3,4-$(CH_3O)_2C_6H_3$-	74

The spectroscopy results obtained on these compounds **3a-d** were satisfactory. The ^{1}H NMR spectra showed the presence of two broad singlets in the regions of 5.63–7.83 and 7.21–7.41 ppm due to the amine protons and a singlet at 7.21–11.05 ppm for the HC proton of the imidazole ring. The ^{13}C NMR spectra were fully consistent with the assigned structures. The infrared spectra confirmed the presence of the NH and C=N stretching vibrations within the region of 3,450–3,180, and 1,640–1,620 cm^{-1} respectively. The infrared spectra also showed a sharp absorption band at 2,300–2,200 cm^{-1} for the C≡N stretching vibration.

The purines **4a-d** were prepared by stirring a suspension of the corresponding 2-(5-amino-1-phenyl or benzyl)-4-(cyanoformimidoyl)-1*H*-imidazoles **3a-d** with a slight excess of dione in acetonitrile at room temperature (Scheme 10.1). The reactions were monitored by TLC (mixture of dry diethyl ether and chloroform 2:3) and reaction times varied between 20 and 24 h. Depending upon the solvent used for the reaction and the rate of precipitation, these purines can be isolated as solids in color from white to green. These were fully characterized by TLC, IR, ^{1}H NMR and mass spectroscopy. The infrared spectra of compounds **4a-d**

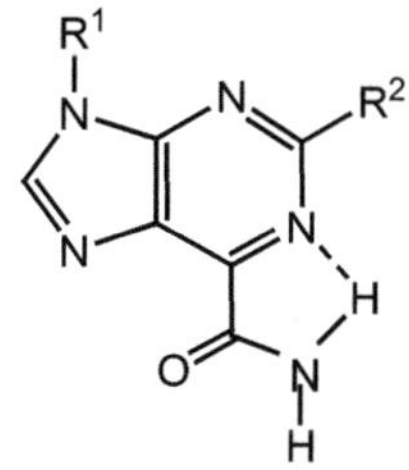

Scheme 10.2 The intra-molecular hydrogen bond in purine

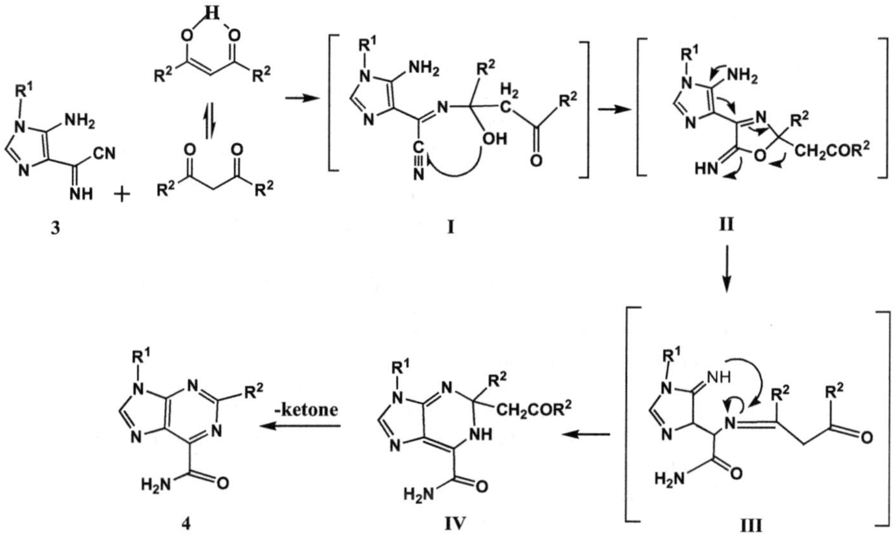

Scheme 10.3 Proposed mechanism for the formation of purine **4**

confirmed the presence of the NH and C=N stretching vibrations within the region of 3,450–3,010, and 1,650–1,560 cm^{-1} respectively. The C=O of the amide group appeared in the region of 1,680–1,615 cm^{-1} as a strong band.

In the ^{1}H NMR spectra of the isolated compounds **4a-d**, the NH_2 protons appeared as two broad singlets in the range δ 6.29–8.56 ppm, because of combined intramolecular hydrogen bonding from an amide N-H to an N atom of the six-member pyrimidine ring and deshielding appears between H of NH_2 and N1 at purine (Scheme 10.2). The H-8 proton of the imidazoles ring was seen as a sharp singlet in the region of 8.18–9.09 ppm and the aromatic protons showed the expected patterns in the range of 7.01–7.70 ppm. The high resolution mass spectra gave a molecular ion peak at 298, 314, 433, 451 $(M + 1)^{+}$ which fit with the expected molecular weight of 297, 313, 432, 450 for the purines **4a-d**. The elemental analysis results of purines **4a-d** were satisfactory.

The proposed mechanism for formation of purine **4** is as follows: the first intermediate **I** was formed by the attack of imine to the carbonyl group of 1,3-dione thereby rearranging it to compound **II** and **III**. At the end purine **4** was formed with removal of the acetone from compound **IV** (Scheme 10.3).

Acknowledgements We are thankful to the Guilan University Research Council for partial support of this work.

References

1. Alves MJ, Booth BL, Freitas AP, Proenca MF (1992) (Z)-N3-(2-Amino-1,2-dicyanovinyl) formamidrazone: a precursor in the synthesis of 1,5-diaminoimidazoles and 6-carbamoyl-1, 2-dihydropurines. J Chem Soc Perkin Trans 1:913–917
2. Alves MJ, Booth BL, Proenca MFP (1987) Some reactions of 5-amino-4-(C-cyanoformimidoyl) imidazoles obtained from nitrilium trifluoromethane sulphonate salts and diaminomaleonitrile. J Chem Soc Perkin Trans 1:1521–1526
3. Alves MJ, Booth BL, Proenca MF (1994) Synthesis of 4- and 5-disubstituted 1-benzylimidazoles, important precursors of purine analogs. J Heterocycl Chem 31:345–351
4. Yahyazadeh A, Pourrostam B (2003) Synthesis of 9-benzyl-6-aminopurines from 5-amino-1-benzyl-4-cyanoimidazoles. Bull Korean Chem Soc 24:1723–1724
5. Booth BL, Alves MJ, Carvalho A, Eastwood PR, Nezhat L (1994) Synthesis of 9-aryl-6-carbamoyl-1,2-dihydropurines and a study of their tautomerism. J Chem Soc Perkin Trans 2:1949–1956
6. Yahyazadeh A (2003) Synthesis of 9-aryl-6-aminopurines from 5-amino-1-aryl-1*H*-imidazole-4-carbonitriles. Russ J Org Chem 39:1649–1651
7. Yahyazadeh A, Sharifi Z (2006) Synthesis of 5-amino-4-cyano formimidoylimidazoles and 5-amino-4-cyanoimidazoles from amidines. Phosphorus Sulfur 181:1339–1343
8. Yahyazadeh A, Booth BL (2001) Synthesis of 5-amino-1-aryl-4-cyanoimidazoles from N-aryl-N-(1,2-dicyanovinyl)-formamidines. Synth Commun 21:3225–3230
9. Al-Azmi A, Elassar AA, Booth BL (2003) The chemistry of diaminomaleonitrile and its utility in heterocyclic synthesis. Tetrahedron 59:2749–2763
10. Assunção LR, Marinho ER, Proença FP (2010) Guanidine: studies on the reaction with ethyl N-(2-amino-1,2-dicyanovinyl)formimidate. Arkivoc 5:82–91
11. Yahyazadeh A, Hadisinea S, Daneshmandi MS (2012) Silica sulfuric acid as an efficient and reusable heterogeneous catalyst for the synthesis of formamidines from the diaminomaleonitrile. Iran J Catal 2:153–156
12. Booth BL, Cabral IM, Dias AM, Freitas AP, Matos-Beja AM, Proenca MF, Ramos-Silva M (2001) A new and efficient approach to the synthesis of 6-amidino-2-oxopurines. J Chem Soc Perkin Trans 1:1241–1248
13. Carvalho MA, Esteves TM, Proenca MF, Booth BL (2004) Efficient conversion of 6-cyanopurines into 6-alkoxyformimidoylpurines. Org Biomol Chem 2:1019–1024
14. Yahyazadeh A, Hossani F (2007) Synthesis and characterization of 6-carbamoyl-1, 2-dihydropurines. E-J Chem 4:376–380
15. Carvalho MA, Alvares Y, Zaki ME, Proenca MF, Booth BL (2004) Synthesis of novel 6-enaminopurines. Org Biomol Chem 2:2340–2345
16. Viktor O, Iaroshenko AM, Dmytro O, Alexander V, Anke S, Peter L (2011) Efficient synthesis of purines by inverse electron-demand Diels–Alder reactions of 1-substituted-1*H*-imidazol-5-amines with 1,3,5-triazines. Tetrahedron 67:8321–8330
17. Hong FL, Liang ZZ, Ding MW, Jun TZ (2011) Microwave assisted synthesis of 2,6-substituted aromatic-aminopurine derivatives. J Heterocycl Chem 48:1140–1144
18. Dilip KT, Francesca D, Khai P, Zhan-Guo G, Tina C, Elizabeth G, John A, Kenneth AJ (2012) Structure-guided design of A3 adenosine receptor-selective nucleosides: combination of 2-arylethynyl and bicyclo[3.1.0]hexane substitutions. J Med Chem 55:4847–4851

Chapter 11
Therapeutic Potential of Common Culinary Herbs and Spices of Mauritius

Jugjeet S. Ramkissoon, Mohamad F. Mahomoodally, Nessar Ahmed, and Anwar H. Subratty

Abstract Compounds with combined anti-glycation and antioxidant properties may offer therapeutic potential for diabetic and oxidative stress related pathologies. We previously demonstrated significant ($p < 0.05$) anti-glycation properties of culinary herbs and spices by an optimized *in vitro* glucose-bovine serum albumin assay. In the present investigation we describe the antioxidant potential of these plants and the therapeutic potential of dietary compounds as possible anti-glycation and antioxidant agents. This study was geared to appraise the total antioxidant and free radical scavenging activities of ten common culinary herbs and spices (*Allium sativum, Zingiber officinale, Thymus vulgaris, Petroselinum crispum, Murraya koenigii* Spreng, *Mentha piperita* L., *Curcuma longa* L., *Allium cepa* L., *Allium fistulosum* and *Coriandrum sativum* L.) commercially available on the Mauritian market. The total antioxidant capacity as measured by the phosphomolybdenum method ranged from 0.76 to 2.49 μg AAE/ml. It was observed that the ethanolic extracts exhibited significant ($p < 0.05$) free radical-scavenging potential as measured by the 2,2′-azinobis-(3-ethylbenzothiazoline-6-sulphonic acid) radical cation assay, ferrous ion chelating assay and the Griess assay. The observed anti-glycation and antioxidant activity of the extracts together with their previously reported total phenolic, flavonoid and tannin contents support the therapeutic value of the plants investigated. Dietary agents interfering in the glycation pathway might offer new lead compounds geared towards glucose-derived and oxidative stress related complications. Our findings showed that the herbs and spices studied possess significant antioxidant and anti-glycation properties and hence can be exploited as functional foods.

J.S. Ramkissoon • M.F. Mahomoodally (✉) • A.H. Subratty
Department of Health Sciences, Faculty of Science, University of Mauritius, Réduit, Mauritius
e-mail: groszacko@yahoo.com; f.mahomoodally@uom.ac.mu; subratty@uom.ac.mu

N. Ahmed
School of Healthcare Science, Manchester Metropolitan University, Manchester, United Kingdom
e-mail: N.Ahmed@mmu.ac.uk

M. Gupta Bhowon et al. (eds.), *Chemistry: The Key to our Sustainable Future*,
DOI 10.1007/978-94-007-7389-9_11,

11.1 Introduction

Plants have historically been an important source of lead molecules in drug discovery [1, 2]. Several plants have been used as dietary adjuvant and in treating and/or managing a number of diseases even without any knowledge of their proper functions and constituents [3]. This practice may be attributed to the cost and side effects of synthetic agents. Medicinal foods are used widely even when their biologically active compounds are unknown, because of their safety, effectiveness, and availability [4]. Various plants have been found to possess significant anti-diabetic properties [5]. Use of these plants may delay the development of diabetic complications. These days great attention is being given to management of diabetes with medicinal plants along with dietary restriction [6]. The World Health Organization (WHO) has recommended the evaluation of traditional plant treatments for diabetes as they are cost-effective, non-toxic, with less or no side effects and are considered to be excellent candidates for oral therapy [7]. There is an increasing demand by patients to use natural products with anti-diabetic activity [8]. Herbal therapy has been used to treat various types of disease including diabetes all over the world successfully [9]. The use of nutraceuticals is an attempt to accomplish desirable therapeutic outcomes with reduced side effects, as compared with other synthetic therapeutic agents [6].

Plants with known antioxidant and anti-glycation properties are receiving a lot of attention as they are believed to have less adverse effects [10]. Some plants contain chemicals that can be used directly as drugs or lead compounds for hypoglycemic properties useful in the management of diabetic complications [2]. For example polyphenolic compounds including tannin are known to possess free radical scavenging properties [10]. Curcumin from turmeric (*Curcuma longa*) reduces blood glucose levels in type 2 diabetic mice [11]. Green tea (*Camellia sinensis*) polyphenols are reported as useful agents to protect against protein oxidation and glycation-induced pathogenesis of diabetic complications [12]. In recent years, garlic (*Allium sativum*) has received attention because of its ability to inhibit formation of advanced glycation endproducts (AGEs) [13, 14]. In this report we emphasize the anti-glycation as well as antioxidant properties from plant material. However, many other active agents obtained from plants have not been well characterized. More investigations must be conducted to evaluate the mechanism of action of medicinal plants with anti-diabetic and antioxidant effects. Consequently, it is necessary to perform toxicological investigation of all plants empirically used in order to avoid the risk of the side effects related to phytotherapy [15].

In searching for new therapeutic agents, our research aimed to evaluate anti-glycation and antioxidant potential of ethanolic extracts of ten selected culinary herbs and spices and to measure the impact of polyphenolic constituents on their effectiveness. We have previously reported the total phenolic, flavonoids and condensed tannins of these herbs [16]. Free radical scavenging activities had also been demonstrated by 2,2-diphenyl-1-picrylhydrazyl (DPPH) and ferric-reducing antioxidant power (FRAP) assays. In this report, we complement those data with additional antioxidant assays such as the total antioxidant capacity (TAC), 2,2′-azinobis-(3-ethylbenzothiazoline-6-sulphonic acid) ABTS assay, nitric oxide (NO) scavenging assay and ferrous ion chelating (FIC) assay.

11.2 Methodology

11.2.1 Chemicals

Bovine serum albumin (BSA; fraction V, fatty acid free, low endotoxin), D-glucose, sodium azide, phosphate buffered saline (PBS), aminoguanidine, ascorbic acid, trichloroacetic acid, gallic acid, anhydrous sodium carbonate, ethanol, methanol 100 %, iron(III) chloride 6-hydrate, potassium ferricyanide, potassium dihydrogen phosphate and dipotassium hydrogen phosphate, $NaNO_2$, $AlCl_3$, 1M NaOH, vanillin methanolic solution, concentrated hydrochloric acid (HCl), potassium persulfate, ABTS 2,2′-azinobis-(3-ethylbenzothiazoline-6-sulphonic acid), naphthyl ethylenediamine dihydrochloride, sodium nitroprusside, sodium phosphate, ammonium molybdate, sulphuric acid, and quercetin were purchased from Sigma-Aldrich (St. Louis, USA).

11.2.2 Plant Materials and Preparation of Extracts

Culinary herbs and spices were purchased from a local market. Ten commercially available herbs and spices were tested namely garlic (*Allium sativum*), ginger (*Zingiber officinale*), thyme (*Thymus vulgaris*), parsley (*Petroselinum crispum*), curry leaves (*Murraya koenigii* Spreng), peppermint (*Mentha piperita* L.), turmeric (*Curcuma longa* L.), onion (*Allium cepa* L.), green onion scallion (*Allium fistulosum*) and coriander (*Coriandrum sativum* L.). The study was limited to products that are widely available to the public and are in routine use for daily food cooking in Mauritius. Fresh culinary herbs and spices were ground into a paste and 5 g of the latter were extracted with ethanol (50 %) at a ratio of 10 ml per gram at room temperature (28 ± 2 °C) for 1 week. The extracts were centrifuged at 1,000 g for 10 min to remove precipitate.

11.2.3 Anti-Glycation Assay

Albumin glycation was determined using fluorometry as described by Matsuura et al. [17]. Briefly, 1 mg/ml of fatty acid-free BSA was incubated with D-glucose (200–400 mM) $\pm$ 100 µl of extracts in 0.2 M potassium phosphate buffered saline (PBS, pH 7.4 containing 0.01 % sodium azide) at 37 °C for defined time periods. Aliquots of the reaction mixture were removed at weekly intervals and fluorescent AGEs were assessed by their emission at 440 nm following excitation at 370 nm using a spectrofluorimeter (F-7000 FL) as described previously [16]. The reactions were stopped by adding 10 µl of 100 % (w/v) trichloroacetic acid and after 10 min the mixture was centrifuged at 1,000 g. The precipitate was re-dissolved in alkaline PBS and quantified for the relative amount of fluorescent AGEs. Aminoguanidine (20 mM) was included as positive control.

11.2.4 Determination of Total Antioxidant Capacity by Phosphomolybdenum Method

The total antioxidant activity was evaluated by the method of Prieto [18]. An aliquot of 0.2 ml of extracts was combined with 2 ml of reagent (0.6 M sulphuric acid, 28 mM sodium phosphate and 4 mM ammonium molybdate). The tubes were capped and incubated in a water bath at 80 °C for 90 min. After cooling the sample to room temperature, the absorbance of the solution was measured at 695 nm against blank using a Perkin-Elmer spectrophotometer. For reference, the appropriate solution of ascorbic acid was used and the reducing capacity of the extracts was expressed as ascorbic acid equivalents.

11.2.5 ABTS: 2,2′-Azinobis(3-ethylbenzothiazoline-6-sulphonic acid) Assay

The ABTS assay according to Chen et al. [19] was used to measure the antioxidant activity of the extracts. ABTS was dissolved in distilled water to a concentration of 7.4 mM, and potassium persulfate was added to a concentration of 2.45 mM. The reaction mixture was allowed to stand at room temperature for 16 h in the dark before use. The resulting intensely colored $ABTS^{\bullet+}$ radical cation was diluted with 0.01 M PBS (phosphate-buffered saline), pH 7.4, to give an absorbance value of ~0.70 at 734 nm. $ABTS^{\bullet+}$ cation solution (2 ml) was mixed with 100 μl extract. Absorbance was measured at 734 nm after reaction for 6 min. The ABTS scavenging activity was calculated as inhibition % = (1-(test sample absorbance/blank sample absorbance)) × 100. Quercetin was used as positive control.

11.2.6 Ferrous Ion-Chelating Ability Assay

The ferrous ion-chelating (FIC) assay was carried out according to the method of Viuda-Martos et al. [20], with some modifications. Solutions of 0.2 mM $FeCl_2{\cdot}4H_2O$ and 0.5 mM ferrozine were diluted 20 times. Briefly, an aliquot (0.5 ml) of different extracts was mixed with 0.5 ml $FeCl_2{\cdot}4H_2O$. After 5 min of incubation, the reaction was initiated by the addition of ferrozine (0.5 ml). The mixture was shaken vigorously and after a further 10 min incubation period, the absorbance of the solution was measured spectrophotometrically at 562 nm. The percentage inhibition of ferrozine–Fe^{2+} complex formation was calculated by using the formula: Chelating effect % $= [(1 - A_S)/A_B] \times 100$, where A_S is the absorbance of a tested sample and A_B is the absorbance of control sample (the control contains $FeCl_2$ and ferrozine complex formation molecules). Gallic acid was included as a standard.

11.2.7 Nitric Oxide Radical Inhibition Assay

Nitric oxide radical inhibition can be estimated by the use of Griess Illosvoy reaction [20]. In this assay, Griess Illosvoy reagent was modified by using naphthyl ethylenediamine dihydrochloride (0.1 % w/v) instead of 1-napthylamine (5 %). The reaction mixture (3 ml) containing sodium nitroprusside (10 mM, 2 ml), phosphate buffer saline (0.5 ml) and extracts (0.25 ml) or standard solution (quercetin) was incubated at 25 °C for 2.5 h. After incubation, 0.5 ml of the reaction mixture was mixed with 1 ml of sulfanilic acid reagent (0.33 % in 20 % glacial acetic acid) and allowed to stand for 5 min for complete diazotization. Then, 1 ml of naphthyl ethylenediamine dihydrochloride was added, mixed and allowed to stand for 30 min at 25 °C. A pink coloured chromophore was formed in diffused light. The absorbance of these solutions was measured at 540 nm against the corresponding blank solutions.

11.2.8 Statistical Analysis

Results were presented as mean $\pm$ S.D. of experiments. Difference between groups was compared using unpaired *t*-test with one-tailed test. Correlations between variables were quantified by the correlation factor "*r*". Correlation and linear regression analysis was performed using Microsoft Excel 2007. In each analysis $p < 0.05$ was considered statistically significant.

11.3 Results and Discussion

AGE formation and accumulation have been implicated as a major factor in the development of diabetic complications, atherosclerosis, Alzheimer's disease, and the normal aging process [21]. Linked to the process of AGE formation, lipoxidation reactions, oxidative and carbonyl stress should also be given as much attention where their tight association is now being increasingly reported [22]. Thus, AGE formation and AGE-protein cross-link formation should not be the only target for therapeutic interventions. Highly reactive carbonyl intermediates responsible for the formation of AGEs and advanced lipoxidation end-products (ALEs) should also be targeted. As formation of these end-products involves a complexity of pathways and reactions, inhibitors that react with each step and intermediate products should be evaluated. Another factor to be considered is the fact that glycoxidized proteins generate reactive oxygen species (ROS) and induce oxidative stress through the reaction with receptor for advanced glycation end-products (RAGEs). Also, ROS are generated by other reactions in the cascade of AGE formation such as methylglyoxal (MGO) and Schiff's base pathways leading to lipoxidation and oxidative damage. Therefore, strategies such as suppression of receptor signaling pathways (e.g. RAGE antagonists), and the use of antioxidants should be targeted. AGEs being complex and having a variety of cross-links, the development of inhibitors and AGE breakers might be difficult. Nonetheless, oxidative and carbonyl stress associated with AGEs can be inhibited by compounds having both metal chelating and carbonyl scavenging properties [23]. Unfortunately, from a large number of naturally occurring and synthetic compounds reported as AGE inhibitors, only the mechanism of action of a few compounds have been studied extensively. Unraveling the mechanism(s) of action of these inhibitors is essential for understanding the roles of AGE in the pathogenesis of a number of age-related chronic diseases and to design more effective therapeutic strategies for these diseases in the future [23].

During the current study, we focused our efforts to evaluate effective anti-glycation as well as antioxidant ability of extracts of culinary herbs and spices with the objective to be incorporated as therapeutic strategy in diabetes control and prevention. Our main approach was to screen a number of natural products. Furthermore, the purpose of this study was also to investigate the antioxidant activity of some local herbs and spices in order to evaluate the scientific basis of their potential therapeutic application as traditional medicine. All the herbs and spices tested in this study showed antioxidant activities. Based on the screening studies, the most promising compounds were selected to conduct further studies. Some of these results are presented below.

Hence, results gathered from our study and a literature search is suggestive of "Formulation" a concept of combining herbs with similar therapeutic activity. There is no doubt that most herbs exhibit their effects owing to a variety of constituents and therefore it becomes important to evaluate the idea of synergy within and between them and to investigate whether a mixture or formulation possesses greater activity than the individual components or whether the overall potency is synergistic in mechanism. However, more detailed pre-clinical and clinical evidences are required to establish these purported potencies.

11.3.1 In vitro Anti-Glycation Activity of Extracts

In vitro glycation assays demonstrated that the ten extracts showed marked inhibition of fluorescent AGEs formation as depicted in Fig. 11.1. When glycation was monitored over 12 weeks the percentage inhibition was found to be: garlic (26.1 %), ginger (25.7 %), thyme (42.3 %), parsley (41.2 %), curry leaves (40.9 %), peppermint (39.8 %), turmeric (39.3 %), onion (11.9 %), scallion (27.9 %) and coriander (38.8 %). The correlation between anti-glycation activities of the ten extracts and their total phenolic contents, DPPH and FRAP activities has been presented elsewhere [16].

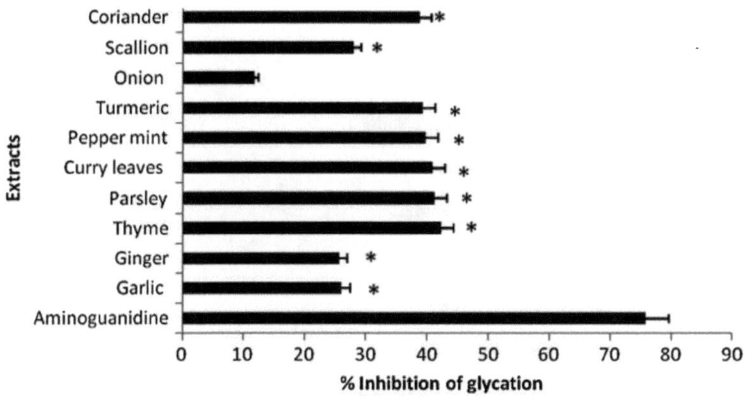

Fig. 11.1 Anti-glycation activity of the ten food plants. Percentage inhibition of fluorescent AGEs following glycation of BSA with 200 mM glucose in the presence of ten extracts in 0.2 M PBS at 37 °C. Fluorescent AGEs were measured after 12 weeks. Results are presented as means ± S.D. (n = 3) (*Values significantly different ($p < 0.05$) from control)

11.3.2 Determination of Total Antioxidant Capacity by Phosphomolybdate Method

The antioxidant activities as measured by the TAC assay are depicted in Fig. 11.2 and are as follows: garlic (0.0025 mg AAE/ml), ginger (0.0015 mg AAE/ml), thyme (0.0011 mg AAE/ml), parsley (0.0009 mg AAE/ml), curry leaves (0.0010 mg AAE/ml), peppermint (0.0012 mg AAE/ml), turmeric (0.0010 mg AAE/ml), onion (0.0015 mg AAE/ml), scallion (0.0010 mg AAE/ml) and coriander (0.0008 mg AAE/ml).

11.3.3 ABTS Assay

All the ten extracts displayed antioxidant activity as they were able to scavenge the $ABTS^{•+}$ radical cation (Fig. 11.3). The ABTS assay involves the oxidation of ABTS to an intensely-coloured nitrogen-centred radical cation, $ABTS^{•+}$. It is thus useful for testing food extracts because $ABTS^{•+}$ has an absorption maxima at 734 nm and most food extracts are highly coloured but do not absorb light at 734 nm. It is also viable for both aqueous and lipophilic systems. As the DPPH assay, ABTS measures the total antioxidant activity of extracts. The antioxidant ability of the extracts expressed as the percentage scavenging activity of the radical cation $ABTS^{•+}$ was as follows: garlic (47 %), ginger (51 %), thyme (95 %), parsley (84 %), curry leaves (98 %), peppermint (97 %), turmeric (95 %), onion (98 %), scallion (69 %) and coriander (38 %).

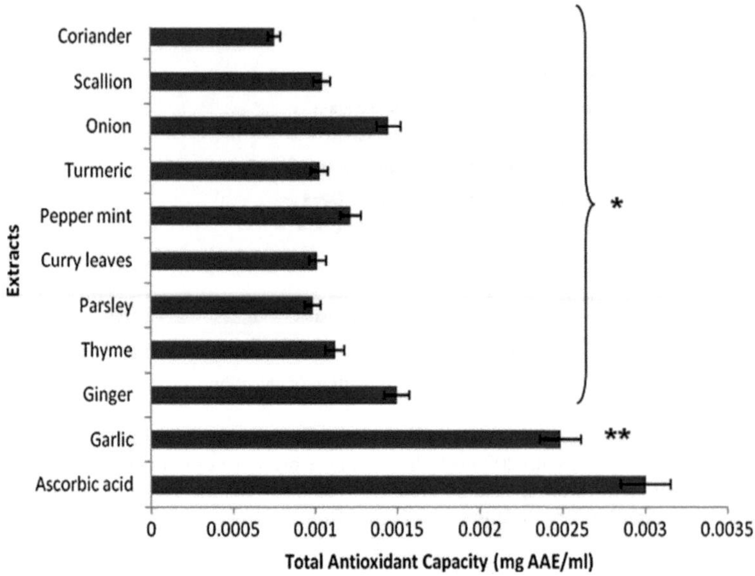

Fig. 11.2 Total antioxidant capacities of the ten extracts. Results are presented as mean ± S.D. (n = 3) (*Values significantly lower ($p < 0.05$) compared to positive control ascorbic acid. **Values not significantly ($p > 0.05$) different from ascorbic acid)

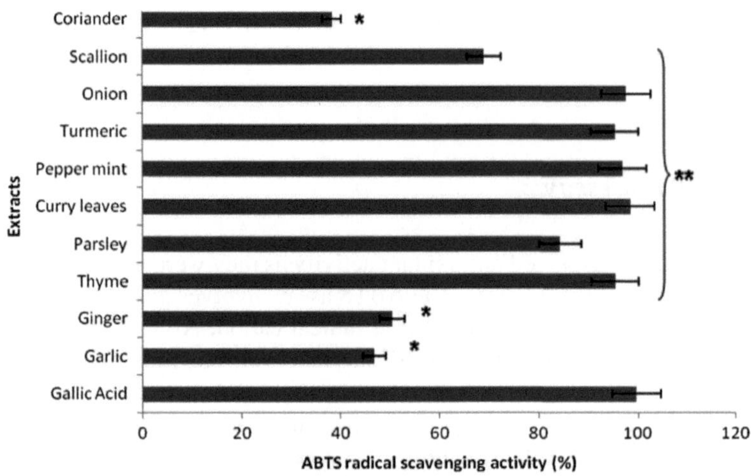

Fig. 11.3 ABTS radical assay. Percentage scavenging activity of the ten extracts. Results are presented as mean ± S.D. (n = 3) (*Values significantly lower ($p < 0.05$) compared to positive gallic acid. **Values not significantly ($p > 0.05$) different from gallic acid)

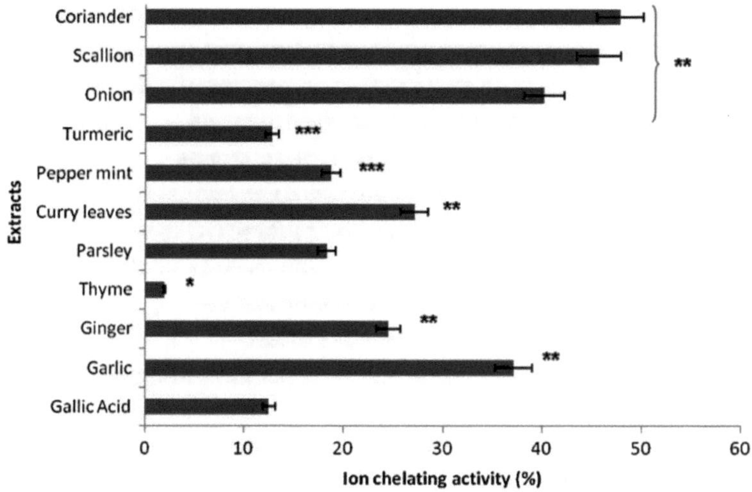

Fig. 11.4 Ferrous ion chelating activity. Percentage inhibition of the formation of the ferrous-ferrozine complex of the ten extracts. Results are presented as mean ± S.D. (n = 3) (*Values significantly lower ($p < 0.05$) compared to positive gallic acid. **Values significantly ($p < 0.05$) higher than positive control. ***Values not significantly ($p > 0.05$) different from gallic acid)

11.3.4 Ferrous Ion Chelating Ability

Addition of the extracts interferes with the ferrous-ferrozine complex and the red colour of the complex decreased with increasing concentrations of the antioxidant agents. All the extracts captured ferrous ions before ferrozine and thus have ferrous chelating ability at varied extent (Fig. 11.4). Among the extracts tested, the chelating activities were as follows: garlic (37 %), ginger (24 %), thyme (2 %), parsley (18 %), curry leaves (27 %), peppermint (19 %), turmeric (13 %), onion (40 %), scallion (46 %) and coriander (48 %).

11.3.5 Nitric Oxide Radical Scavenging Assay

The extracts effectively reduced the generation of nitric oxide from sodium nitroprusside (Fig. 11.5). The nitric oxide radical scavenging activity expressed as percentage inhibition was as follows: garlic (62 %), ginger (67 %), thyme (84 %), parsley (81 %), curry leaves (79 %), peppermint (73 %), turmeric (71 %), onion (86 %), scallion (85 %) and coriander (83 %).

A wide range of antioxidants from both natural and synthetic origin have been proposed for use in the treatment of various human diseases [23]. Medicinal plants are used as a source of phytochemicals to cure various illnesses in many parts of the world [24]. Being the backbone of traditional medicine, medicinal plants as a good source of biologically active compounds known as phytochemicals have been

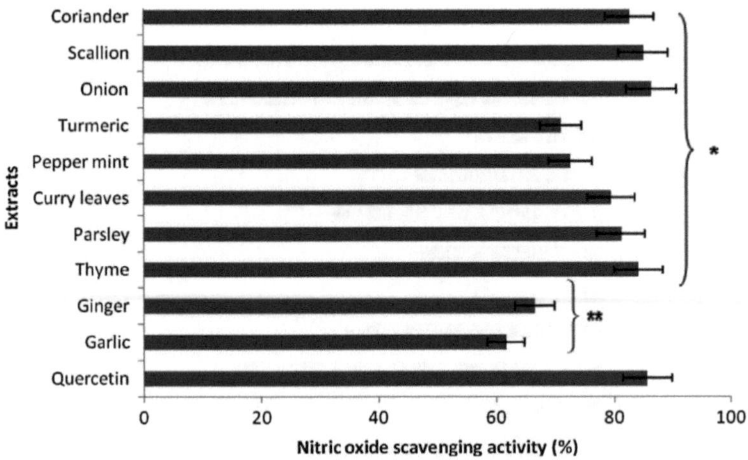

Fig. 11.5 Nitric oxide radical scavenging assay. Percentage scavenging activity of the ten extracts. Results are presented as mean ± S.D. (n = 3) (*Values not significantly ($p > 0.05$) different from positive control quercetin. **Values significantly lower ($p < 0.05$) compared to quercetin)

found to act as antioxidants by scavenging free radicals and having therapeutic potential for free radical associated disorders [25] as it is well known that free radicals are the major cause of various chronic and degenerative diseases such as coronary heart disease, inflammation, stroke, diabetes mellitus and cancer [26]. ROS have been found to play an important role in the initiation and/or progression of various diseases [27]. Thus, recent studies have investigated the potential of plant products as antioxidants against various diseases induced by free radicals [28]. Additionally, it has been determined that the antioxidant effect of plant products is mainly attributed to phenolic compounds, such as flavonoids, phenolic acids, tannins, and phenolic diterpenes [29]. In this regard, during our study we investigated the glycation inhibitory capacity which was correlated with total phenolic, flavonoids and tannins contents of the extracts [16]. As AGEs are related to reactive species, antioxidants have therapeutic potential by scavenging free radicals but also suppress glycation. Antioxidant compounds such as vitamin C, vitamin E and the carotenoids have been shown to reduce *in vitro* and *in vivo* protein glycation [30]. For example, treatment of diabetic rats with vitamin E resulted in a decrease of plasma lipid peroxidation [31]. Therefore, antioxidant activity of the plants used in herbal medicine should be assessed either to elucidate the mechanism of their pharmacological action or to provide information on antioxidant activity of these herbal plants. Dietary intake of antioxidants can help scavenge free radicals and oxidants and protect against diseases as suggested by intervention trial studies [32]. Research on natural antioxidants has become increasingly active in various fields. Accordingly, numerous articles on natural antioxidants, including polyphenols, flavonoids, vitamins, and volatile chemicals, have been published [15]. Phenolics are ubiquitous secondary metabolites in plants and possess a wide range of therapeutic uses such as antioxidant, anti-mutagenic, anti-carcinogenic, free radical scavenging activities and also decrease cardiovascular complications [33]. The scavenging ability of the phenolics is mainly due to the presence of hydroxyl groups [34]. Flavonoids and other phenolic compounds of plant origin have been reported as scavengers of free radicals [35]. Flavonoids are a group of polyphenolic compounds, which exhibit several biological effects such as anti-inflammatory, anti-hepatotoxic, anti-ulcer, anti-allergic, antiviral and anticancer activities [36]. Hence, nowadays search for natural antioxidant sources is gaining much importance.

Medicinal plants have been reported to have not only antioxidant but also anti-inflammatory as well as anti-glycation effects [37]. Natural products are an obvious place to search for potential anti-diabetic regimens. Several plant extracts have been reported to exhibit anti-diabetic activity [38]. Among the various medicinal and culinary plants, some endemic species are of particular interest because they may be used for producing raw materials or preparations containing phytochemicals with significant antioxidant capacities and health benefits [39]. The harmful effects of diabetes have been found to be mediated through increased oxidative stress playing a critical role which leads to increased formation and accumulation of AGEs contributing to diabetic complications [15].

In addition to those endogenously formed, AGEs can also be introduced in the body from exogenous sources [40]. Studies have provided evidence that diet is a significant exogenous source of highly reactive AGEs and as environmental risk factors for diabetic complications [41]. To some extent this can be prevented by dietary AGE restriction and incorporation of edible plants with demonstrated anti-glycation and antioxidant properties suggesting that modulation of food-AGE content could become an important ingredient for the therapeutic management of diabetic patients. Until effective and safe drugs become available, physicians and dietitians can, for instance, advise increased reliance on foods with anti-glycation and antioxidant activity. As temporary normalization of serum glycemia is not sufficient, dietary modification to reduce AGEs and prevent glycation can be exploited as a means for reducing serum AGEs for extended periods of time e.g. months or years [42]. Ethnomedical literature contains a large number of plants that can be used against diseases, in which ROS and free radicals play an important role [43]. There is a plethora of plants that have been found to possess strong antioxidant activity. Recent reports indicate that there is an inverse relationship between the dietary intake of antioxidant rich foods and the incidence of human diseases [44]. Interest in finding naturally occurring antioxidants in foods or medicines to replace synthetic antioxidants has increased considerably, given that synthetic antioxidants have restricted use due to their side effects [6]. The potential toxicity of synthetic antioxidants (e.g. butylated hydroxyanisole, butylated hydroxytoluene, tertiary butylhydroquinone, esters of 3,4,5-trihydroxybenzoic acid, etc.) [45] has aroused an increased interest and scientists have focused on isolation and characterization of natural antioxidants from natural sources such as herbs, spices, seeds, cereals, fruits and vegetables by extraction, fractionation and purification [46].

Table 11.1 Active chemicals isolated from the culinary herbs and spices studied

Plant	Active compounds
Garlic	S-allylcysteine, S-ethylcysteine, S-propylcysteine, S-methylcysteine, kaempferol, allixin and saponins [11, 50]
Ginger	6-Gingerdiols, 6-shogao, ascorbic acid, beta-carotene, beta-sitosterol, caffeic acid, capsaicin, chlorogenic acid, delphinidin, ferulic acid, gamma-terpinene, kaempferol, lauric acid, palmitic acid, quercetin [51, 52]
Thyme	Thymol, 4-terpineol, anethole, apigenin, ascorbic acid, beta-carotene, caffeic acid, camphene, carvacrol, chlorogenic acid, chrysoeriol, eriodictyol, eugenol, ferulic acid, gallic acid, gamma-terpinene, isochlorogenic acid, kaempferol, labiatic acid, luteolin, *p*-coumaric acid, *p*-hydroxy-benzoic acid, rosmarinic acid, tannin [51, 52]
Parsley	Flavonoids: apiin, apigenin, crisoeriol, luteolin, myristicin, limonene, eugenol, alpha-thujene [53]
Curry leaves	Carbazoles, mahanimbine, murrayanol, mahanine [53]
Peppermint	Perillyl alcohol, rosmarinic acid, tetramethyl-2-hexadecen-1-ol, Octadecadienoic acid, pulgeone, piperitenone [54]
Turmeric	Diferuloylmethane, ascorbic acid, beta-carotene, caffeic acid, eugenol, *p*-coumaric acid, protocatechuic acid, syringic acid, vanillic acid [52, 55]
Onion	Quercetin, arachidonic acid, catechin, epicatechin, thiosulfinates [56]
Scallion	S-allyl cysteine, S-allyl-mercaptocystein, furostane saponins, quercetin, isorhametin, thiosulfinates [57, 58]
Coriander	Apigenin, beta-carotene, beta-sitosterol, caffeic acid, camphene, chlorogenic acid, gamma-terpinene, isoquercitrin, myristic acid, *p*-hydroxybenzoic acid, protocatechuic acid, quercetin, rhamnetin, rutin, tannin [52]

It is also clear that most of the dietary antioxidants have low or minimal toxicity, and that intake can be increased without adverse effects [47]. Many fresh fruits and vegetables have been found to contain natural antioxidants, mainly phenolic compounds such as ferulic acid, catechins, as well as ascorbic acid [48]. Polyphenols are the biggest group of phytochemicals, and many of them have been found in plant-based foods. Polyphenol-rich diets have been linked to many health benefits. Dietary polyphenols have received tremendous attention among nutritionists, food scientists and consumers due to their roles in human health [49]. Some of the reported active compounds of the extracts studied are given in Table 11.1.

Our observation that high values of antioxidant activity and phenolic content were found in ethanol extracts is in line with reported data [16]. In general, there was a good positive correlation between the TPC and antioxidant activity as assessed by FRAP and DPPH-scavenging assays suggesting that polyphenols are important contributors to the antioxidant and free-radical scavenging activities. However, the lack of correlation between TPC and antioxidant activity reported in some studies could be due to the different antioxidant activity parameters determined in these studies, different responses of phenolic compounds in different bioassays used to determine the antioxidant activity and/or the chemical structure of the phenolic compounds present in different plants. Moreover, the total phenolic content estimated by the Folin-Ciocalteu reagent may overestimate TPC because it is known to react with other components present in the plant extracts such as minerals and ascorbic acid [59, 60]. Consequently, the antioxidant/free-radical scavenging activities of the plant extracts cannot be predicted on the basis of their TPC alone, but also requires proper characterization of the individual phenolic components [61].

The results of the current study have shown that most of the studied plants are potentially good sources of free-radical scavenging compounds and support the traditional medicinal application of some of the tested plants. During this study, we systematically screened extracts of medicinal and edible plants used as culinary herbs and spices for their anti-glycation activities through biochemical assay. BSA was incubated with reducing sugar at physiological pH for various durations. The extent of glycation of proteins was measured mainly by the spectrofluorimetric method. As a result, the ten extracts under study were identified as having anti-glycation potential and their activities were compared with aminoguanidine. The total phenolic, flavonoids, tannins contents and antioxidant activity as determined by the DPPH and FRAP assays were correlated with the observed anti-glycation activity. Details of this study have been published elsewhere [16]. The phenolic compounds make a significant contribution to the antioxidant activity in these extracts as evidenced by the positive correlation between phenolic contents and antioxidant activities. Therefore, ingestion of extracts from these plants may help to prevent *in vivo* oxidative damage associated with diseases and illnesses.

11.4 Conclusions

The present study was aimed at investigating the antioxidant activities of ethanolic extracts of selected culinary herbs and spices. It is reasonable to consider that good glycemic control, in combination with a careful diet in terms of reduced AGE consumption and increased intake of natural products with anti-glycation and antioxidant properties should be among the new goals for optimal management of diabetic patients. Addressing dietary habits from a new perspective, while difficult, could achieve the best long-term effects as novel drug interventions become available for clinical use in the future.

Comparison of nutritional regime and values of AGEs shows that the higher intake of fructose in alternative nutrition of healthy subjects may cause an increase in plasma AGE values. Protective effect of regular consumption of vegetables and fruits dominantly concerns the prevention of free-radical diseases. The risks of alternative nutrition may be reduced by better choice of foodstuffs.

Further studies to evaluate the *in vivo* potential of these dietary agents in various animal models and the isolation and identification of the antioxidant principles are needed. Nonetheless these studies provide a database as a valuable instrument for available information on natural products with reported therapeutic ability for guiding food choices to reduce AGE and oxidative stress.

Acknowledgements We are grateful to the University of Mauritius and the Tertiary Education Commission, Mauritius for financial support.

References

1. Patwardhan B, Vaidya ADB, Chorghade M (2004) Ayurveda and natural products drug discovery. Curr Sci 86:789–794
2. Grover JK, Yadav S, Vats V (2002) Medicinal plants of India with antidiabetic potential. J Ethnopharmacol 81:81–100
3. Jung M, Park M, Lee HC, Kang YH, Kang ES, Kim SK (2006) Anti-diabetic agents from medicinal plants. Curr Med Chem 3:1203–1218
4. Dewanjee S, Das AK, Sahu R, Gangopadhyay M (2009) Anti-diabetic activity of *Diospyros peregrina* fruit: effect on hyperglycemia, hyperlipidemia and augmented oxidative stress in experimental type 2 diabetes. Food Chem Toxicol 47:2679–2685
5. Patel DK, Kumar R, Laloo D, Hemalatha S (2012) Diabetes mellitus: an overview on its pharmacological aspects and reported medicinal plants having antidiabetic activity. Asian Pac J Trop Biomed 2:411–420
6. Singh U, Singh S, Kochhar A (2012) Therapeutic potential of antidiabetic nutraceuticals. Phytopharmacol 2:144–169
7. Shokeen P, Anand P, Murali YK, Tandon V (2008) Antidiabetic activity of 50% ethanolic extract of *Ricinus communis* and its purified fractions. Food Chem Toxicol 46:3458–3466
8. Wadkar KA, Magdum CS, Patil SS, Naikwade NS (2008) Anti-diabetic potential and Indian medicinal plants. J Herb Med Toxicol 2:45–50
9. Bnouham M, Ziyyat A, Mekhfi H, Tahri A, Legssyer A (2006) Medicinal plants with potential anti-diabetic activity-A review of ten years of herbal medicine research (1990–2000). Int J Diabetes Metab 14:1–25
10. Elosta A, Ghous T, Ahmed N (2012) Natural products as anti-glycation agents: possible therapeutic potential for diabetic complications. Curr Diabetes Rev 8:92–108
11. Nishiyama T, Mae T, Kishida H, Tsukagawa M, Mimaki Y, Kuroda M, Sashida Y, Takahashi K, Kawada T, Nagagawa K, Kitahara M (2005) Curcuminoids and sesquiterpenoids in turmeric (*Curcuma longa* L.) suppress an increase in blood glucose level in type 2 diabetic KK-Ay mice. J Agric Food Chem 53:959–963
12. Frei B, Higdon JV (2003) Antioxidant activity of tea polyphenols *in vivo*: evidence from animal studies. J Nutr 133:3275S–3284S
13. Morihara N, Nishihama T, Ushijima M, Ide N, Takeda H, Hayama M (2007) Garlic as an anti-fatigue agent. Mol Nutr Food Res 51:1329–1334
14. Sheikh N, Safari M, Mani Kashani K, Araghchian M, Zeraati F (2004) Study on the effect of garlic on the *in vitro* albumin glycation reaction. Acta Med Iran 42:113–116
15. Patel DK, Kumar R, Laloo D, Hemalatha S (2012) Natural medicines from plant source used for therapy of diabetes mellitus: an overview of its pharmacological aspects. Asian Pac J Trop Med 42:239–250
16. Ramkissoon JS, Mahomoodally MF, Nessar A, Subratty AH (2012) Relationship between total phenolic content, antioxidant potential and antiglycation abilities of common culinary herbs and spices. J Med Food 15:1116–1123
17. Matsuura N, Aradate T, Sasaki C, Kojima H, Ohara M, Hasegawa J (2002) Screening system for the Maillard reaction inhibitor from natural product extracts. J Health Sci 48:520–526
18. Prietto P, Pineda M, Aquilar M (1999) Spectrophotometric quantification of antioxidant capacity through the formation of phosphomolybdenum complex: specification application to the determination of vitamin E. Anal Biochem 269:337–341
19. Chen YF, Roan HY, Li CK, Huang YC, Wang TS (2011) Relationship between antioxidant and antiglycation ability of saponins, polyphenols, and polysaccharides in Chinese herbal medicines used to treat diabetes. J Med Plants Res 5:2322–2331
20. Viuda-Martos M, Navajas YR, Zapata ES, Fernández-López J, Pérez-Álvarez JA (2010) Antioxidant activity of essential oils of five spice plants widely used in a Mediterranean diet. Flavour Frag J 5:13–19

21. Bandyopadhyay U, Das A, Bannerjee RK (1999) Reactive oxygen species: oxygen damage and pathogenesis. Curr Sci 77:658–666
22. Rahbar S, Figarola JL (2003) Novel inhibitors of advanced glycation endproducts. Arch Biochem Biophys 419:63–79
23. Cuzzocrea S, Riley DP, Caputi AP, Salvemini D (2001) Antioxidant therapy: a new pharmacological approach in shock, inflammation and ischemia/reperfusion injury. Pharmacol Rev 53:135–159
24. Rice-Evans CA, Miller NJ, Paganga G (1997) Antioxidant properties of phenolic compounds. Trends Plant Sci 2:152–159
25. Povichit N, Phrutivorapongkul A, Suttajit M, Chaiyasut C, Leelapornpisid P (2010) Phenolic content and *in vitro* inhibitory effects on oxidation and protein glycation of some Thai medicinal plants. Pak J Pharm Sci 23:403–408
26. Smirin P, Taler D, Abitbol G, Brutman-Barazani T, Kerem Z, Sampson SR, Rosenzweig T (2010) *Sarcopoterium spinosum* extract as an antidiabetic agent: *in vitro* and *in vivo* study. J Ethnopharmacol 129:10–17
27. Halliwell B (1997) Antioxidants and human diseases: a general introduction. Nutr Rev 55: S44–S52
28. Hou WC, Lin RD, Cheng KT, Hung YT, Cho CH, Chen CH, Hwang SY, Lee MH (2003) Free radical-scavenging activity of Taiwanese native plants. Phytomedicine 10:170–175
29. Pietta PG (2000) Flavonoids as antioxidants. J Nat Prod 63:1035–1042
30. Cameron NE, Cotter MA (1993) Potential therapeutic approaches to the treatment or prevention of diabetic neuropathy: evidence from experimental studies. Diabetes Med 10:593–605
31. Kinalski M, Sledziewski A, Telejko B, Zarzycki W, Kinalska N (1999) Antioxidant therapy and streptozoticin-induced diabetes in pregnant rats. Acta Diabetol 36:113–117
32. Frei B (1994) Natural antioxidants in human health and disease. Annu Rev Nutr 16:33–50
33. Yen GH, Chen HY (1995) Antioxidant activity of various tea extracts in relation to their antimutagenicity. J Agric Food Chem 43:27–32
34. Huang D, Ou B, Proir RL (2005) The chemistry behind the antioxidant capacity assays. J Agric Food Chem 53:1841–1856
35. Formica JV, Regelson W (1995) Review of the biology of quercetin and related bioflavonoids. Food Chem Toxicol 33:1061–1080
36. Cao G, Sofic E, Prior RL (1997) Antioxidant and pro-oxidant behaviour of flavonoids: structure activity relationships. Free Radic Biol Med 22:749–760
37. Chua MT, Tung YT, Chang ST (2008) Antioxidant activities of ethanolic extracts from the twigs of *Cinnamomum osmophloeum*. Bioresour Technol 99:1918–1925
38. Xu A, Wang H, Hoo RL, Sweeney G, Vanhoutte PM, Wang Y, Wu D, Chu W, Qin G, Lam KS (2009) Selective elevation of adiponectin production by the natural compounds derived from a medicinal herb alleviates insulin resistance and glucose intolerance in obese mice. Endocrinology 150:625–633
39. Exarchou V, Nenadis N, Tsimidou M, Gerothanassis IP, Troganis A, Boskou D (2002) Antioxidant activities and phenolic composition of extracts from Greek oregano, Greek sage, and summer savory. J Agric Food Chem 50:5294–5299
40. Koschinsky T, He CJ, Mitsuhashi T, Bucala R, Liu C, Buenting C, Heitmann K, Vlassara H (1997) Orally absorbed reactive glycation products (glycotoxins): an environmental risk factor in diabetic nephropathy. Med Sci 94:6474–6479
41. Ling X, Nagai R, Sakashita N, Takeya M, Horiuchi S, Takahashi K (2001) Immunohistochemical distribution and quantitative biochemical detection of advanced glycation endproducts in fetal to adult rats and in rats with streptozotocin-induced diabetes. Lab Invest 81:845–861
42. Peppa M, Uribarri J, Vlassara H (2003) Glucose, advanced glycation end products, and diabetes complications: what is new and what works. Clin Diabetes 21:186–187
43. Huxley RR, Neil H (2003) The relationship between dietary flavonol intake and coronary heart disease mortality: a metaanalysis of prospective cohort studies. Eur J Clin Nutr 57:904–908

44. Kumar DS, Muthu AK, Smith AA, Manavalan R (2010) *In vitro* antioxidant activity of various extracts of whole plant of *Mucuna pruriens* (Linn). Int J PharmTech Res 2:2063–2070
45. Johnson IT, Fenwick GR (2000) Dietary anticarcinogens and antimutagens: chemical and biological aspects. Royal Society of Chemistry, London, pp 92–95
46. Umamaheswari M, Chatterjee TK (2008) *In vitro* antioxidant activities of the fractions of *Coccinia grandis* L. leaf extract. Afr J Tradit Complement Altern Med 5:61–73
47. Batool F, Sabir SM, Rocha JBT, Shah AH, Saify ZF, Ahmed SD (2010) Evaluation of antioxidant and free radical scavenging activities of fruit extract from *Zanthoxylum alatum*: a commonly used spice from Pakistan. Pak J Bot 42:4299–4311
48. Re R, Pellegrini N, Proteggente A, Pannala A, Yang M, Rice-Evans C (1999) Antioxidant activity applying an improved ABTS radical cation decolourisation assay. Free Radic Biol Med 26:1231–1237
49. Tsao R (2010) Chemistry and biochemistry of dietary polyphenols. Nutr 2:1231–1246
50. Ahmad MS, Ahmed N (2006) Antiglycation properties of aged garlic extract: Possible role in prevention of diabetic complications. J Nutr 80:796–799
51. Moon JK, Shibamoto T (2009) Antioxidant assays for plant and food components. J Agric Food Chem 57:1655–1666
52. Suhaj M (2006) Spice antioxidants isolation and their antiradical activity: a review. J Food Compos Anal 19:531–537
53. Haidari F, Keshavarz A, Shahi MM, Mahboob SA, Rashidi MR (2011) Effects of parsley (*Petroselinum crispum*) and its flavonol constituents, kaempferol and quercetin, on serum uric acid levels, biomarkers of oxidative stress and liver xanthine oxidoreductase a activity in oxonate-induced hyperuricemic rat. Iran J Pharm Res 10:811–819
54. Das AK, Rajkumar V, Dwivedi DK (2001) Antioxidant effect of curry leaf (*Murraya koenigii*) powder on quality of ground and cooked goat meat. Int Food Res J 18:563–569
55. Wu CH, Huang SM, Lin JA, Yen GC (2011) Inhibition of advanced glycation end-product formation by foodstuffs. Food Funct 2:224–234
56. Rice-Evans C, Miller N, Bolwell P, Bremley P, Pridham J (1995) The relative antioxidant activities of plant-derived polyphenolic flavonoids. Free Rad Res 22:375–383
57. Naidu JR, Ismaill RB, Yeng C, Sasidha S, Kumar P (2012) Chemical composition and antioxidant activity of the crude methanolic extracts of *Mentha spicata*. J Phytol 4:13–18
58. Leelarungrayub N, Chanarat N, Rattanapanonel V (2004) Potential activity of Thai shallot (*Allium ascalonicum* L.) extract on the prevention of hemolysis and glutathione depletion in human erythrocyte from oxidative stress. CMU J 3:225–234
59. Matthaus B (2002) Antioxidant activity of extracts obtained from residues of different oil seeds. J Agric Food Chem 50:3444–3452
60. Deepa N, Kaur C, Singh B, Kapoor HC (2006) Antioxidant capacity in some red sweet pepper cultivars. J Food Compos Anal 19:572–578
61. Molan AL, Faraj AM, Mahdy AS (2012) Antioxidant activity and phenolic content of some medicinal plants traditionally used in northern Iraq. Phytopharmacol 2:224–233

Chapter 12
Metal Burden as Template for Assessing the Quality of Raw Water Sourced from Two Rivers by Lagos State Water Corporation, Nigeria

Adeleke Adeniyi, Olawale Osifeko, Olabisi Owoade, Yusuf Omotayo, A. Emelia, Aminah Ibrahim, and Raheemot Balogun

Abstract Water and surface sediment samples from three different points of River Owo at Isashi and Ogun River at Akute Odo were collected and quantitatively analyzed for metals burden. The two rivers serve as sources of raw water processed by the Lagos State Water Corporation for domestic and industrial consumption. The levels of lead, cadmium, copper, calcium and zinc in the samples were determined by atomic absorption spectrometry. The results obtained revealed high mean values of calcium in both water and sediment samples. Metals concentration (mg/l) in water samples are within the Nigeria's background values (NBV): ND–0.46 (Pb), ND–0.35 (Cd), ND–0.78 (Cu), 8.47–330.00 (Ca) and ND–1.89 (Zn). However, cadmium and calcium levels in sediment samples exceeded the NBV (μg/g) with ND–2.02 and 0.24–87.50 respectively. Similarly, with the exception of lead and cadmium, all other metals in the water samples are found to be within the World Health Organization (WHO) limit set for drinking water. Correlation coefficient and statistical *t*-test were used to test for statistical significant differences at 95 % confidence level. The levels of cadmium and lead especially in the water samples give cause for concern considering their potential health risk.

12.1 Introduction

Lagos is a state surrounded by water and as such it is assumed that water would be easily accessible to the populace. The raw waters in the lagoons surrounding Lagos are not fit for human consumption. The large expanse of freshwater around

A. Adeniyi (✉) • O. Osifeko • O. Owoade • Y. Omotayo • A. Emelia • A. Ibrahim • R. Balogun
Department of Chemistry, Lagos State University, Ojo P.M.B 0001, LASU Post Office, Badagry Expressway, Ojo, Lagos, Nigeria
e-mail: lekeadeniyi@yahoo.com; osifekoo@yahoo.com; owoadeb@yahoo.com; yomi_boyce@yahoo.co; emi_solar@yahoo.com; lekeadeniyi@yahoo.com; adewunmibalogun32@yahoo.com

M. Gupta Bhowon et al. (eds.), *Chemistry: The Key to our Sustainable Future*,
DOI 10.1007/978-94-007-7389-9_12,

Lagos is equally not potable since it does not meet the World Health Organization standards [1].

Changes in any component of the environment may cause discomfort and affect normal life, consequently protection of our environment must be given the highest priority [2]. The river and other water bodies is the ultimate repository of man's wastes [3, 4].

Water is one of the basic requirements for human daily life; the problem of water pollution is increasingly becoming acute as a result of increasing population, urbanization and industrialization [1, 4, 5].

Metals such as cadmium, lead and mercury are known to be toxic with no beneficial effects to man and wildlife [6]. Cadmium and lead are stable and persistent environmental pollutants and are known to exhibit extreme toxicity even at low levels under certain conditions [7]. Meanwhile interests in metals like copper and zinc which are required for metabolic activity in organisms lie in the narrow "window" between their essentiality and toxicity [8]. Calcium is involved in many chemical cycles that take place in water bodies [9]. Contamination with heavy metals may have devastating effect on the ecological balance of the aquatic environment and the diversity of aquatic organism becomes limited [10].

The nature of metals present in rivers depends on ore-bearing deposits in the catchment area, the discharges of human wastes and discharges when the river passes through urban and industrial areas [11, 12].

As a result of adsorption and accumulation, the concentration of metals in bottom sediments is expected to be higher than in the water above and this sometimes can cause secondary pollution problems, therefore bottom sediments are repository of heavy metals [13]. Trace elements are easily influenced by environmental factors such as surface run-off, atmospheric deposition and anthropogenic pollutants. Hence trace metals may be sensitive indicators for monitoring changes in an aquatic environment [14].

There are global concerns that the direct usage of water above permitted limits could cause adverse health effects [15–17]. These concerns are manifested in the significantly large number of freshwater metal contamination studies across different countries. Representative global studies reviewed are from Nigeria, South Africa, USA, India, Bangladesh, Egypt, Turkey, Macedonia, Pakistan and Greece [18–24].

In Lagos State Waterworks fed by the two rivers used for this study, harvested raw water is dosed with alum in the sedimentation tank. It then goes into a filtration bed; calcium hypochlorite is added to the filtered water, allowed to stand for a while before hydrated lime ($Ca(OH)_2$) is used to readjust the pH. The treated water is then released for consumption. It is apparent that this traditional water treatment protocol used by the Water Corporation may not be able to adequately treat raw water with high levels of contaminants. The purpose of this study is to use metal burden as template to assess the quality of raw water sourced from two rivers by the Lagos State Water Corporation, Nigeria. No previous work on these locations has been reported.

12.2 Methodology

12.2.1 Study Area

The study areas are the Ogun River at Akute Odo in Ifo Local Government Area of Ogun State and River Owo at Isashi in Ojo Local Government Area of Lagos State, Nigeria (Fig. 12.1). The two rivers are the sources of raw water processed by Iju, Adiyan and Isashi Water Works respectively.

Three sampling stations were established to cover possible impacted areas. The locations of various representative sampling points are: Point A, Entrance point; Point B, Middle point; Point C, Point of raw water collection.

The metals, cadmium, lead, copper, zinc and calcium were selected for this study. Previous studies have implicated these metals as the major contaminants associated with rivers flowing through urban cities and agricultural settlements [9, 18, 25].

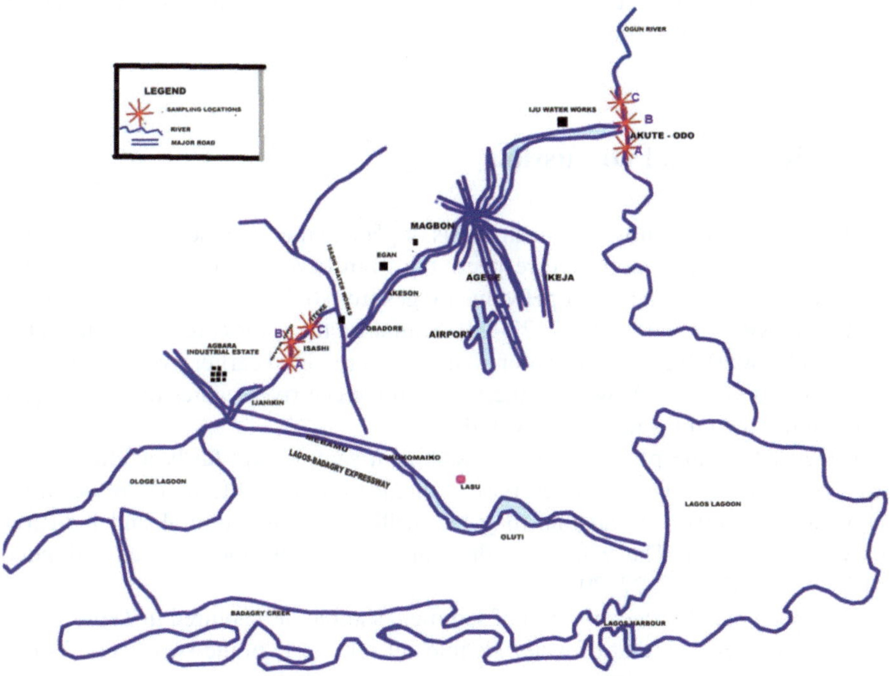

Fig. 12.1 Sketch map showing sampling locations

12.2.2 Sampling and Analyses

Surface water and sediment samples were collected at 2 weeks intervals for a period of 4 months (May – August, 2011).

Water samples were collected using plastic containers while the sediment samples were collected with the help of divers using plastic containers. The pH of the water samples was determined on site. The water samples were then acidified with conc. HNO_3 and stored in a refrigerator prior to metal analyses [26].

The sediments samples were air dried and passed through a 2 mm sieve [7]. The pH of the sediment was measured by taking about 5 g of sediment samples and adding 20 ml of distilled water using a Jenway 3505 pH meter.

Analysis of metals was based on the method previously reported for metal analysis by Adeniyi et al. [7]. Five metals, lead (Pb), cadmium (Cd), copper (Cu), calcium (Ca) and zinc (Zn) were determined in an air-acetylene flame using a Buck Scientific 200A Atomic Absorption Spectrophotometer.

12.2.3 Data Analysis

Data collected from each parameter were subjected to computation and statistical analysis followed by the correlation coefficients between the metals concentrations in water and the sediments.

12.3 Results and Discussion

Tables 12.1 and 12.2 show the mean levels of pH and metals concentration in water and sediment samples collected respectively from River Owo at Isashi and Ogun River at Akute Odo. The water pH value ranges from 6.78 to 7.38, the values falling within the NBV of 5.70–7.79 [7, 27–29]. Similarly, the values are within the WHO limits (6.50–8.50) for drinking water, and 6.1–8.9 reported earlier for Umtata River, South Africa [18, 30]. However, the northern Greece river values of 7.7–8.6 [23] were higher than the values in this study.

Nevertheless, the pH values of the sediment samples are fairly acidic ranging from 5.93 to 6.87. These values pose a threat to aquatic life as decrease in pH aggravates toxicity [7] and this may be attributed to agricultural and industrial activities upstream. The values for this study fall within the reported sediments NBV of 4.02–8.73 [7, 27–29].

The mean metals concentration of sediment samples as captured in Table 12.2 is generally higher than that of water samples (Table 12.1) in the two rivers. This is due to the fact that sediment serves as a bottom sink for all the activities in the aquatic ecosystem [3, 6]. Calcium has the highest levels of the metals determined.

Table 12.1 Mean levels of pH and metals concentration (mg/l ± SD) in Isashi and Akute Odo water samples

Sampling location	pH	Pb	Cd	Cu	Ca	Zn
A	6.94 ± 0.42	0.34 ± 0.10	0.03 ± 0.01	0.02 ± 0.01	122.18 ± 39.64	0.28 ± 0.12
	(7.28 ± 0.40)	(0.31 ± 0.06)	(0.04 ± 0.01)	(0.08 ± 0.02)	(16.26 ± 16.72)	(0.18 ± 0.06)
	0.23[a]	0.33[a]	0.002[a]	0.01[a]	0.002[a]	0.03[a]
B	6.96 ± 0.32	0.32 ± 0.08	0.03 ± 0.01	0.02 ± 0.01	133.68 ± 63.61	0.27 ± 0.09
	(7.38 ± 0.24)	(0.27 ± 0.05)	(0.04 ± 0.01)	(0.08 ± 0.02)	(8.96 ± 4.10)	(0.18 ± 0.07)
	0.014[a]	0.17[a]	0.05[a]	0.024[a]	0.01[a]	0.002[a]
C	6.78 ± 0.13	0.31 ± 0.10	0.03 ± 0.01	0.02 ± 0.01	186.30 ± 70.02	0.31 ± 0.11
	(7.38 ± 0.17)	(0.33 ± 0.08)	(0.08 ± 0.07)	(0.08 ± 0.02)	(11.03 ± 5.38)	(0.20 ± 0.05)
	0.001[a]	0.36[a]	0.13[a]	0.08 [a]	0.003[a]	0.01[a]
NBV	5.70–7.79	ND – 0.46	ND – 0.35	ND – 0.78	8.47–330.00	ND – 1.98
WHO limit	6.50–8.50	0.05	0.005	5.00	75.0–200.0	5.00

Notes: A: Entrance point; B: Middle point; C: Point of raw water collection; Values in parentheses are for Akute Odo samples; [a]*t*-test values; *NBV* Nigeria's background values for some rivers; *WHO* World Health Organization limits for drinking water; *SD* standard deviation; *ND* not detected; t_{tab} at $p = 0.05$ (2.31)

Table 12.2 Mean levels of pH and metals concentration (μg/g ± SD) in Isashi and Akute Odo sediment samples

Sampling location	pH	Pb	Cd	Cu	Ca	Zn
A	6.85 ± 0.39	26.16 ± 6.9	1.79 ± 0.57	182.64 ± 249.57	2616.87 ± 2194.29	26.29 ± 4.61
	(6.37 ± 0.74)	(11.88 ± 1.90)	(0.66 ± 0.10)	(4.18 ± 2.03)	(23.26 ± 20.98)	(27.27 ± 7.36)
	0.13[a]	0.01[a]	0.01[a]	0.113[a]	0.04[a]	0.43[a]
B	6.41 ± 0.62	30.23 ± 2.64	2.05 ± 0.58	123.38 ± 121.82	4893.28 ± 4848.68	24.95 ± 12.67
	(6.20 ± 0.56)	(14.60 ± 2.91)	(0.84 ± 0.17)	(10.57 ± 6.26)	(51.50 ± 35.87)	(30.71 ± 6.44)
	0.36[a]	0.0002[a]	0.013[a]	0.07[a]	0.06[a]	0.28[a]
C	5.93 ± 0.26	27.95 ± 7.86	1.85 ± 0.47	71.82 ± 55.59	3835.44 ± 3109.74	14.00 ± 8.29
	(6.36 ± 0.59)	(17.10 ± 5.50)	(0.83 ± 0.09)	(6.87 ± 1.74)	(90.38 ± 117.31)	(31.53 ± 4.78)
	0.15[a]	0.09[a]	0.01[a]	0.04[a]	0.04[a]	0.01[a]
NBV	4.02–8.73	ND – 405.0	ND – 2.02	ND – 249.5	0.24–87.50	0.04–429.10

Notes: A: Entrance point; B: Middle point; C: Point of raw water collection; Values in parentheses are for Akute Odo samples; [a]*t*-test values; *NBV* Nigeria's background values for some rivers; *SD* standard deviation; *ND* not detected; t_{tab} at $p = 0.05$ (2.31)

The trend in metals distribution is of the order of Ca > Pb > Zn > Cd > Cu for water and Ca > Cu > Pb > Zn > Cd for sediments respectively. Similar river sediment and water metal distribution trends have been observed for River Buriganga, Dhaka, Bangladesh and the Cochin estuary and Periyar River, in the southwest coast of India [19, 23].

The concentrations of the metals are higher in River Owo at Isashi when compared with those of Ogun river at Akute Odo water and sediment samples (Tables 12.1 and 12.2). These higher values are not unexpected because of the proximity of Agbara Industrial Estate to River Owo.

All the elements analyzed in the water samples for both rivers are within the NBV. Pb (0.27–0.34 mg/l) and Cd (0.03–0.08 mg/l) exceed the WHO limit of 0.05 and 0.005 mg/l respectively. However, levels of Cu, Ca and Zn are within the WHO limit set for drinking water. The elevated levels of lead and cadmium in the river water samples are of toxicological and ecotoxicological concern. This may result in adverse health effects arising from bioaccumulation in tissues of living organisms [31, 32].

Metals in river water samples reported for rivers outside Nigeria were as follows (mg/l): 0.24–1.11 (Pb), 0.01–0.26 (Cd), 0.10–0.53 (Cu) and 0.07–0.12 (Zn) for Umtata River, South Africa; 1.10–3.90 (Pb), 13.0–51.0 (Cu) and 5.0–142.0 (Zn) for Bindal River, Dehradum, India; 0.14–0.59 (Pb), 0.09–0.24 (Cd), 1.22–2.76 (Cu) and 0.19–0.42 (Zn) for River Buriganga, Dhaka, Bangladesh; <0.005–0.38 (Pb), 0.002–0.003 (Cd), 0.002–0.041 (Cu) and 0.01–0.02 (Zn) for Nile Delta in Egypt; 0.0003–0.0190 (Pb), 0.0001–0.0012 (Cd) and 0.01–0.02 (Zn) for Avsar Dam Lake, Turkey. The values for Pakistan and Greece were as follows (mg/l): 0.01–2.43 (Pb), 0.001–3.129 (Cd), Not detected–3.0 (Cu) and 0.01–0.70 (Zn) for Indus River, Sindh province, Pakistan; 1.0–16.0 (Pb), 0.10–0.60 (Cd), 2.0–7.0 (Cu) and 20.0–157.0 (Zn) in northern Greece rivers. These global values, many of which exceed the WHO limits for drinking water, are an indication of the level of freshwater contamination arising from increasing urbanization and industrialization [15, 18, 20–23, 25, 27, 33]. National, trans-national and local authorities must therefore brace up to address the issue of water contamination in a sustainable manner.

Pb, Cu and Zn in River Owo sediment samples are all within the reported NBV (μg/g): ND–405.00 (Pb), ND–249.50 (Cu) and 0.04–429.10 (Zn), whereas Cd (1.79–2.05 μg/g) and Ca (2616.87–4893.28 μg/g) exceed the NBV (μg/g) of ND–2.02 and 0.24–87.50. On the other hand, it is only the levels of Ca (23.26–90.38 μg/g) in Ogun river at Akute Odo sediment samples that exceed the NBV of 0.24–87.50 μg/g.

The global sediment metals background values as captured from earlier studies are: 0.64–2740.0 μg/g (Pb), 0.33–17.26 μg/g (Cd), 13.00–255.82 μg/g (Cu) and 7.5–3050.0 μg/g (Zn). These values along with those of this study reveal substantial global contamination of river sediments [19–22, 24, 34]. The United States Environmental Protection Agency (USEPA) classifies natural sediments with >6 μg/g (Cd), >50 μg/g (Cu) and >60 μg/g (Pb), as being heavily contaminated [12].

Table 12.3 Correlation coefficients between metals in water and sediment of Isashi and Akute Odo samples

Sampling location	pH	Pb	Cd	Cu	Ca	Zn
A	0.78^{ns}	-0.04^{ns}	-0.41^{ns}	-0.59^{ns}	0.59^{ns}	0.59^{ns}
	$(-0.34)^{ns}$	$(0.41)^{ns}$	$(0.69)^{ns}$	$(-0.32)^{ns}$	$(-0.56)^{ns}$	$(-0.70)^{ns}$
B	-0.31^{ns}	0.52^{ns}	-0.41^{ns}	-0.35^{ns}	-0.39^{ns}	0.84*
	$(-0.26)^{ns}$	$(0.12)^{ns}$	$(0.29)^{ns}$	$(-0.48)^{ns}$	$(0.19)^{ns}$	$(-0.58)^{ns}$
C	0.06^{ns}	0.58^{ns}	0.06^{ns}	-0.63^{ns}	0.19^{ns}	0.87*
	$(0.17)^{ns}$	$(0.11)^{ns}$	$(0.02)^{ns}$	$(0.23)^{ns}$	(0.81)*	$(0.41)^{ns}$

Notes: A: Entrance point; B: Middle point; C: Point of raw water collection; Values in parentheses are for Akute Odo (water vs. sediment) samples; ns, non significant differences at $p = 0.05$; *significant differences at $p = 0.05$, t_{tab} at $p = 0.05$ (0.81)

The statistical *t*-test analysis comparing results obtained from the two rivers (Tables 12.1 and 12.2) reveal non-significant differences at 95 % level of confidence. This shows that the metals burden in the two rivers are not statistically different. Generally, the correlation obtained in the metals concentration (Table 12.3) between the water and sediment samples in both rivers is small and negative giving a non-significant difference at 95 % level of confidence. Coefficients close to zero or negative are indications that the concentration of the metals in the sediment samples is not directly affecting those in the water samples. However, the correlation coefficient of Zn (Isashi) and Ca (Akute Odo) show a significant difference, which reveals that the amount of Zn and Ca in water is directly dependent on the level of contamination of the sediment.

12.4 Conclusions

The occurrence of metals (Pb, Cd, Cu, Ca and Zn) at relatively elevated concentrations in water and sediment samples in both rivers give cause for concern. Most probable sources of the contaminants are municipal run-offs, agricultural and industrial activities around the rivers' catchments and the discharge of untreated and partially treated domestic effluents into the river bodies. Lead and cadmium occur at levels which exceed the WHO limit set for drinking water. The Lagos State Water Corporation should be encouraged to improve the method of treating raw water sourced from the two rivers with emphasis on continuous raw water quality monitoring. Metals at elevated levels have been implicated as risk to human health and the "health" of the aquatic system. Direct disposal of effluent from domestic and industrial sources into water bodies should be discouraged by Municipal authorities.

References

1. Lagos State Water Corporation (LSWC) (2007) Organization history. http://www.lagoswater.org
2. Manahan SE (1984) Environmental chemistry. Brooks/Cole Publishing Company, California
3. Davutluoglu OI, Seckin G, Ersu CB, Tilmaz I, Sari B (2011) Assessment of metal pollution in water and surface sediments of the Seyhan River, Turkey, using different indexes. Clean Soil Air Water 39:185–194
4. Kucuksezgin F, Kacar A, Kucuksezgin C, Uluturhan E (2010) Monitoring metal contamination levels and fecal pollution in clam (*Tapes decussatus*) collected from Izmir bay. Environ Monit Assess 162:407–415
5. Zvinowanda CM, Okonkwo JO, Shabalala PN, Agyei NM (2009) A novel absorbent for heavy metal remediation in aqueous environments. Int J Environ Sci Tech 6:425–434
6. Awofolu OR, Mbolekwa Z, Mtshemla V, Fatoki OS (2005) Levels of trace metals in water and sediments from Tyume River and its effects on an irrigated farmland. Water SA 31:87–94
7. Adeniyi AA, Owoade OJ, Shotonwa IO, Okedeyi OO, Ajibade AA, Sallu AR, Olawore MA, Ope KA (2011) Monitoring metals pollution using water and sediments collected from Ebute Ogbo river catchments, Ojo, Lagos, Nigeria. Afr J Pure Appl Chem 5:219–223
8. Fatoki OS, Mathabatha S (2001) An assessment of heavy metal pollution in East London and Port Elizabeth harbours. Water SA 27:233–240
9. Adeniyi AA, Yusuf KA, Okedeyi OO (2008) Assessment of the exposure of two fish species to metals pollution in the Ogun river catchments, Ketu, Lagos, Nigeria. Environ Monit Assess 137:451–458
10. Oshisanya KI, Unyimadu JP, Shelle ROD, Nubi OA, Ladigbolu IA, Oguguah NM, Olumodeji OO, Adeleye AO, Fashade AO (2011) Seasonal variation of heavy metals in sediments and water of Lagos lagoon. J Am Sci 7:384–387
11. Varol M, Sen B (2012) Assessment of nutrient and heavy metal contamination in surface water and sediments of the upper Tigris River, Turkey. CATENA 92:1–10
12. Gabarino JR, Hayes HC, Roth DA, Antweiler RC, Brinton TI, Taylor HE (1995) Heavy metals in the Mississippi River. In: Meade RH (ed) Contaminants in the Mississippi River. U.S. Geological Survey Circular, Reston, p 15
13. Emoyan OO, Ogban FE, Akarah E (2006) Evaluation of heavy metals loading of river Ijana in Ekpan – Warri, Nigeria. J Appl Sci Environ Manag 10:121–127
14. Papafilippaki AK, Kotti ME, Stavroulakis GD (2008) Seasonal variations in dissolved heavy metals in the Keritis River, Chania, Greece. Glob NEXT J 10:320–325
15. Varol M (2011) Assessment of heavy metal contamination in sediments of the Tigris River (Turkey) using pollution indices and multivariate statistical techniques. J Hazard Mater 195:355–364
16. Hejabi AT, Basavarajappa HT, Karbassi AR, Monavari SM (2011) Heavy metal pollution in water and sediments in the Kabini River, Karnataka, India. Environ Monit Assess 182:1–13
17. Chen CY, Stemberger RS, Klaue B, Blum JD, Pickhart PC, Folt CL (2000) Accumulation of heavy metals in food web components across a gradient of lakes. Limnol Oceanogr 45:1525–1536
18. Fatoki OS, Lujiza N, Ogunfowokan AO (2002) Trace metal pollution in Umtata River. Water SA 28:183–189
19. Mohan M, Augustine T, Jayasooryan KK, Chandran MSS, Ramasamy EV (2012) Fractionation of selected metals in the sediments of Cochin estuary and Periyar River, southwest coast of India. Environmentalist 32:383–393
20. Mohiuddin KM, Ogawa Y, Zakir HM, Otomo K, Shikazono N (2011) Heavy metals contamination in water and sediments of an urban river in a developing country. Int J Environ Sci Tech 8:723–736
21. El Bouraie MM, El Barbary AA, Yehia MM, Motawea EA (2010) Heavy metal concentrations in surface river water and bed sediments at Nile Delta in Egypt. Suosera Finish Peatl Soc Hels Suo 61:1–12

22. Spasovski O, Ristova E (2012) Heavy metals in sediments along the River Zletovska in eastern Macedonia. Zastita Mater 53:238–242
23. Simeonov V, Stratis JA, Samara C, Zachariadis G, Kouimtzis T (2003) Assessment of the surface water quality in northern Greece. Water Res 37:4119–4124
24. Ozturk M, Ozozen G, Minareci O, Minareci E (2009) Determination of heavy metals in fish, water and sediments of Avsar Dam Lake in Turkey. Iran J Environ Health Sci Eng 6:73–80
25. Kumar A, Bisht BS, Joshi VD (2010) Chemical report of sediments, phytoplanktons and fishes of Bindal River, Dehradin. J Biol Environ Sci 4:53–58
26. America Public Health Association (1998) Standard methods for the examination of water and wastewater, APHA, AWWA, and WPCE, 20th edn. Byrd Progress, Springfield
27. Ekeanyanwu CR, Ogbuinyi CA, Etienajirhevwe OF (2011) Trace metals distribution in fish tissues, bottom sediments and water from Okumeshi River in Delta State Nigeria. Environ Res J 5:6–10
28. Aremu MO, Atolaiye BO, Gav BL, Opaluwa OD, Sangari DU, Madu PC (2011) Metal concentrations in sediments and water from rivers Doma, Farinruwa and Mada in Nasarawa State, Nigeria. J Environ Chem Ecotoxicol 3:244–251
29. Ladigbolu IA, Balogun KJ (2011) Distribution of heavy metals in sediments of selected streams in Ibadan metropolis, Nigeria. Int J Environ Sci 1:1186–1191
30. World Health Organization (WHO) (1996) Guidelines for drinking water quality, vol 2, 2nd edn. World Health Organisation, Geneva
31. Katzantzis G (1987) The mutagenic and carcinogenic effect of cadmium. An update. J Toxicol Environ Chem 15:83–100
32. Vinodhini R, Narayanan M (2008) Bioaccumulation of heavy metals in organs of fresh water fish *Cyprinus carpio* (Common carp). Int J Environ Sci Tech 5:179–182
33. Farooq MA, Shaukat SS, Zafar MU, Abbas Q (2012) Variation pattern of heavy metal concentrations during pre- and post-monsoon seasons in the surface water of river Indus (Sindh Province). World Appl Sci J 19:582–587
34. Gaur VK, Gupta SK, Pandey SD, Gopal K, Misra V (2005) Distribution of heavy metals in sediment and water of River Gomti. Environ Monit Assess 102:419–433

Chapter 13
Adsorption of Selected Ions on Ferro-Precipitates from Aqueous Solutions

Roman Marsalek

Abstract The main aim of the paper is to evaluate ferro-precipitate as an immobilization agent for various ions from aqueous solutions. Heavy metals, namely lead, copper and chromium were adsorbed as well as arsenate and phosphate. In addition, the adsorption of surfactants, sodium dodecyl sulfate (SDS) and cetyltrimethylammonium bromide (CTAB) was carried out. Linear as well as non-linear regression of adsorption isotherms confirmed the Langmuir model applicability. Adsorption capacities (a_m) were calculated. The highest adsorption capacity was found for arsenate and phosphate respectively (a_m = 1.36 mmol g^{-1} for arsenate and 0.70 mmol g^{-1} for phosphate). Lower but still enough adsorption capacity was found for heavy metals (a_m = 0.76 mmol g^{-1} for lead, 0.58 mmol g^{-1} for copper, 0.38 mmol g^{-1} for chromium, and the lowest values were shown for surfactants (a = 0.38 mmol g^{-1} for CTAB and 0.21 mmol g^{-1} for SDS).

13.1 Introduction

Many pollutants are released into the environment due to human activity (industry, agriculture, etc.). Heavy metals, surfactants, or even arsenic and phosphorus compounds are among the most serious. These substances can be removed from waste water in different ways, including also adsorption.

Heavy metals in waste water cause a serious environmental problem. High toxicity and the ability to accumulate in living organisms are included among their negative qualities. Recently, efforts have occurred to eliminate the content of heavy metals by means of adsorption on various types of adsorption materials. Apart from the most common active coal, cheaper options are analyzed. The following natural materials are used – various kinds of coal (lignite coals, charcoal,

R. Marsalek (✉)
Department of Chemistry, Faculty of Science, University of Ostrava, Ostrava, Czech Republic
e-mail: roman.marsalek@osu.cz

M. Gupta Bhowon et al. (eds.), *Chemistry: The Key to our Sustainable Future*,
DOI 10.1007/978-94-007-7389-9_13,

brown coal, bituminous coal), humic acids, peat, algae, woolen and rice materials [1–5]. The second category is waste materials, e.g. various kinds of sludge. Other interesting adsorbents offering good prospects include ferro-compounds such as goethite, hematite, magnetite and others [6–10]. These compounds are very often artificially synthesized and used as nano-adsorbents.

Nitrogen and phosphorus compounds are often used as fertilizers. These substances penetrate into the groundwater and then further into the aquatic environment where they cause a number of problems, among the most important of which is eutrophication. Arsenic compounds are highly toxic to living organisms. Compounds of nitrogen and phosphorus are removed by microorganisms in the biological purification stage in a wastewater treatment plant. An alternative to this could be adsorption, which also applies for those substances that can be removed using a variety of iron oxides [11–14].

Surfactants are commonly used in many industrial and commercial products and processes throughout the world. For example, surfactants are in demand for industrial processes requiring colloid stability, metal treatment, mineral flotation, oil production, emulsion polymerization, pesticides, and pharmaceutical formulation. This is a positive role of surface active agents, but their applications can produce environmental pollution and increase a series of problems for wastewater treatment plants [15]. Many materials have been studied as surfactant adsorbents, e.g. carbon black, chitosan, latex, chrysotile, montmorillonite, kaolinite, shungite, titania, and bentonite [16–23]. Oxidic iron compounds such as hematite, may be used for the adsorption of surfactants similarly as in the case of heavy metals and phosphorus compounds [24].

These oxy-ferro-compounds also occur as waste products in particular manufacturing processes; e.g. in the Bochemie Company (Czech Republic), zinc chloride is manufactured particularly from recycled acid, which represents waste from zinc works.

The problem of this production is the thick and almost unsettling sediment of iron which, moreover, retains rather large amounts of zinc chloride. The sludge consists of approximately one third water, one third of zinc chloride, and the remnant ferro-compounds. Due to the presence of zinc chloride, the sludge cannot be used to charge a blast furnace. The solution is to find a suitable process of precipitating iron in solutions, e.g. using the goethite process. The goethite process is successfully used in the electrolytic galvanization industry to turn the iron in a solution (mostly sulfide or chloride) into a sediment which is easy to filter [25, 26]. The Electrolytic Zinc Company of Australasia developed a continuous process, during which all ferrous ions in a solution are at first oxidized, and subsequently the solution continuously flows into a reactor at a speed which eliminates the solid phase [27].

This method was also used to precipitate iron from the waste acid used to produce zinc chloride in the Bochemie Company. The process was carried out under varied conditions (temperature, pH, time) and resulted in sediments with varied composition and texture parameters. Sediments high in zinc, the further use of which is rather limited, were analyzed with regard to their possible use as sorbents to eliminate heavy metals from aqueous solutions.

When being adsorbed, solid anions exhibit a surface charge. Zeta potential is one of a few effective techniques for characterization of surface charge, as well as the surface chemical properties of solids in a solution, and for understanding changes on the solid surfaces. Zeta potential and its changes were also observed during adsorption on iron compounds [28–33].

The main goal of this study is to judge waste materials on the basis of ferrous oxide as an adsorption material to eliminate selected ions from aqueous solutions. Further goals were to describe the adsorption process by means of some adsorption theory (Langmuir, Freundlich) and to calculate the adsorption capacity of the sorbent. By means of the zeta potential, changes on the surface of adsorption particles were monitored during the adsorption process.

13.2 Methodology

13.2.1 General

The chemicals used in the goethite process were zinc chloride ($ZnCl_2$, Sigma-Aldrich), zinc oxide (ZnO, Sigma-Aldrich) and hydrogen peroxide (H_2O_2, Sigma-Aldrich). The chemicals used for preparation of stock solutions for adsorption were lead nitrate ($PbNO_3$, Sigma-Aldrich), copper nitrate ($CuNO_3.3H_2O$, Sigma-Aldrich), chromium nitrate ($Cr(NO_3)_3.9H_2O$, Sigma-Aldrich), sodium arsenate ($NaAsO_2$, Sigma-Aldrich), sodium phosphate (NaH_2PO_4, Sigma-Aldrich), sodium dodecyl sulfate (SDS, Sigma-Aldrich) and cetyltrimethylammonium bromide (CTAB, Sigma-Aldrich), Fig. 13.1. Hydrochloric acid (HCl, Sigma-Aldrich) and sodium hydroxide (NaOH, Sigma-Aldrich) were used for adjusting the pH during zeta potential measurements. All the solutions were made in distilled water. Analytical grade chemicals were used.

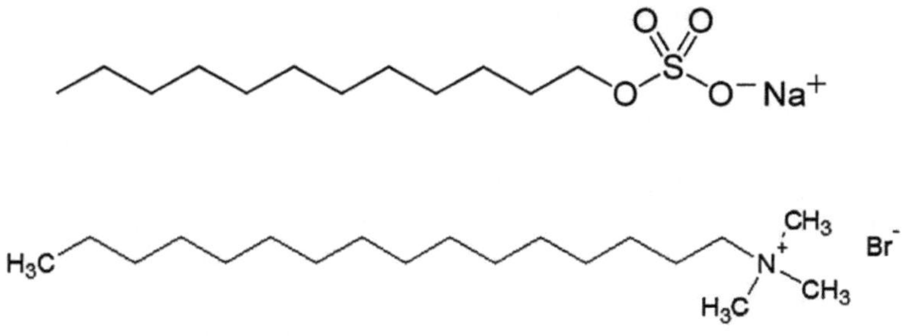

Fig. 13.1 Structures of SDS and CTAB

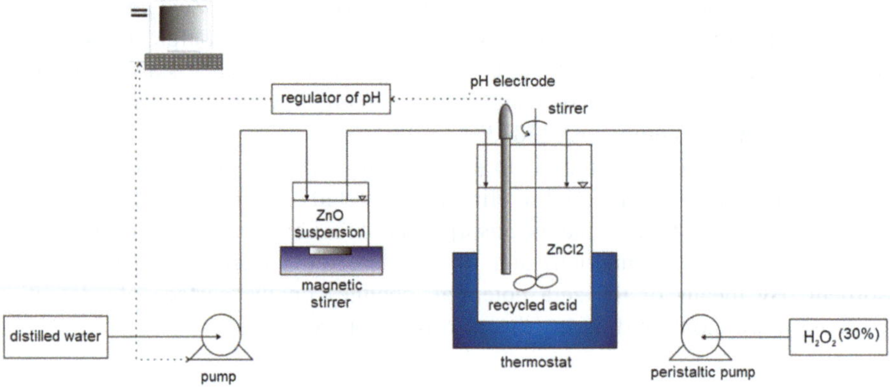

Fig. 13.2 Scheme of apparatus for precipitation of iron

13.2.2 Goethite Process

A continuous flow reactor (Fig. 13.2) was used in this process. An 800 g measure of zinc chloride solution was put into a reaction vessel (filtering flask). The solution had been prepared by dissolving 240 g of solid zinc chloride in 560 g of distilled water. Recycled acid was rid of solid impurities by filtering it over a textile filter. Subsequently, all ferrous ions were oxidized into ferric ions by slowly adding hydrogen peroxide (30 %) and continuously stirring the substance. Intermittent checks were carried out to find out whether all the iron had oxidized. The discharge of oxidized acid into the reaction vessel was set on a peristaltic pump. A pH electrode, a stirrer and hoses from both pumps (with oxidized recycled acid and a zinc oxide suspension) were immersed into the reaction vessel. The stirrer was switched on. After having reached the desired temperature and pH, both pumps and a computer program for pH were switched on. This way, the experiment began.

After finishing the experiments we obtained precipitates (sediments) with varied composition and texture parameters (see Results and Discussion).

13.2.3 Adsorption: Batch Experiments

About 0.2 g of dried sample was added to 50 ml of a given ion (Pb^{2+}, Cu^{2+}, Cr^{3+}, $C_{12}H_{25}OSO_3^-$, $(C_{16}H_{33})N(CH_3)_3^+$, AsO_2^-, $H_2PO_4^-$) solutions of known concentration varying in the range 10–0.1 mmol/l and occasionally shaken. Adsorption equilibration took about 3 days, as deduced from previous measurements. The zeta potential of the adsorption suspensions was measured. The samples were then removed by filtering through a paper filter. The equilibrium concentration of the filtered solutions was determined by inductively coupled plasma atomic emission spectroscopy (heavy metals) and by UV/VIS spectroscopy (arsenic, phosphorus and surfactants).

The amount of ion adsorbed (a) was determined from the change in the solution concentration before and after equilibrium, according to:

$$a = \frac{(c_0 - c_e)V}{m} \tag{13.1}$$

where c_0 is the initial concentration of adsorbate, c_e the concentration of adsorbate at the adsorption equilibrium, V the volume of adsorbate solution and m the mass of the sample.

13.2.4 The Zeta Potential Measurements

The zeta potential was measured by analyzing 0.2 g of precipitate in 50 ml of solution using a Zetasizer Nano ZS (Malvern Instruments Ltd., GB). This machine uses Laser Doppler Velocimetry to determine the electrophoretic mobility. The zeta potential was obtained from the electrophoretic mobility by the Smoluchowski equation:

$$\zeta = \frac{\mu.\eta}{\varepsilon} \tag{13.2}$$

ζ is the zeta potential (V), η represents dynamic viscosity (Pa.s), and ε stands for the dielectric constant. The fixed conditions of measuring were the following: temperature (298 K), electric field (15 V), frequency (500 Hz), and the properties of the samples – viscosity (0.0089 kg/m.s), refraction index (1.333), and dielectric constant (78.36).

We had two main reasons for the measurement of zeta potential. Firstly, we measured the zeta potential of the precipitate particles, depending on the pH. The aim was to find the differences between precipitates and determine the isoelectric points. Secondly, the zeta potential of the sample was measured after a running adsorption, i.e. in the equilibrium. The objective was to describe the changes on the particle surface due to adsorption of ions.

13.3 Results and Discussion

13.3.1 Characterization of Samples

Table 13.1 shows the conditions of precipitation for selected samples 1–3.

The temperature was maintained at 80 °C for these selected samples. This is the optimal temperature. There is still a high content of zinc in the solid phase at a lower

Table 13.1 Conditions of precipitation and composition of precipitates and solutions

	Conditions of precipitation			Sludge composition (% wt.)				Composition of solutions	
					Solid				
Sample	pH	T (°C)	Contact time (min)	% wet	Fe	Zn	Cl^-	$ZnCl_2$ (%)	Fe (ppm)
1	2.2	80	360	62.3	55.2	7.4	0.7	23.1	8
2	2.2	80	160	63.9	55.1	6.8	2.3	21.9	1
3	1.0	80	160	35.6	55.4	0.8	7	20.8	3

temperature and the higher temperature is energy consuming. The pH was adjusted to between 1 and 2.2 and the contact time was 160 min whereas for sample 1, 360 min was used. The resulting composition was not significantly affected by time. The most important parameter is pH and we could say that the decrease in pH leads to a reduction of zinc.

Samples were subjected to further analysis, namely the determination of specific surface area, X-ray diffraction, particle size distribution and zeta potential measurements. The results are summarized in Table 13.2.

The surface area of the sample from the adsorption isotherms of nitrogen (SBET) was evaluated according to BET theory, using a nitrogen molecular area of 0.162 nm^2 for the calculation.

The normal powder XRD patterns were recorded between 5.0° and 50.0° 2θ using CuKα radiation with an INEL CPS 120 diffractometer equipped with a curved position-sensitive detector, in transmission mode with a germanium monochromator.

Particle size distribution was determined as the mean value of D_{50}. Measurements were performed using a Mastersizer 2000 machine (Malvern Instruments Ltd. GB).

The isoelectric point (IEP) is the pH at which the zeta potential is zero. The pH was adjusted by adding a solution of hydrochloric acid and sodium hydroxide.

Data from RTG diffraction showed that samples 1 and 2 contain a significant amount of amorphous phase; the following two ferrous compounds are present – ferrihydrite $Fe_5O_7(OH).4H_2O$ (30 %) and feroxyhyte δ-FeO(OH) (20 %). The relatively high content of zinc makes disposal of this material (waste) in a blast furnace impossible. On the other hand sample 3 contains only 0.8 % zinc and the mineral composition is also different. This sample contains mainly akaganeite β-FeO(OH) (54.5 %). Samples 1 and 2 have approximately seven times higher specific surface area in comparison with sample 3 (see Table 13.2). Particle size corresponds to the size of the surface. In addition we followed the influence of pH on the zeta potential of our samples. Figure 13.3 shows the effect of pH on the zeta potential.

The isoelectric point of sample 1 lies around pH 6. The changes of zeta potential in sample 2 depend on whether the pH was very similar. The zeta potential of sample 3 is shifted to positive values. The isoelectric point is about pH 8.5. These data are very important for the adsorption of ions, especially by surfactants.

Table 13.2 Specific surface area, mineral composition, particle size and isoelectric points of samples

Sample	S_{BET} (m^2/g)	RTG diffraction	D_{50} (μm)	IEP
1	220.45	ferrihydrite + feroxyhyte	3.25	6
2	250.21	ferrihydrite + feroxyhyte	2.49	6.2
3	30.04	akaganeite	4.75	8.5

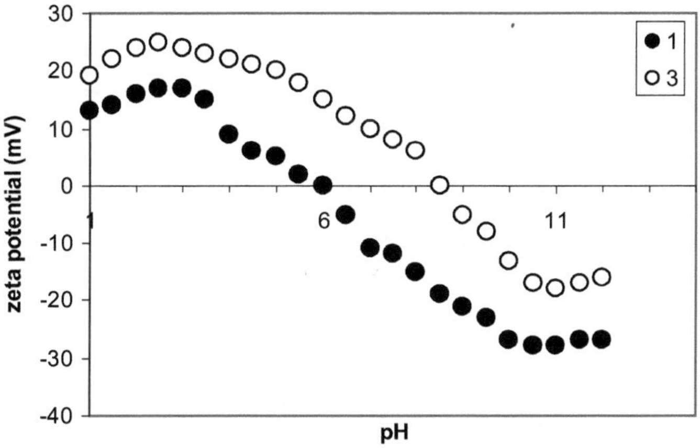

Fig. 13.3 Isoelectric points of samples 1 and 3, influence of pH on zeta potentials

For all adsorptions the pH was maintained at pH 6, at which the zeta potential of samples 1 and 2 showed values around 0 mV, and the zeta potential of sample 3 approximately +17 mV.

13.3.2 Adsorption of Selected Ions

Heavy Metals

Figure 13.4 depicts typical adsorption isotherms obtained from the experimental data.

The shape of isotherms indicates that the adsorption data could be well suited to the Langmuir adsorption model of monolayer coverage. In a linear form, the Langmuir equation is given by:

$$\frac{c_e}{a} = \frac{c_e}{a_m} + \frac{1}{a_m b} \tag{13.3}$$

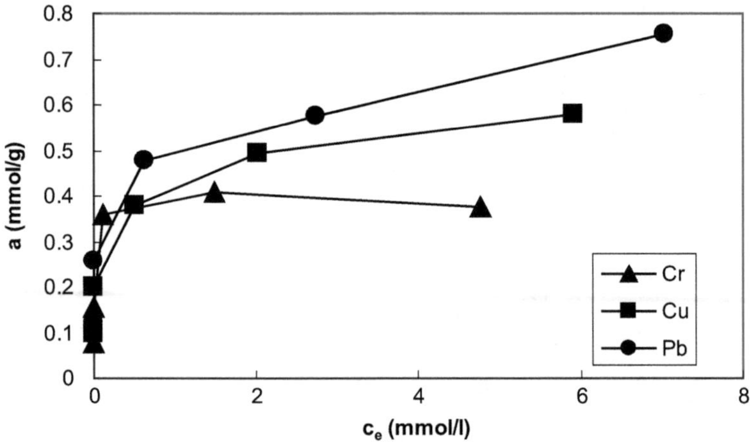

Fig. 13.4 Adsorption isotherms of heavy metals on sample 1

Table 13.3 Results obtained from the Langmuir and Freundlich equations for heavy metals adsorptions

	Langmuir isotherm			Freundlich isotherm		
Heavy metal	a_m *(mmol/g)*	b *(l/mmol)*	r^2	k	n	r^2
Pb	0.756	4.001	0.985	0.517	5.487	0.983
Cu	0.584	7.112	0.996	0.430	5.060	0.956
Cr	0.376	46.296	0.999	0.373	7.000	0.859

where a is the amount of the surfactant adsorbed, c_e is the equilibrium concentration of the surfactant in solution, b represents a monolayer binding constant and a_m is the monolayer adsorption capacity.

All adsorption isotherms proved to be consistent with the Langmuir model as deduced from calculated r-squared values close to 1 (minimal r^2 value was 0.985). The most informative parameter in the Langmuir equation is a_m, providing information on the adsorbed amount at monolayer surface coverage. These values are compiled in Table 13.3 as found for studied metals on sample 1. There are also comparable parameters of Freundlich adsorption isotherms:

$$a = k.{c_e}^{1/n} \tag{13.4}$$

where k and n are dimensionless Freundlich constants.

The Langmuir adsorption theory is based on the adsorption of gases and it is not apparently valid without any differences for adsorption from the solutions. Thus, the experimentally obtained data were also analyzed in terms of non-linear regression. Four adsorption models (Langmuir, Freundlich, Langmuir-Freundlich and Temkin) were used and compared. The non-linear regression was calculated by means of the OPstat program using various algorithms – genetic algorithm, simplex, and Levenberg-Marquardt algorithm. Akaic criterion was chosen to evaluate the experimental data correspondence with a given theory.

Table 13.4 Akaic criteria for adsorption theories, adsorption of heavy metals on sample 1

	Akaic criterion			
Heavy metal	Langmuir	Langmuir-Freundlich	Freundlich	Temkin
Pb	−41.82	−39.74	−26.16	−26.66
Cu	−43.36	−40.63	−34.04	−23.84
Cr	−52.47	−48.98	−45.65	−25.22

The lowest value of Akaic criterion was found for Langmuir model, Table 13.4. In other words non-linear regression also confirmed Langmuir model applicability.

From Fig. 13.4 and Table 13.3 it is apparent that the maximum adsorption capacity was found in lead ions, followed by copper and the lowest affinity was in chromium ions. The reason for this behavior may be less hydration of lead ions in comparison with copper and chromium ions. Lead ions are thus better available as an adsorbent surface. Similar results were also found for the second sample. Conversely adsorption was practically not realized on the third sample. Sample 3 has a much smaller surface area. It seems that the specific surface area plays a decisive role for the adsorption of heavy metals on this type of material. In addition to physical adsorption on the surface, the ion exchange is probably the second way of heavy metals immobilization. Zinc ions are replaced with the appropriate heavy metal. Samples 1 and 2 contain relatively significant amounts of zinc (see Table 13.1). This fact is also demonstrated by an increased amount of zinc in the solution after adsorption.

Surfactants

Adsorption of surfactants was different from the adsorption of heavy metal ions. Molecules of surfactants are much larger and contain a hydrophobic chain and a hydrophilic head. Formation of micelles at the critical micelle concentration is also one of the important traits of these compounds. The critical micelle concentration was determined for both surfactants by conductivity measurements. The critical micelle concentrations were 8.5 mmol/dm^3 for SDS and 1 mmol/dm^3 for CTAB respectively. Figure 13.5 shows the adsorption isotherms for SDS samples 1 and 3.

The shape of adsorption isotherms of anion active surfactant SDS suggests that it can be described by Langmuir adsorption theory. The parameter r^2 is closer to one in comparison with the Freundlich theory and it is possible to calculate concrete adsorption capacities. The data are summarized in Table 13.5.

From Table 13.5 it is obvious that SDS adsorption on sample 3 is approximately two times higher in comparison with sample 1. These results are surprising because sample 1 has a markedly larger specific surface area (see Table 13.2). It seems that the specific surface area does not play the most important role in the immobilization process and specific interactions between iron minerals and surfactant molecules probably exist. The reason for the higher determined adsorption capacity of sample 3

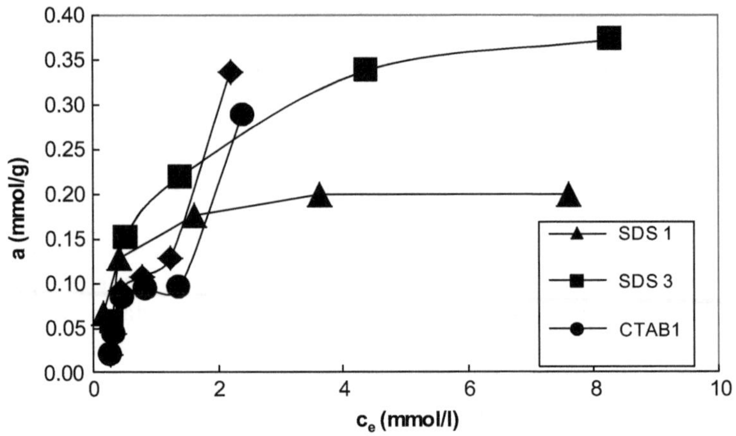

Fig. 13.5 Adsorption isotherms of SDS and CTAB on samples 1 and 3

Table 13.5 Results obtained from the Langmuir and Freundlich equations for SDS adsorptions

	Langmuir isotherm			Freundlich isotherm		
Sample	a_m *(mmol/g)*	b *(l/mmol)*	r^2	k	n	r^2
1	0.207	3.706	0.999	0.135	3.605	0.860
3	0.443	0.674	0.987	0.152	1.968	0.856

is in its different zeta potential value. At pH 6 (value for all studied systems), the surface of sample 3 has zeta potential around +18 mV; on the other hand the surface of sample 1 is without charge. The positively charged surface of sample 3 attracts and subsequently reacts with the negatively charged SDS molecules. So from the zeta potential measurements, as well as from the surface characteristics of the precipitate, it was deduced that the sorption mechanism of SDS on the adsorbent surface is connected mainly with interactions between the hydrophilic part of the surfactant and the precipitate surface. A similar mechanism was also described in the SDS adsorption on hematite [24]. In all cases further measurement is necessary to explain the mechanism of the impact of the surfactant.

The situation is different for adsorption of CTAB. A lower solubility and lower critical micelle concentration (1 mmol/l) made it impossible to measure the whole concentration range as like in the previous cases, and the default concentration is in the range of 0.1–3 mmol/l. It first reached maximum before a dissociation of some micelles occurred, and these free molecules again adsorbed on the surface. The system is not in equilibrium. At these initial concentrations the system has not reached equilibrium, and it was not possible to determine the adsorption capacity using the Langmuir theory. The adsorption quantity calculated from the differences between the initial and equilibrium concentrations was shown to be higher than 0.35 mmol/g.

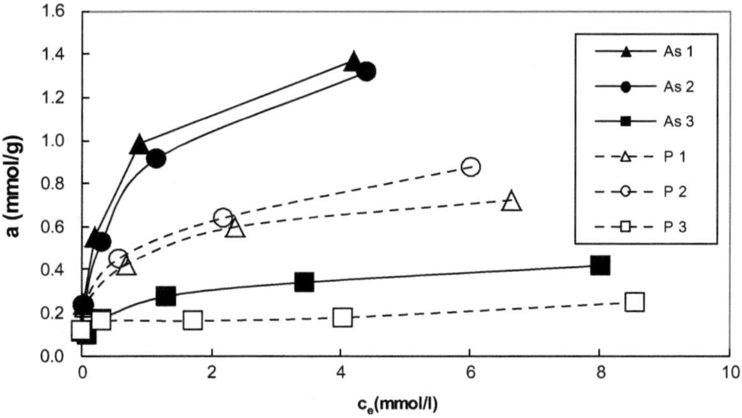

Fig. 13.6 Adsorption isotherms of arsenate and phosphate on samples 1, 2 and 3

Table 13.6 Results obtained from the Langmuir and Freundlich equations for arsenate and phosphate adsorptions

		Langmuir isotherm			Freundlich isotherm		
Adsorbate	Sample	a_m *(mmol/g)*	b *nl/mmol)*	r^2	k	n	r^2
Arsenate	1	1.420	4.949	0.994	0.930	2.703	0.984
	2	1.398	2.895	0.989	0.801	2.393	0.985
	3	0.438	1.682	0.993	0.230	3.125	0.981
Phosphate	1	0.736	4.013	0.992	0.482	5.270	0.990
	2	0.889	3.244	0.979	0.552	4.874	0.992
	3	0.245	1.931	0.962	0.177	17.292	0.758

Arsenate and Phosphate

A third group (Fig. 13.6) of substances to be adsorbed on the surface of iron precipitates was inorganic anions: arsenate $(AsO_2)^-$ and phosphate $(H_2PO_4)^-$.

Samples 1 and 2 showed a high adsorption capacity for these inorganic anions, especially for arsenate. Sample 3 also adsorbed the followed inorganic anions. The course of adsorption was similar to the previous cases and the results are summarized in Table 13.6.

In the case of the adsorption of arsenate and phosphate, we assume there was a complex formation between iron atoms on the adsorbents surface (Fig. 13.7). The formation of such complexes was also confirmed using IR-FTIR during the adsorption of phosphate on akageinete and goethite [12, 14]. The higher adsorption capacity of samples 1 and 2 compared with sample 3 is again attributed to the significantly greater surface area of samples 1 and 2.

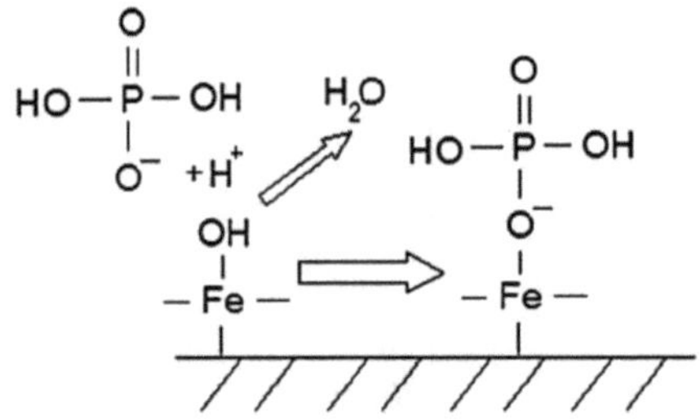

Fig. 13.7 Adsorption of phosphate ions by inner sphere complexes

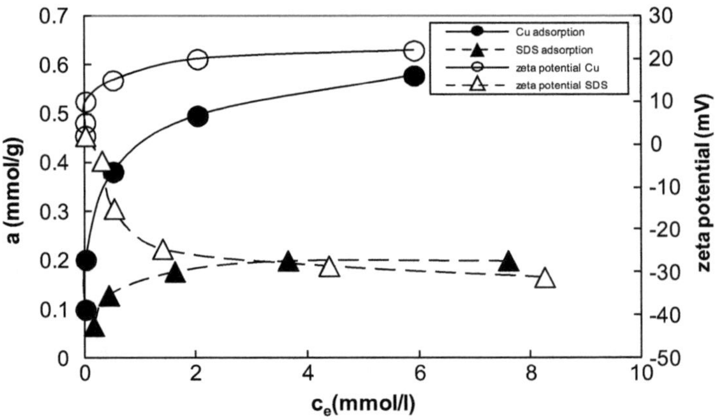

Fig. 13.8 Dependence of zeta potential (blank symbols) and adsorption of Cu and SDS (full symbols) on sample 1 on the equilibrium concentration of the Cu and SDS solution (c_e)

13.3.3 Changes of Zeta Potential

Figure 13.8 shows the noteworthy similarity of the adsorption isotherms with the changes of zeta potential of the adsorption systems. In other words the adsorption of ions causes a change of the surface charge. The same results were found for other adsorption systems [34–37].

The figure shows the adsorption of copper on sample 1, and SDS surfactant adsorption on the same sample, to be both solid symbols. Adsorption took place at pH 6, the value of the zeta potential of sample 1 in water at this pH was equal to 0 mV. After the addition of copper ions and similarly surfactant SDS, the ions were adsorbed and caused a change to the zeta potential. In the case of adsorption of copper, but also other heavy metals (Pb and Cr), the zeta potential increased. In the case of copper, the resulting value of the zeta potential was about 20 mV in the equilibrium. In contrast, in the case of adsorption of anionic surfactants (and inorganic anions arsenate $(AsO_2)^-$ and phosphate $(H_2PO_4)^-$) zeta potential is shifted into negative values. After adsorption of SDS the final value of zeta potential was approximately −33 mV. Total changes of the zeta potential correspond to the amount adsorbed at the highest initial concentration of the substances. The more ions on the surface to be adsorbed, the greater the change in zeta potential appeared.

13.4 Conclusions

Adsorption of selected heavy metals, surfactants and inorganic anions on ferrous-precipitates was monitored. We changed conditions of precipitation which resulted in a wide spectrum of precipitates. They varied in composition and textural parameters. Our materials contained various iron compounds such as akageinete, ferrihydrite and feroxyhyte. These materials, which represent unusable waste, show a relatively high adsorption capacity for ions, especially for inorganic anions. The adsorption process can be described by the "classic" Langmuir theory and the calculated parameters of the Langmuir equation can be used for potential particular applications. The adsorption process was studied as well with regard to changes of surface charge. Measuring proved a connection between adsorption and the change of the zeta potential of the particles. Particularly due to its adsorption capacity and price, this material ranks among many cheap adsorption materials possibly usable to treat waste water containing heavy metals.

Acknowledgements This article has been done in connection with the Institute of Environmental Technologies project, reg. no. CZ.1.05/2.1.00/03.0100 supported by the Research and Development for Innovations Operational Programme financed by European Union Structural Funds and from the means of the state budget of the Czech Republic. It was also supported by the Ministry of Industry and Trade of the Czech Republic (FR-T11/246).

References

1. Taraba B, Marsalek R (2007) Immobilization of heavy metals and phenol on altered bituminous coals. Energ Source A 29:885–894
2. Wang YH, Lin SH, Juang RS (2003) Removal of heavy metal ions from aqueous solutions using various low-cost adsorbents. J Hazard Mater B 102:291–302
3. Klucakova M, Omelka L (2004) Sorption of metal ions on lignite and humic acid. Chem Pap Chem Zvesti 58:170–175
4. Cox M, El-Shafey E, Pichugin AA, Appleton Q (2000) Removal of mercury(II) from aqueous solution on a carbonaceous sorbent prepared from flax shive. J Chem Technol Biot 75:427–435
5. Meena AK, Mishra GK, Kumar S, Rajagopal C, Nagal NP (2004) Low-cost adsorbents for the removal of mercury(II) from aqueous solution – a comparative study. Defence Sci J 54:537–548
6. Chen YH, Li FA (2010) Kinetic study on removal of copper(II) using goethite and hematite nano-photocatalysts. J Colloid Interf Sci 347:277–281
7. Mustafa S, Khan S, Zaman MI (2010) Effect of Ni^{2+} ion doping on the physical characteristics and chromate adsorption behavior of goethite. Water Res 44:918–926

8. Campos V (2009) The sorption of toxic elements onto natural zeolite, synthetic goethite and modified powdered block carbon. Environ Earth Sci 59:737–744
9. Li W, Zhang SZ, Shan XQ (2006) Effect of phosphate on the adsorption of Cu and Cd on natural hematite. Chemosphere 63:1235–1241
10. Chowdhury SR, Yanful EK (2010) Arsenic and chromium removal by mixed magnetite-maghemite nanoparticles and the effect of phosphate on removal. J Environ Manage 91:2238–2247
11. Kolbe F, Weiss H, Morgenstern P, Wennrich R, Lorenz W, Schurk K, Stanjek H, Daus B (2011) Sorption of aqueous antimony and arsenic species onto akaganeite. J Colloid Interf Sci 357:460–465
12. Deliyanni EA, Peleka EN, Lazaridis NK (2007) Comparative study of phosphates removal from aqueous solutions by nanocrystalline akaganeite and hybrid surfactant-akaganeite. Sep Purif Technol 52:478–486
13. Peleka EN, Deliyanni EA (2009) Adsorptive removal of phosphates from aqueous solutions. Desalination 245:357–371
14. Chitrakar R, Tezuka S, Sonoda A, Sakane K, Ooi K, Hirotsu T (2006) Phosphate adsorption on synthetic goethite and akaganeite. J Colloid Interf Sci 298:602–608
15. Wu HS, Pendleton PJ (2001) Adsorption of anionic surfactant by activated carbon: effect of surface chemistry, ionic strength, and hydrophobicity. J Colloid Interf Sci 243:306–315
16. Ridaoui H, Jada A, Vidal L, Donnet JB (2006) Effect of cationic surfactant and block copolymer on carbon black particle surface charge and size. Colloid Surf A 278:149–159
17. Chang MY, Juang RS (2005) Equilibrium and kinetic studies on the adsorption of surfactant, organic acids and dyes from water onto natural biopolymers. Colloid Surf A 269:35–46
18. Vale HM, McKenna TF (2005) Adsorption of sodium dodecyl sulphate and sodium dodecyl benzene sulphonate on poly(vinyl chloride) latexes. Colloid Surf A 268:68–72
19. Valentim IB, Joekes I (2006) Adsorption of sodium dodecyl sulphate on chrysotile. Colloid Surf A 290:106–111
20. Qi L, Liao W, Bi Z (2007) Adsorption of conventional and gemini cationic surfactants in nonswelling and swelling layer silicite. Colloid Surf A 302:568–572
21. Sineva AV, Parfenova AM, Fedorova AA (2007) Adsorption of micelle forming and non-micelle forming surfactants on the adsorbents of different nature. Colloid Surf A 306:68–74
22. Kosmulski M, Prochniak P, Rosenholm JB (2009) Electrokinetic studies of adsorption of ionic surfactants on titania from organic solvents. Colloid Surf A 348:298–300
23. Atia AA, Farag FM, El-Fatah A, Youssef M (2006) Studies on the adsorption of dodecyl benzene sulphonate and cetylpyridinium bromine at liquid/air and bentonite/liquid interfaces. Colloid Surf A 278:74–80
24. Gao XD, Chorover J (2010) Adsorption of sodium dodecyl sulphate (SDS) at ZnSe and α-Fe_2O_3 surfaces: Combining infrared spectroscopy and bath uptake studies. J Colloid Interf Sci 348:167–176
25. Davey PT, Scott TR (1976) Removal of iron from leach liquors by the goethite process. Hydrometallurgy 2:25–33
26. Ismael MRC, Carvalho JMR (2003) Iron recovery from sulphate leach liquors in zinc hydrometallurgy. Miner Eng 16:31–39
27. Davey PT, Scott TR (1975) Formation of β-FeOOH and α-Fe_2O_3 in the goethite process. T I Min Metall C 84:83–86
28. Jolsterå R, Gunneriusson L, Forsling W (2010) Adsorption and surface complex modeling of silicates on maghemite in aqueous suspensions. J Colloid Interf Sci 342:493–498
29. Rica RA, Jiménez ML, Delgado AV (2010) Electric permittivity of concentrated suspensions of elongated goethite particles. J Colloid Interf Sci 343:564–573
30. Kumpulainen S, von der Kammer F, Hofmann T (2008) Humic acid adsorption and surface charge effects on schwertmannite and goethite in acid sulphate waters. Water Res 42:2051–2060

31. Zhang JS, Stanforth R, Pehkonen SO (2007) Proton–arsenic adsorption ratios and zeta potential measurements: implications for protonation of hydroxyls on the goethite surface. J Colloid Interf Sci 315:13–20
32. Zhang JS, Stanforth R, Pehkonen SO (2007) Effect of replacing a hydroxyl group with a methyl group on arsenic(V) species adsorption on goethite (α-FeOOH). J Colloid Interf Sci 306:16–21
33. Madigan C, Leong YK, Ong BC (2009) Surface and rheological properties of as-received colloidal goethite (α-FeOOH) suspensions: pH and polyethylenimine effects. Int J Miner Process 93:41–47
34. Gallardo-Moreno AM, Gonzáles-Garcia CM, González-Martín M, Bruque JM (2004) Arrangement of SDS adsorbed layer on carbonaceous particles by zeta potential determinations. Colloid Surf A 249:57–62
35. Basar CA, Karagunduz A, Keskinler B, Cakici A (2003) Effect of presence of ions on surface characteristics of surfactant modified powdered activated carbon (PAC). Appl Surf Sci 214:169–174
36. Juang RS, Wu WL (2002) Adsorption of sulfate and copper(II) on goethite in relation to the changes of zeta potentials. J Colloid Interf Sci 249:22–29
37. Marsalek R, Navratilova Z (2011) Comparative study of CTAB adsorption on bituminous coal and clay mineral. Chem Pap Chem Zvesti 65:77–84

Chapter 14
Stochastic Approach for Enzyme Reaction in Nano Size via Different Algorithms

Farid Taherkhani and Shahram Ranjbar

Abstract Stochastic simulations have been done for enzyme kinetics reaction with Michaelis-Menten mechanism in low population number. Gillespie and Poisson algorithms have been used for investigation of population number and fluctuation population around their mean values as a function of time. Our result shows that equilibrium time for population dynamics via Poisson algorithm is smaller than Gillespie algorithm. Variations of average population number versus time for all species have the following order: deterministic approach (mean fields) $>$ Gillespie $>$ Poisson. There is asymptotic limit for fluctuation population as a function of time via Poisson algorithm but there is not such trend for fluctuation population via Gillespie algorithm. There is a maximum for fluctuation population for all species for kinetics reaction with Michaelis-Menten mechanism as a function of time via Gillespie algorithm. The stochastic approach has also been used for horse liver alcohol dehydrogenase which catalyses the NAD^+ (nicotinamide heterocyclic ring) oxidation of ethanol to acetaldehyde and three kinds of third order reactions. Probability distribution function and fluctuation population for reactants are calculated as a function of time. Increasing a variety of species for third order reactions leads to decrease of coefficient variation.

14.1 Introduction

Within its host cell, a complex coupling of transcription, translation, genome replication, assembly, and virus release processes determines the growth rate of a virus. Mathematical models that account for these processes can provide insights into the understanding as to how the overall growth cycle depends on its constituent

F. Taherkhani (✉) • S. Ranjbar
Department of Physical Chemistry, Razi University, Kermanshah, Iran
e-mail: f.taherkhani@razi.ac.ir; ranjbar@razi.ac.ir

M. Gupta Bhowon et al. (eds.), *Chemistry: The Key to our Sustainable Future*,
DOI 10.1007/978-94-007-7389-9_14,

reactions [1]. A virus infection may be initiated by a single virus particle that delivers its genome, a single molecule of DNA or RNA, to its host cell [2]. Under such conditions, the inherently stochastic nature of the ensuing processes may give rise to dynamics that differ significantly from those predicted by deterministic models. At the molecular level, random fluctuations are inevitable, with their effect being most significant when molecules are at low numbers in the biochemical system. This typically occurs in the regulation of gene expression where transcription factors interact with DNA binding sites in the gene's regulatory sequences. These intrinsic fluctuations have recently been measured using fluorescent probes [3, 4]. Deterministic approach for studying the kinetics of small systems is not appropriate [5]. To investigate the chemical kinetics, the stochastic approach is more consistent than deterministic approach for small systems [6–9]. Deterministic approach does not give any information regarding the fluctuation of concentration as a function of time [10–12]. McQuarrie et al. [10, 11] worked on the irreversible first and second order elementary reactions and compared the average concentration obtained from the master equation (ME) with that of the deterministic approach. In addition, Rose and Zheng [12] solved numerically the cubic Schlogl model with a single steady state. Erdi and Toth [13] also considered simple enzyme kinetics, ligand migration kinetics and membrane noise, to compare the stochastic results with those of deterministic. Stochastic simulations of homogeneous chemically reacting systems have been done by Fabio and Stefano [14] for Lotka–Volterra mechanism. Kramers theory of the rates of chemical reaction was reviewed by Gomes [15]. Many studies have been done on enzyme kinetics reaction with Michaelis-Menten mechanism [16–18].

The stochastic version of the enzyme kinetics predicts that catastrophic bottlenecks in the system are more likely than one would expect from deterministic theory for Michaelis-Menten mechanism [17]. Many biochemical reactions occurring in human are catalyzed by enzymes. On the other hand all biological reactions occur in low population number. Stochastic simulation yields a correct average population number for all reactants species as a function of time. There is no information regarding fluctuation via deterministic approach. In present work development of stochastic simulation for enzyme kinetics has been done via two different algorithms namely Gillespie and Poisson.

$$E + S \underset{k_2}{\overset{k_1}{\rightleftarrows}} ES \xrightarrow{k_3} E + P \tag{14.1}$$

Rate constants for k_1, k_2 and k_3 are 10 ($\text{molecule}^{-1}.\text{s}^{-1}$), 0.1 ($\text{s}^{-1}$) and 0.14 ($\text{s}^{-1}$) respectively. The mentioned rate constants can be applied for Chymotrypsin enzyme [19]. Fluctuation populations for all species in enzyme reaction have been calculated via the two mentioned algorithms. Many biological reactions contain some elementary reactions, therefore we investigate probability distribution, coefficient variation, average number of particles, and discrepancy for the number of particles with regard to the stochastic and deterministic approaches as a function of time for reversible second order and three kinds of third order reaction namely:

$$A + B \underset{k_2}{\overset{k_1}{\rightleftarrows}} C + D \tag{14.2}$$

$$3\mathrm{A} \rightarrow \mathrm{P} \tag{14.3}$$

$$2\mathrm{A} + \mathrm{B} \rightarrow \mathrm{P} \tag{14.4}$$

$$\mathrm{A} + \mathrm{B} + \mathrm{C} \rightarrow \mathrm{P} \tag{14.5}$$

It is worthwhile to notice that McQuarrie et al. [10, 11] did mention that the master Eq. 14.2 can be solved exactly by using the method of separation of variables and the ordinary differential equation. However the evaluation of Fourier type coefficients is very difficult and it seems unlikely that numerical results could be easily found for the coefficients [20]. It is noteworthy that the master equation was exactly solved only at the equilibrium state by Darvey et al. [21].

14.2 Methodology

14.2.1 Master Equation for General Chemical Reaction

A Markov process, which satisfies the Markov property, is defined by the following relation:

$$P(y_n, t_n | y_{n-1}, t_{n-1}, \ldots y_1, t_1) = P(y_n, t_n | y_{n-1}, t_{n-1}) \tag{14.6}$$

where

$$t_1 < t_2 < \ldots < t_n \tag{14.7}$$

The Markov property merely expresses that, for a Markov process, the probability of a transition at time t_{n-1} from a y_{n-1} value to a y_n value at time t_n (Eq. 14.6) depends only on the value of y_{n-1} at time t_{n-1} and not to the previous history of the system [22]. For a discrete set of states, the ME may be given as [23]:

$$\frac{\mathrm{d}P_n(t)}{\mathrm{d}t} = \sum_j W_{j,n}P_j(t) - W_{n,j}P_n(t) \tag{14.8}$$

$W_{n,j}$ is the conditional probability that n reactant molecules exist in the system at time $t + \Delta t$, assuming that j reactant molecules existed at time t [23].

14.2.2 Stochastic Algorithm and Simulation

Another way to investigate the kinetics of a small system is stochastic algorithm. Up to now several authors have applied the stochastic algorithms [24–29]. In recent years, stochastic modeling has emerged as a physically more realistic alternative for modeling of the *vivo* reactions [2]. Let consider X as the time of the event. By a constant hazard we mean that:

$$P(X \in (t, t + dt]|X > t) = \alpha dt \tag{14.9}$$

where $\alpha > 0$ is a constant whose value may be calculated as, $\alpha = \sum_{i=1}^{M} W_{j,n} = \sum_{i=1}^{M} \alpha_i$, where $\alpha_i = W_{j,n} = k_i \frac{j!}{n!(j-n)!}$, k is a rate constant and it can be obtained via density functional approach [30, 31]. For a small δt we will have:

$$P(X \in (t, t + dt]|X > t) = \alpha \delta t \tag{14.10}$$

Considering a time $t > 0$, and a large integer N, dividing the interval $(0, t]$ into N subintervals of the form $((i - 1)\delta t, i\delta t]$, i $= 1, 2, \ldots$, N where $\delta t = \frac{t}{N}$, then we have:

$$P(X > t) = P[(X \notin (0, t])] = P(\{(X \notin (0, \delta t])\} \cap \{X \notin (\delta t, 2\delta t]\} \cap \ldots \\ \{X \notin (N - 1)\delta t, t]\}) \tag{14.11}$$

Hence

$$\begin{aligned} P(X > t) &= P(X \notin (0, \delta t])P(X \notin (\delta t, 2\delta t]|X > \delta t) \\ &\quad \ldots P(X \notin ((N - 1)\delta t, t]|X > (N - 1)\delta t) \\ &\approx (1 - \alpha\delta t) \times (1 - \alpha\delta t) \times \ldots (1 - \alpha\delta t) \\ &= (1 - \alpha\delta t)^N \\ &= \left(1 - \frac{\alpha t}{N}\right)^N \end{aligned} \tag{14.12}$$

If $N \to \infty$ and $\delta t \to 0$, therefore Eq. 14.12 will convert to $\exp(-\alpha t)$ then $P(X \leq t) = (1 - \exp(-\alpha t))$. Consequently, whenever we consider a time dependent event with constant hazard α, in Gillespie algorithm [24–27], we can conclude that the time distribution is an exponential function. By choosing two uniform random numbers and within the interval $[0, 1]$ and by definition of two following expressions, we may write:

$$\tau = \frac{1}{\alpha} \ln\left(\frac{1}{r_1}\right) \tag{14.13}$$

$$\sum\nolimits_{i=1}^{\mu-1} \alpha_i < r_2\alpha \leq \sum\nolimits_{i=1}^{\mu} \alpha_i \tag{14.14}$$

There are three loops for algorithm as follows:

1. Calculating $\alpha_i = k_i \frac{j!}{n!(j-n)!}$, whereas k is a rate constant ($1 \leq \mu \leq M$).
2. Generating two uniform random numbers r_1 and r_2 and calculating τ and μ according to Eqs. 14.13 and 14.14.
3. Increasing t by τ and adjusting population of reactants for reaction μ.

In Poisson algorithm, simulation time which is needed for the phenomenon to take place is [26, 31]:

$$\Pr(P(\lambda t = n)) = \frac{e^{-\lambda t}(\lambda t)^n}{n!} \tag{14.15}$$

and

$$\lambda t = \sum_i k_i \frac{j!}{n!(j-n)!} \tag{14.16}$$

where k_i, j, and n are rate constants, initial population, number of particles which are created or destroyed on the basis of stoichiometry coefficient of reaction i and $\lambda = \alpha$.

14.3 Results and Discussion

14.3.1 Stochastic Simulation for Average Number of Particles via Gillespie Algorithm for Michaelis-Menten Reaction

It is possible to study enzymatic reactions at the level of a single molecule via fluorescence correlation spectroscopy [32]. At low population of reactant, the usual description of such reactions via rate equations breaks down, so more appropriate stochastic models of single-molecule Michaelis-Menten kinetics have been developed recently [16, 32]. Stochastic simulations have been done for Eq. 14.1 for enzyme (E), substrate (S), intermediate (ES) and product (P) species. Stochastic simulation result has been shown for substrate via Gillespie algorithm in Fig. 14.1. Initial populations for substrate, enzyme, intermediate and product species are 100, 10, 0, and 0 respectively. It is worthwhile to notice that rate constant k_1, k_2 and k_3 of 10 (molecule^{-1}.s^{-1}), 0.1 (s^{-1}), 0.14 (s^{-1}) can explain a real dynamics of Chymotrypsin enzyme. On the basis of Fig. 14.1, substrate population decreases as a function of time and finally it approaches zero.

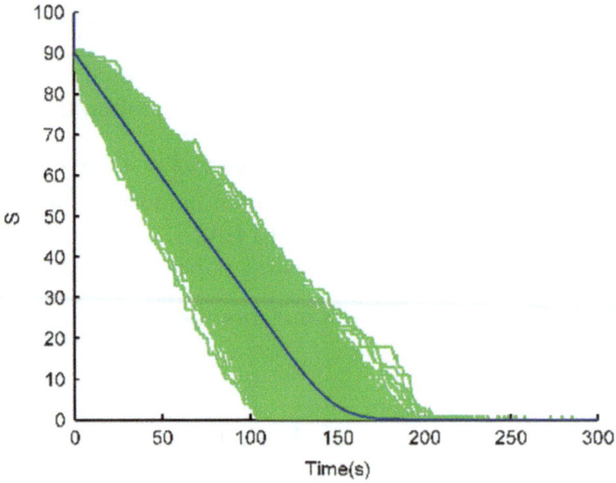

Fig. 14.1 1,000 stochastic simulations for population number of product (P) species and its average as a function of time via Gillespie algorithm

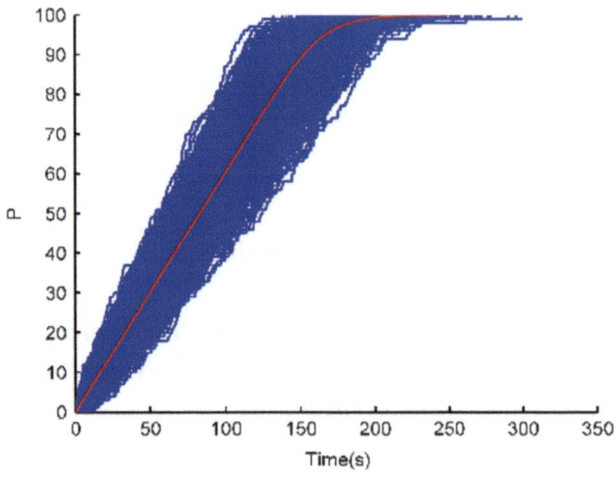

Fig. 14.2 Gillespie algorithm for 1,000 stochastic simulations of substrate population number and its average as a function of time

Figure 14.1 shows the average of 1,000 stochastic simulations via Gillespie algorithm for substrate species as a function of time.

Figure 14.1 indicates that fluctuation of substrate population is not small. On the basis of Eq. 14.1, 1,000 stochastic simulations have been done for product species (P). The result of stochastic simulation is shown in Fig. 14.2. On the basis of Fig. 14.2, product population increases as a function of time. It is important to notice that there is a fluctuation around mean population number of product species. The obtained simulation result is consistent with that observed in the literature [33].

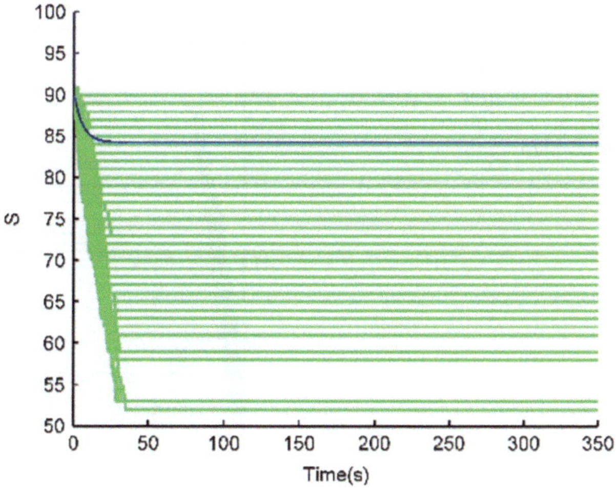

Fig. 14.3 1,000 stochastic simulations via Poisson algorithm for substrate population as a function of time

14.3.2 Poisson Algorithm for Stochastic Simulation

Stochastic simulations have been done for substrate species via Poisson algorithm. Similar to Gillespie algorithm, 1,000 stochastic simulations have been done for average population number of substrate species via Poisson algorithm. Result of Poisson algorithm for substrate species is shown in Fig. 14.3. Figure 14.3 indicates that substrate population approaches the equilibrium as a function of time quickly. Stationary population for substrate species via Poisson algorithm is 85 particles which is greater than zero particle for substrate species from Gillespie algorithm.

Stochastic simulation is investigated for population number of product species as a function of time. The result of simulation for product species is shown in Fig. 14.4. Average population number of product species from stochastic simulation via Poisson algorithm shows that stationary population for product species is 5 particles. On the basis of Fig. 14.4, average population number for product species is very much smaller than average population number of product from Gillespie algorithm. If 100 particles as an initial population for substrate are considered, the population of substrate species is predicted to be 100 particles in stationary state. As a result, Poisson algorithm does not predict the correct population number for stationary state of Michaelis-Menten kinetics. It is worthwhile to notice that equilibrium time for substrate and product species via Poisson algorithm is less than Gillespie algorithm. Variation of population for substrate and product as a function of time via Poisson algorithm is smaller than Gillespie algorithm (see Figs. 14.1, 14.2, 14.3, and 14.4).

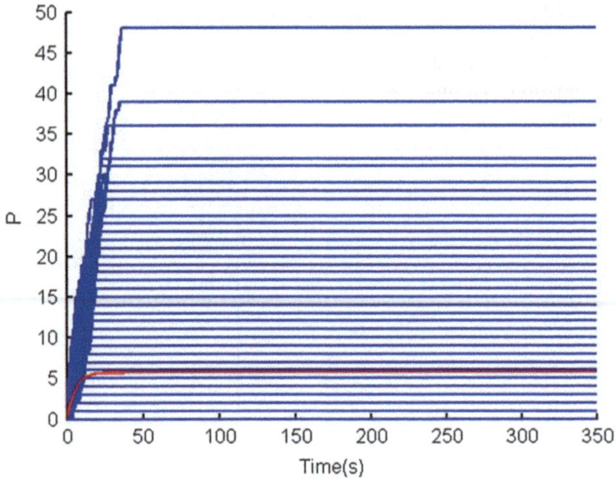

Fig. 14.4 1,000 stochastic simulations for population number and its average as a function of time for product species via Poisson algorithm

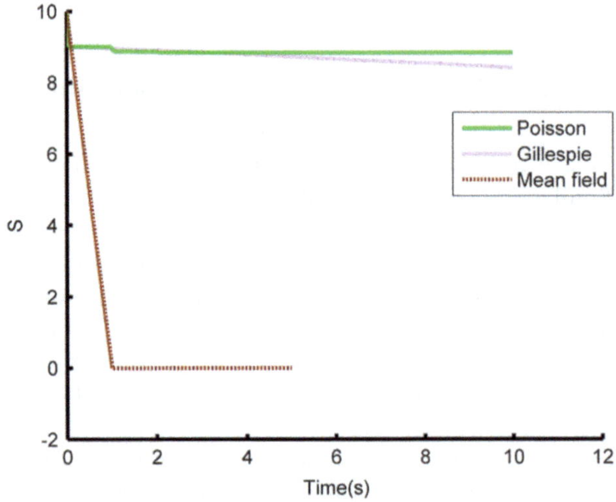

Fig. 14.5 Average population number for substrate species as a function of time via Gillespie, Poisson algorithms and mean field approach

14.3.3 Comparison of Gillespie and Poisson Algorithm with Mean Field Approach

Comparison of Gillespie and Poisson algorithms with mean field approach (deterministic differential equation) has been done for substrate population as a function of time. 1,000 stochastic simulations have been done to obtain average number of substrate. Initial population for substrate, enzyme, intermediate, and product of 10, 1, 0, 0 is taken respectively.

On the basis of Fig. 14.5, substrate population from mean field approach, stochastic simulations via Gillespie and Poisson algorithms are shown as a function

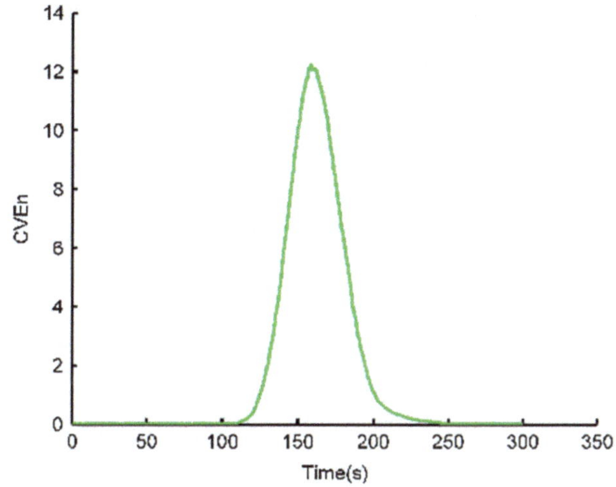

Fig. 14.6 Fluctuation population number as a function of time for enzyme (CVEn) from averaging of 1,000 simulations via Gillespie algorithm

of time. Population variation of substrate species via mean field approach is greater than Poisson and Gillespie algorithm. On the other hand, substrate population approaches equilibrium in lower time via mean field approach. There is a little variation for substrate population as a function of time via Poisson and Gillespie algorithms. As a result, equilibrium time for substrate population via Poisson and Gillespie is greater than mean field approach.

14.3.4 Fluctuation of Population for Substrate, Enzyme, Intermediate, Product Species as a Function of Time

Mean field approach does not represent fluctuation population $CV(t)$

$$CV(t) = < N^2(t) > - < N(t)>^2 \tag{14.17}$$

during the dynamics, where N(t) is number of particles. It is worthwhile to note that there is a significant fluctuation for the number of particles in low population level. Stochastic simulations have been done for investigation of fluctuation of population for all species in Michaelis-Menten reaction as a function of time via Poisson and Gillespie algorithms. Average result of 1,000 stochastic simulations for fluctuation population of enzyme (CVEn) and substrate (CVS) versus time have been shown in Figs. 14.6 and 14.7 via Gillespie algorithm respectively. Initial population for substrate, enzyme, intermediate, product of 100, 10, 0, 0 is considered respectively.

On the basis of Fig. 14.6, fluctuation population for enzyme species is zero initially and it increases after elapsing time, then it approaches zero finally. Result of stochastic simulation for fluctuation population number for intermediate (CVES), product (CVP) species are shown versus time in Figs. 14.8 and 14.9 respectively.

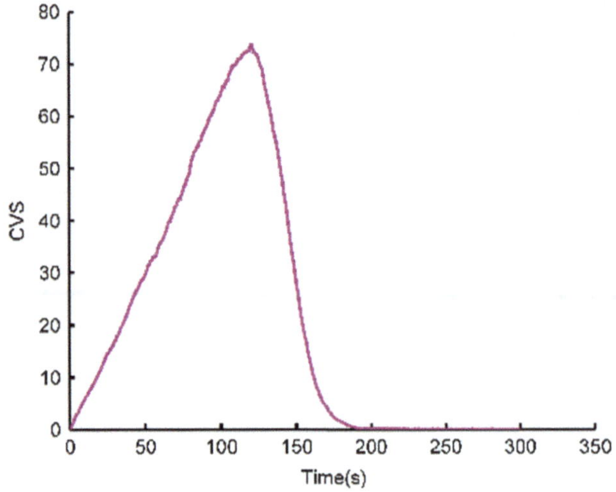

Fig. 14.7 Average of 1,000 stochastic simulations of substrate species (CVS) versus time via Gillespie algorithm

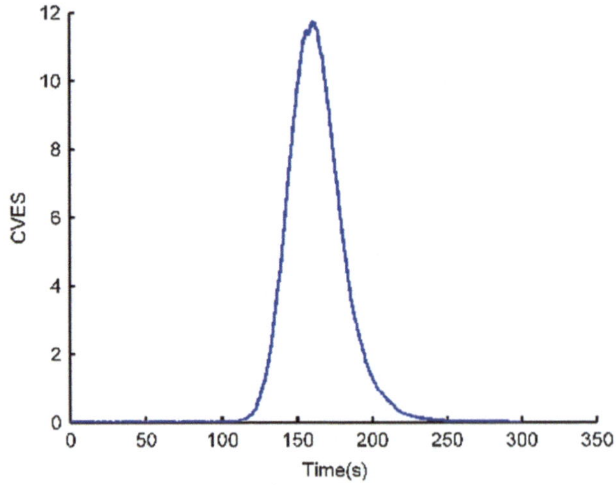

Fig. 14.8 Fluctuation population of intermediate species (CVES) as a function of time via Gillespie algorithm

14.3.5 Comparison of Gillespie and Poisson Fluctuation of Population

Gillespie and Poisson algorithms have been used for investigation of fluctuation population of product species as a function of time. Result of average of 1,000 stochastic simulations is shown in Fig. 14.10. It indicates fluctuation population for product species via Gillespie algorithm is greater than Poisson algorithm. Fluctuation population for product species approaches a constant value as a function of time. On the other hand fluctuation population for product species via Gillespie algorithm first increases, after some time it approaches its maximum value, then it decreases and approaches zero finally.

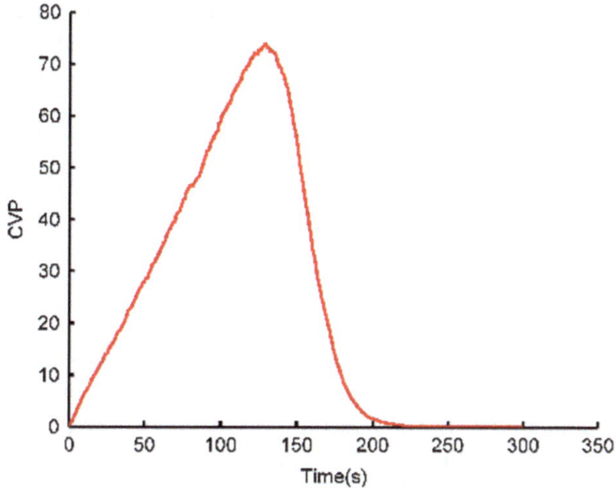

Fig. 14.9 Gillespie algorithm for fluctuation population of product species (CVP) as a function of time

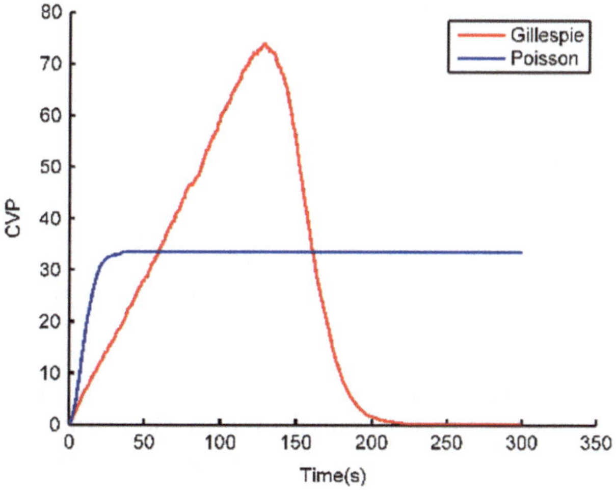

Fig. 14.10 Comparison of Gillespie and Poisson algorithms for fluctuation of product population as a function of time

14.3.6 Numerical Solution of Master Equation of Some Elementary Reactions

Master Equation for Second Order Reversible Reaction

An example of a second order reversible reaction is the horse liver alcohol dehydrogenase which catalyses the NAD^+ (nicotinamide heterocyclic ring) for oxidation of ethanol to acetaldehyde. Incubation of the enzyme with deuterated ethanol CH_3CD_2OH followed by re-isolation of oxidized NAD^+ revealed no deuterium incorporation into the oxidized cofactor, showing that the deuterium atom transfer to the cofactor is stereospecific, which is removed in the reverse reaction [34].

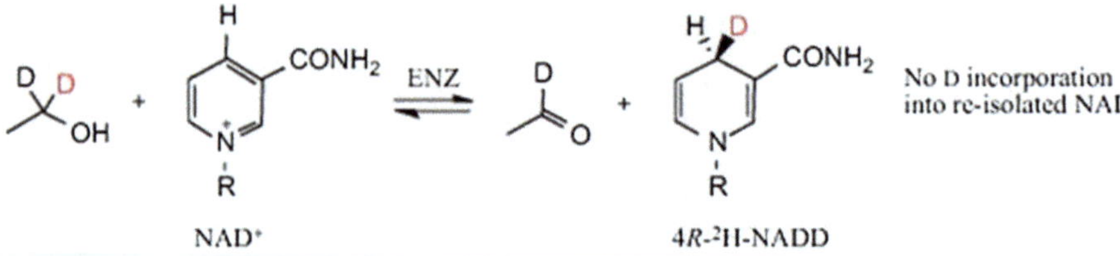

(14.18)

In order to derive the master equation for the second order reversible reaction, Eq. 14.2, we suppose *a*, *b*, *c* and *d*, with their values achieved randomly, as the number of A, B, C, D species, respectively. The possible states of a system at time *t* which could lead to state specified by *a*, *b*, *c*, *d* at time $t + \Delta t$ are $(a + 1, b + 1, c\text{-}1, d\text{-}1)$ and (a, b, c, d). Consequently, we may choose $\Delta t \to 0$. If the initial concentrations of A, B, C and D are a_0, b_0, c_0 and d_0, respectively at $t = 0$, then due to the fact that the total number of species is constant, we may get a conclusion as below:

$$a_0 - a = b_0 - b = c - c_0 = d - d_0 \tag{14.19}$$

Then

$$\frac{\mathrm{d}P_a(t)}{\mathrm{d}t} = k_1(a+1)(b_0 - a_0 + a + 1)P_{a+1}(t) + k_2(c_0 + a_0 - a + 1)(d_0 + a_0 - a + 1)P_{a-1}(t) - [k_1 a(b_0 - a_0 + a) + k_2(c_0 + a_0 - a)(d_0 + a_0 - a)]P_a(t) \tag{14.20}$$

This is the master equation for the reversible second order reaction. This equation has been already derived by McQuarrie without giving any solution for it [20]. Later, we will discuss its numerical solution.

Master Equation for Three Kinds of Third Order Reaction

The solution of deterministic reaction rate for Eq. 14.3 can be shown:

With a similar argument which led to Eq. 14.20, we may find the ME as:

$$\frac{\mathrm{d}P_x(t)}{\mathrm{d}t} = \frac{1}{6}P_{x+3}(t) \times k \times (x+3) \times (x+2) \times (x+1) - \frac{1}{6}P_x(t) \times k \times (x) \times (x-1) \times (x-2) \tag{14.21}$$

To obtain master equation Eq. 14.4, let concentrations of A and B at time t be $X(t)$ and $Y(t)$, respectively. We may define

$$Z_0 \equiv 2Y(t) - X(t) \tag{14.22}$$

According to both Eqs. 14.8 and 14.22 we can write:

$$\begin{aligned}\frac{dP_x(t)}{dt} =& \frac{1}{4}P_{x+2}(t) \times (x+2) \times (x+1) \times (x+Z_0+2) \\ & - \frac{1}{4}P_x(t) \times k \times (x) \times (x-1) \times (x+Z_0)\end{aligned} \tag{14.23}$$

In Eq. 14.23, coefficient $\frac{1}{4}$ is derived from Eq. 14.22 and combinational term.

To obtain master equation Eq. 14.5, let the concentrations of A, B and C be $x(t)$, $y(t)$, $z(t)$, respectively, at time t. We define the time independent variables Z_1 and Z_2 as:

$$Z_1 \equiv B(t) - A(t) \tag{14.24}$$

$$Z_2 \equiv C(t) - A(t) \tag{14.25}$$

With a similar discussion which led to Eq. 14.20 we find the master equation as:

$$\begin{aligned}\frac{dP_x(t)}{dt} =& P_{x+1}(t) \times k \times (x+1) \times (x+Z_1) \times (x+Z_2) \\ & - P_x(t) \times k \times (x) \times (x+Z_1) \times (x+Z_2)\end{aligned} \tag{14.26}$$

Numerical Solution of Master Equation

Number of particles given by deterministic approach, A_d, can be easily obtained from the solution of the rate equation (Note that A_d is the same as A). However, the average number of particles obtained from the solution of Master equation, $\langle A_m \rangle$, given by the stochastic approach may be easily calculated as:

$$A_m = \sum_A AP_A(t) \tag{14.27}$$

where $A(t)$ is the number of A and $P_A(t)$ is the probability of having A particles at time t. Consider the following equation:

$$\Delta A(t) = A_m - A_d \tag{14.28}$$

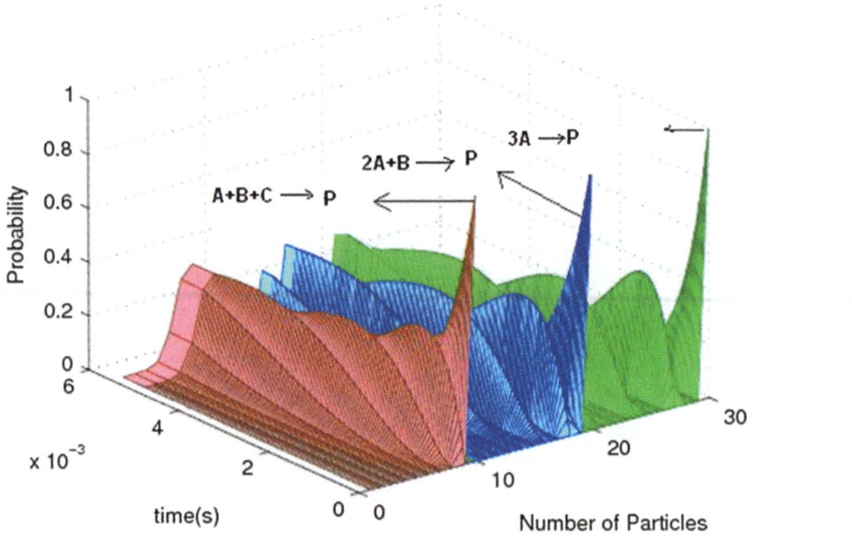

Fig. 14.11 Probability of having A molecules as a function of time and number of A particles for Eqs. 14.3, 14.4, and 14.5, when total number of molecules is 30

where A_m and A_d are the average number of A species obtained from the Master equation and deterministic approaches, respectively (Note that the probability distribution function for A in deterministic approach is a delta function). At $t = 0$ the initial conditions are:

$$P_A(0) = 1 \text{ if } A = A_0 \tag{14.29}$$

$$P_A(0) = 0 \text{ if } A \neq A_0 \tag{14.30}$$

$$\text{then } A_m = A_0 \text{ and}$$

$$\Delta A(0) = A_m - A_d = 0 \tag{14.31}$$

For numerical solution of Master equation, we have chosen the value of $dt = 0.00005$ s for each time step, for smaller time steps all quantities become unchanged. Besides that, we have calculated all values of $P_A(t)$ and then the average number of A molecules. Result of probability distribution of Eqs. 14.3, 14.4, and 14.5 is shown in Fig. 14.11, right to left, respectively. In the same way, the mean value of $\langle A^2 \rangle$ and $\langle A \rangle^2$ can be calculated, from which we may calculate the coefficient variation, CF, as follows:

$$CF = \frac{\langle A^2 \rangle - \langle A \rangle^2}{\langle A \rangle^2} \tag{14.32}$$

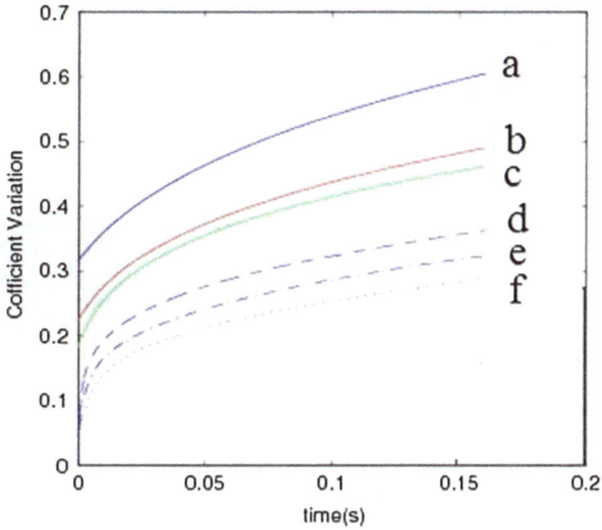

Fig. 14.12 The comparison of *CF* obtained from equilibrium statistical mechanics (*a, b, c*) and the Master equation (*d, e, f*) for three different third order reactions Eqs. 14.3, 14.4, and 14.5 respectively (Initial number of reactant in all of the reactions is 90 and $Z_1 = 0, Z_2 = 0$)

It is worthwhile to note that the value of CF versus time for three different types of third order reactions are shown in Fig. 14.12 and are compared to those obtained from equilibrium statistical mechanics $\frac{1}{\sqrt{A}}$. As shown in Fig. 14.12, the value of CF obtained from the solution of ME is smaller than $\frac{1}{\sqrt{A}}$ of the equilibrium statistical mechanics for all reactions. By using Markov assumption via the mentioned time step, the probability of having A molecules at time $t, P_A(t)$, may be calculated numerically. The results of such calculations for Eq. 14.2 are given in Fig. 14.13. Also the CF for the second order reversible reaction is calculated, for which the results are shown in Fig. 14.14.

14.4 Conclusions

We have simulated enzyme kinetics reaction via stochastic simulation. Two different algorithms namely Gillespie and Poisson have been used for stochastic simulation. Our results show that equilibrium time via Poisson algorithm is smaller than Gillespie algorithm. Fluctuation population via Poisson algorithm is lower than Gillespie algorithm as a function of time. There is an asymptotic limit for fluctuation population for product species as a function of time via Poisson algorithm. Fluctuation population via Gillespie algorithm is zero at the first time, and it increases as a function of time, then it approaches its maximum value. Finally it decreases and goes to zero. Average population number via Gillespie algorithm is less than Poisson algorithm. Also Master equations have been solved for three different third order reactions, namely Eqs. 14.3, 14.4, and 14.5, and for the reversible second order

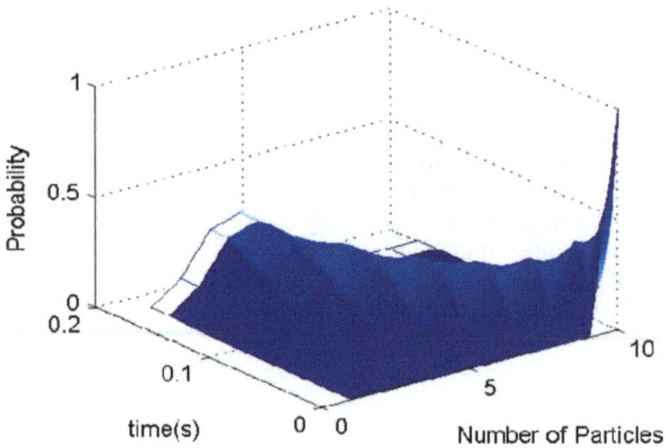

Fig. 14.13 Probability of having A molecules for Eq. 14.2 as a function of time and the number of A molecules

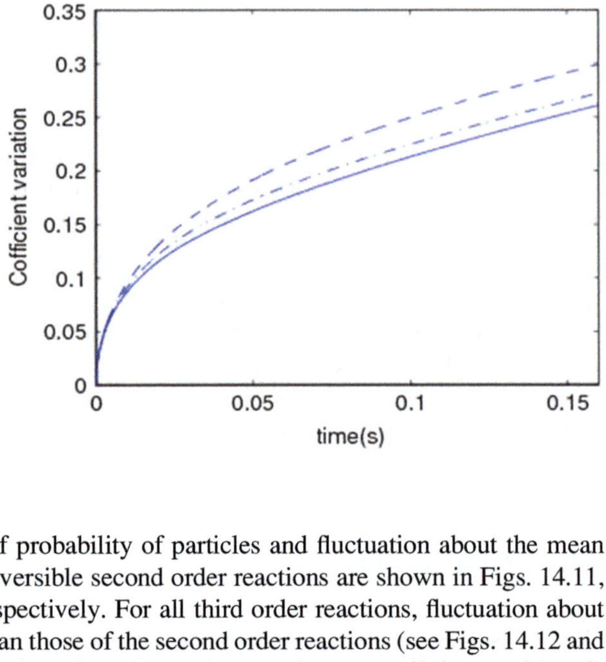

Fig. 14.14 Calculated coefficient variation for Eq. 14.2 when A = 10 (*dash* –), 20 (*dash dot* -.-.), 30 *solid* -) as a function of time

reaction Eq. 14.2. Graph of probability of particles and fluctuation about the mean value for third order and reversible second order reactions are shown in Figs. 14.11, 14.12, 14.13, and 14.14 respectively. For all third order reactions, fluctuation about the mean value is greater than those of the second order reactions (see Figs. 14.12 and 14.13). For elementary kinetics of small systems, maximum coefficient variance is smaller than $\frac{1}{\sqrt{N}}$ (Fig. 14.12) (N is number of particles). Besides that, by increasing a variety of species, coefficient variation decreases (see Fig. 14.12).

Acknowledgements We are thankful to the Computational and Theoretical Physical Chemistry Research Center of Razi University and we wish to acknowledge Mr. Mohammad Reza Poopari for useful suggestion and continued support on this work.

References

1. Jilbert AR, Miller DS, Scougall CA, Turnbull H, Burrell C (1996) Kinetics of duck hepatitis B virus infection following low dose virus inoculation: one virus DNA genome is infectious in neonatal ducks. Virology 226:338–345
2. Srivastava R, You L, Summersy J, Yin J (2002) Stochastic vs. deterministic modeling of intracellular viral kinetics. J Theor Biol 218:309–321
3. Blake WJ, Kærn M, Cantor CR, Collins JJ (2003) Noise in eukaryotic gene expression. Nature 422:633–637
4. Elowitz MB, Levine J, Siggia ED, Swain PS (2002) Stochastic gene expression in a single cell. Science 297:1183–1186
5. Gillespie T (2000) The chemical Langevin equation. J Chem Phys 113:297–306
6. Peeters P, Nicolis G (1992) Intrinsic fluctuations in chaotic dynamics. Phys A 188:426–435
7. Geysermans P, Nicolis G (1993) Thermodynamic fluctuations and chemical chaos in a well-stirred reactor: a master equation analysis. J Chem Phys 99:8964–8969
8. Matias MA, Güémez J (1995) On the effects of molecular fluctuations on models of chemical chaos. J Chem Phys 102:1597–1606
9. Wang H, Li Q (1998) Master equation analysis of deterministic chemical chaos. J Chem Phys 108:7555–7559
10. McQuarrie DA (1963) Kinetics of small systems I. J Chem Phys 38:433–436
11. McQuarrie DA, Jachimowski CJ, Russell ME (1964) Kinetics of small systems II. J Chem Phys 40:2914–2921
12. Zheng Q, Ross J (1991) Comparison of deterministic and stochastic kinetics for nonlinear systems. J Chem Phys 94:3644–3648
13. Erdi P, Toth J (1989) Mathematical models of chemical reactions: theory and applications of deterministic and stochastic models. Princeton University Press, Princeton
14. Mavelli F, Piotto S (2006) Stochastic simulations of homogeneous chemically reacting systems. J Mol Struc Theochem 771:55–64
15. Gomes JANF (1984) Classical dynamics stochastic theories of chemical reactions. J Mol Struc Theochem 107:139–155
16. Qian J, Elson EL (2002) Single-molecule enzymology: stochastic Michaelis–Menten kinetics. Biophys Chem 101:565–576
17. Armbruster D, Nagy JD, van de Rijt EAF, Rooda JE (2009) Dynamic simulations of single-molecule enzyme networks. J Phys Chem B 113:5537–5544
18. Kou SC, Cherayil BJ, Min W, English BP, Xie XS (2005) Single-molecule Michaelis-Menten equations. J Phys Chem B 109:19068–19081
19. Mathews CK, van Holde KE, Ahern KG (1999) Biochemistry, 3rd edn. Prentice Hall Publ. Co., San Fransisco
20. McQuarrie DA (1967) Stochastic approach to chemical kinetics. Methuen, London
21. Darvey IG, Ninham BW, Staff PJ (1996) Stochastic models for second-order chemical reaction kinetics. The equilibrium state. J Chem Phys 45:2145–2155
22. Van Kampen NG (1992) Stochastic processes in physics and chemistry. Elsevier, Amsterdam
23. Steinfeld JI, Francisco JS, Hase WL (1998) Chemical kinetics and dynamics. Upper Saddle River, Prentice Hall Publ. Co., San Fransisco
24. Gillespie DT (1976) A general method for numerically simulating the stochastic time evolution of coupled chemical reactions. J Comput Phys 22:403–434
25. Gillespie DT (1992) A rigorous derivation of the chemical master equation. Phys A 188:404–425
26. Gillespie DT (2007) Stochastic simulation of chemical kinetics. Annu Rev Phys Chem 58:35–55
27. Gillespie DT (2001) Approximate accelerated stochastic simulation of chemically reacting systems. J Chem Phys 115:1716–1733

28. Rao V, Arkin P (2002) Stochastic chemical kinetics and the quasi-steady-state assumption: application to the Gillespie algorithm. J Chem Phys 118:4999–5010
29. Gibson MA, Bruck J (2000) Efficient exact stochastic simulation of chemical systems with many species and many channels. J Phys Chem A 104:1876–1889
30. Li W, Zhao Y, Wang Z-Q, Ju C-G, Feng Y-L, Zhang J-L (2010) Theoretical study and rate constants calculations of hydrogen abstraction reactions CF_3CHCl_2 + F and CF_3CHClF + F. J Mol Struc Theochem 959:101–105
31. Lu RH, Li QS, Zhang SW, Kang LJ (2004) Direct DFT dynamics study of the addition reaction $CF_2 = CHF + H \rightarrow$ Product. J Mol Struc Theochem 685:147–154
31. Resat H, Wiley HS, Dixon DA (2001) Probability-weighted dynamic Monte Carlo method for reaction kinetics simulations. J Phys Chem B 105:11026–11034
32. Rigler R (1995) Fluorescence correlations, single molecule detection and large number screening. Applications in biotechnology. J Biotechnol 41:177–186
33. Turner TE, Schnell S, Burrage K (2004) Stochastic approaches for modeling in vivo reactions. Comput Biol Chem 28:165–178
34. Bugg T (1997) An introduction to enzyme and coenzyme chemistry. Blackwell Science, London

Chapter 15
Enhancing Conceptual Understanding of the "Chemistry of Life" at the 'A'-Level Through Use of Computer Animations

Ummeh W. Ahsun and Fawzia Narod

Abstract "Chemistry of Life" is a topic that forms part of the "Applications of Chemistry" section introduced as a compulsory part in the Cambridge International Examinations (CIE) Advanced Level ('A'-Level) chemistry syllabus in 2007. In the present study, an attempt was made to enhance students' conceptual understanding of the "Chemistry of Life" through use of computer animations. The research was carried out in a girls' private secondary school, located in the central region of Mauritius. The sample consisted of 14 girls of age 18–19 years, who were preparing to sit for the CIE 'A'-Level examinations in November 2011. The study was carried out through an action research involving three cycles; each cycle included three lessons during which computer animations were used. Data were collected through observation checklists, achievement tests, and a students' questionnaire. The findings have clearly revealed that use of computer animations has impacted positively on students' motivation, interest, and engagement, leading to enhanced conceptual understanding, as evidenced by data gathered through observation checklists and students' questionnaire. These results were further supported by an improvement in students' performance in the achievement tests administered at the end of the second and third cycles, even though the concepts taught became increasingly more difficult from Cycles 1 to 2 and 3. Further, it was found that "cues and labeling", proper design in terms of colour and graphics, as well as embedded "voice questions" can all play a crucial role in helping learners to retrieve information from computer animations, and, to develop understanding of the concepts under study. Last but not least, results have also indicated that learner-controlled interactive computer animations are most effective in enhancing students' motivation, interest, and conceptual understanding as compared to non-interactive ones.

U.W. Ahsun
Aleemiah College (Girls), Phoenix, Mauritius

F. Narod (✉)
Department of Science Education, Mauritius Institute of Education, Réduit, Mauritius
e-mail: zia373@eudoramail.com

M. Gupta Bhowon et al. (eds.), *Chemistry: The Key to our Sustainable Future*,
DOI 10.1007/978-94-007-7389-9_15,

15.1 Introduction

"Chemistry of Life" is a topic that forms part of the "Applications of Chemistry" section introduced as a compulsory part in the Cambridge International Examinations (CIE) 'A'-Level chemistry syllabus in 2007, following phasing out of the option topics "Phase Equilibria", "Transition Elements", "Biochemistry", "Spectroscopy" and "Environmental Chemistry". This has been specially a challenge for almost all chemistry educators in Mauritius, who were used to teaching the options "Phase Equilibria" and "Transition Elements". It has indeed been reported by Cambridge examiners [1] that "Phase Equilibria" and "Transition Elements" were the most popular options among Mauritian candidates. This can be attributed to the fact that "Phase Equilibria" and "Transition Elements" were the most common options taught in the chemistry classrooms in the years 1990–2006. Thus, most chemistry educators were confident in, and familiar with the teaching of "Phase Equilibria" and "Transition Elements". The sudden phasing out of option topics, and the incorporation of the "Chemistry of Life" as a compulsory integral part in the CIE 'A'-Level syllabus in 2007, have proved to be a challenge not only for chemistry educators, but also for students who have not opted to study Biology at 'A'-level. Chemistry educators were required to prepare for the new curriculum content, and to review their teaching strategies and teaching resources. Furthermore, analysis of CIE Examiners' Reports (2007–2010) [2] has revealed that many chemistry students encountered difficulties when dealing with questions pertaining to the "Chemistry of Life".

The topic "Chemistry of Life" is very closely related to biochemistry, which deals with the study of chemical substances, and chemical reactions and processes occurring in living organisms. Thus, "Chemistry of Life" as per CIE 'A'-Level syllabus, comprises entirely biochemistry-related concepts, namely, proteins and amino acids, enzymes, nucleic acids, ATP and metal ions in biological systems. In addition, just like chemistry, biochemistry is a science that is investigated within the macroscopic, microscopic, and particularly, the submicroscopic (molecular) levels of organization [3, 4]. Accordingly, for proper and meaningful understanding of biochemistry or "Chemistry of Life", students are required to readily translate between these three levels of organization, a condition that can be rather difficult and confusing for them. Not surprisingly, several authors have highlighted that students' difficulties and misconceptions in learning biochemical concepts and processes emerge at the molecular level due to the abstract nature of the concepts [5, 6]. Another reported difficulty in learning biochemistry includes the visualisation of the three dimensional orientation of molecules, which play a critical role in determining reactivity and properties [7].

According to CIE Examiners' Reports [2], many students encountered difficulties in answering questions pertaining to "Chemistry of Life". Some of the problems encountered by students relate to: deducing the sequence of amino acids condensing and drawing correct polypeptide chain; reading the sequence of codons given; explaining the roles of RNA in protein synthesis; understanding the sequence of events occurring during protein synthesis; and sketching curves to illustrate the effect of competitive and non-competitive inhibitors on the rate of enzyme-catalyzed reactions amongst others. It has also been reported by CIE examiners that some students tend to answer questions of the "Chemistry of Life" from a biological perspective rather that from a chemical perspective, and students made little efforts to apply chemical knowledge when answering the questions. Furthermore, "Chemistry of Life" is a topic that involves many microscopic chemical concepts that cannot be easily viewed or demonstrated.

On the other hand, it is acknowledged that computer technologies offer potential means to promote molecular understanding and to enhance the teaching and learning of chemistry concepts [8]. Computer technologies are also reported to allow simultaneous representation of molecular and macroscopic views of the same phenomena [9]. Further, findings from research in education have revealed that, use of ICT coupled with the necessary pedagogical strategies, can engage students in higher-order thinking [10], and can allow students to work at their own pace [11]. Findings from Abdoolatiff and Narod [8] have also shown that use of computer simulations in classroom teaching has a positive impact on students, both in terms of enhancing acquisition of concepts, and of increasing their motivation, interest and engagement in the lessons. Modern computer aids and ICT tools thus not only represent an innovative model on which chemistry teaching can be based, but also offer a wide range of potential uses in classroom teaching and learning of chemistry, ranging from powerpoint presentations, use of short videos, computer simulations to the use of 3D-animations, and data-logging amongst others.

In the literature, numerous advantages have been linked with the use of computer animations in the teaching and learning of chemistry concepts. Computer-generated animations are series of still computer generated pictures that are presented in succession so that the illusion of motion is developed, much like a picture flip-book [12]. Animations differ in that they offer two unique attributes that still pictures do not, namely trajectory and motion [13]. Further, O'Day [14] claimed that animations can address the different learning styles of students, and can assist long-term memory retention. It has also been reported that students enjoy learning with animations [13]. Findings from several studies have also shown that students who receive instructions including computer animations of chemical processes at the molecular level are better able to answer conceptual questions about particulate phenomena representation [12, 15, 16]. Computer animations also allow dynamic phenomena to be visualized and are represented more easily; research has indeed shown that computer animations can enhance understanding of dynamic molecular processes [17]. Further, according to Gardner et al. [18] learning is multi-dimensional with computer animations.

In addition, Muñoz-Repiso and Tejedor [19] claimed that in this technological era and information age, students are required to restructure their conceptions on the characteristics of learning environments, while Selwyn and Husen [20] mentioned that the ability to make good use of ICT, in this highly globalised world, is seen as an essential 'life skill' for individuals. On the national scale, it needs to be highlighted that there has been a concerted endeavour and emphasis on integrating use of ICT in classroom teaching in Mauritius over the past years. It has been clearly specified [21] that to be able to function effectively in the fast evolving socio-economic context, and to respond to challenges and opportunities of the modern world, learners should be able to use ICT tools. In fact, it is clearly stipulated that technology be embedded in the education system as from the pre-primary to secondary levels in the Education and Human Resource Strategy Plan 2008–2020 [22]. It is also recommended that ICT be used as a support for learning [21], and that pre-primary and primary schools be encouraged to use ICT as a tool in the teaching/learning process in view of exposing young learners to modern technology [22]. It has also proposed that ICT facilities would be provided to all secondary schools to ensure that teachers use ICT on a regular basis for teaching and learning, and it is envisaged that all students leaving secondary schools are equipped with ICT skills to adapt to the requirements of future needs of independent learning [22]. In the same context, the National Curriculum Framework for Secondary Education has also called upon teachers to exploit ICT tools in their practice to empower students to cope with our technology-oriented society [23]. Thus, the incorporation of ICT as a subject at primary level with the provision of a computer room with computers in primary schools, and the provision of ICT equipment in state secondary schools provide clear evidence of the commitment of the Government in supporting and facilitating the use of ICT in classroom teaching.

In light of the above discussions pertaining to the difficulties encountered in teaching and learning of the topic “Chemistry of Life”, and taking into consideration the apt need of incorporating modern technology in the teaching/learning process, we have planned this study to investigate the use of computer animations in the teaching of the afore-mentioned topic through an action research. Further, being aware of the reported underpinning benefits of the use of computer animations in the teaching of abstract and complex concepts brought forward by other research studies, the present action research aimed at enhancing students’ conceptual understanding of the “Chemistry of Life” by making use of computer animations.

15.2 Methodology

15.2.1 Participants

The action research was carried out in a girls’ private secondary school during the second school term of year 2011, over a period of 9 weeks in the months of May and June. The participants were 14 students, girls in the range of 18–19 years, who have opted to study chemistry as a main subject at the CIE ‘A’-Level. The participating students were enrolled in an Upper Six class, which is the second and final preparatory year for the CIE ‘A’-Level examinations (students are usually prepared for the CIE ‘A’-Level examinations over a period of 2 years). The participants were preparing to sit for the CIE 'A'-Level examinations in November 2011. While all the participants have studied biology at the CIE Ordinary-Level ('O'-Level), only five of them had opted to study biology as a main subject at the 'A'-level together with chemistry.

15.2.2 Data Collection Tools

For the purpose of the study, a multi-method approach known as triangulation [24] has been used to obtain both qualitative and quantitative data. The data-gathering instruments employed during the study include:

a. **Observation Checklists**
   While implementing the lessons during the action research, the students were observed at work to gather information about their motivation, interest, participation, interactions and understanding when computer animations were used. For this purpose, detailed observation checklists were used. The checklists were carefully designed and included criteria like students' motivation, interest, participation, interactions and understanding of concepts being addressed. In addition, the observation checklists also included criteria related to the use of computer animations, types of computer animations used as well as the different ways in which animations were presented to students during the lessons.
b. **Students' Achievement Tests**
   Students' achievement tests were designed and administered to the participants to gather information regarding their understanding of concepts taught through use of computer animations. Three tests were set, namely Tests 1, 2 and 3 which were administered to all participant students, after Cycles 1, 2 and 3 respectively to find out about their understanding of concepts dealt with in each cycle. All the tests set included multiple choice items, short answer questions and structured open-ended questions. The maximum number of marks achievable for each test was 40 and students were allocated 80 min to complete each test.
c. **Students' Questionnaire**
   After completing the three cycles of the action research, a questionnaire was administered to all participants in an attempt to determine whether use of computer animations had an impact on students' motivation and interest during the lessons. The questionnaire also sought to answer other questions related to students' understanding of concepts taught through use of computer animations, as well as their perceptions, attitudes and personal opinions regarding use of computer animations during the lessons. In addition, the questionnaire also included items that helped the researchers to find out about students' preferences for the different types of computer animations used during the study. Further, it allowed the researchers to get an insight into features and certain aspects that could not be observed during the lessons.

The questionnaire was designed in line with the objectives of the study, and consisted of both open-ended and close-ended questions. For the close-ended questions, the participants were required to select the answer they found most appropriate from given answers. On the other hand, the open-ended questions ensured qualitative dimensions to the participants' views and opinions.

15.2.3 Research Design

The study was carried out during May and June 2011, through an action research comprising three cycles, each of which included three lessons with a duration of either 70 min (2 periods of 35 min) or 105 min (3 periods of 35 min). During all the nine lessons of the action research, computer animations have been used to teach the "Chemistry of Life" topic. The section "Computer Animations Used" gives an overview of the types of computer animations used during the study, and the concepts addressed.

Computer Animations Used

As mentioned in Sect. 15.1, computer animations are series of still computer-generated pictures that are presented in succession so that the illusion of motion is developed, much like a picture flip-book [12]. Computer animations used for the purpose of the present study have been broadly classified into the following three groups depending on their sources and interactivity:

1. ***Teacher-prepared PowerPoint 2D computer animations***
   These 2D-animations were designed and prepared by the teacher-researcher using Microsoft PowerPoint. The animations were non-interactive. Thirteen such animations were prepared and used during the study to teach the following concepts:

   - The generalized structure of an amino acid;
   - Peptide link formation;
   - Apoenzyme, holoenzyme, cofactors, coenzymes and prosthetic group;
   - Chemical denaturation of enzymes;
   - ATP and its role in metabolism; and
   - Role of metal ions (Fe^{2+}, Na^{+} and K^{+}) in biological systems.

2. ***2D and 3D flash animations***
   2D and 3D flash animations were selected from the web in line with concepts pertaining to the "Chemistry of Life" and were downloaded freely. These downloaded 2D and 3D-animations were also non-interactive. A total of nine such animations were used to teach the following concepts:

   - The different levels of protein structure (primary, secondary and tertiary);
   - Components and structure of DNA and RNA;
   - The genetic code and codons; and
   - Protein synthesis – transcription and translation.

3. ***Interactive online animations***
   The third type of animation included interactive flash animations which could be used freely online. These animations were selected in line with concepts pertaining to the "Chemistry of Life". A total of nine interactive flash animations were selected from the web and used in the teaching of the following concepts:

   - Enzyme specificity;
   - Enzyme inhibition (competitive and non-competitive);
   - Complementary base pairing in DNA;
   - Semi-conservative replication of DNA;
   - DNA replication;
   - Mutations and types of mutations; and
   - Genetic diseases (sickle cell anaemia and cystic fibrosis).

   Thus, a total of 31 computer animations were used in the present study. All the computer animations were either prepared by the teacher-researcher, or selected from the web in line with the aims and objectives of the CIE 'A'-Level chemistry syllabus for the topic "Chemistry of Life". The lessons were carefully planned, and appropriate worksheets were designed in line with the syllabus requirements and the animations' contents, in an attempt to incorporate use of the selected computer animations during the lessons.

15.3 Results and Discussion

For the purpose of the present study, both quantitative and qualitative data-collection tools have been used. Qualitative data were collected mostly through observation checklists and open-ended questions in the questionnaire, while quantitative data were collected through achievement tests and close-ended questions in the questionnaire. The data gathered during the research are presented and discussed in this section.

15.3.1 Observation Checklists

In Cycle 1, it was observed that the use of PowerPoint 2D animations did facilitate students' understanding of concepts, like 'condensation polymerization of amino acids' and 'peptide bond formation'. However, it was also noted that some students encountered difficulties to retrieve information from the animations. Based on observation and reflection, these difficulties were attributed to the fact that the designed animations lacked verbal cues and labeling that would have facilitated retrieval of appropriate information. Indeed, Mautone and Mayer [25] reported that when using computer animations, an important aspect to consider is the "*detection of information by the learners, especially if the latter are novices*". Further, students' participation level was generally low; they passively and silently watched the PowerPoint animations. It was also noted that a few students were not showing signs of motivation as evident from their facial expressions; this might be due to the fact that students were not actively engaged but were passively viewing the animations. The lack of motivation could be related to the fact that the teacher-designed animations were not appealing to students, in terms of colour or design, in line with Grabe and Grabe [26] who mentioned that attractive integrated features in animations such as colour may tap on students' interest. Based on observations during lessons of Cycle 1, amendments and improvements were made for the design of the PowerPoint animations, as well as regarding strategies to be used during the lessons in Cycle 2.

Observations during the lessons in Cycle 2 have revealed that students' motivational level has increased compared to Cycle 1; they were found to view the animations with greater zeal. This was attributed to the fact that the computer animations designed for Cycle 2 included more colour schemes as well as accompanying verbal cues and labeling which were absent in the teacher-designed animations used in Cycle 1. It was also observed that the use of the designed computer animations promoted understanding of complex and abstract concepts, like cofactors, prosthetic groups and coenzymes. In addition, verbal cues and labels integrated in the animations were found to facilitate retrieval of information, and to promote conceptual understanding. The verbal cues directed students' attention to the relevant key information as claimed by Mautone and Mayer [25]. Observations made during Cycle 2 have clearly pointed out that use of questioning (both oral and written) together with computer animations not only helped to increase classroom interactions (teacher-students and student-student) but also enhanced students' understanding of the concepts. It was clearly evident in Cycle 2 that students' participation and engagement during the lessons were improved when the computer animations included appropriate colour schemes, verbal cues and labeling.

During the present study, two separate animations were designed by the teacher-researcher to explain what coenzymes are, and what prosthetic groups are. A very interesting finding of the study was that students were found to have difficulties to differentiate between these two terms without teacher's scaffold. These difficulties could be attributed to students' inability to relate information presented through the two animations, in concurrence with Ainsworth [27] who claimed that translation between two representations can influence students' understanding; the author also emphasized on the need of supporting translation across representations to maximize learning outcomes.

Observations from Cycle 3 have further revealed that the designed computer animations helped to promote conceptual understanding and that students were successful in retrieving relevant information from the animations. For example, they could easily identify the components of ATP as 'adenine', 'ribose', and 'phosphate groups' from the relevant animation. They were also able to develop understanding of abstract concepts such as 'role of ATP in metabolism', and 'hydrolysis of ATP', through use of the designed PowerPoint animations. It was also observed that integrated 'audio-questioning' in the designed animations enhanced students' conceptual understanding. Students were also found to be fully and actively engaged in the lessons including use of PowerPoint animations; they viewed the animations attentively and answered the oral and written questions. It was also noted that students' motivational level and engagement were highest in Cycle 3 as compared to Cycles 1 and 2. This could be attributed to improvements made in the PowerPoint animations in terms of colour design, graphics and integration of "audio" questions as the action research moved from cycle to cycle. Thus, we report here that designed PowerPoint animations can promote students' understanding of concepts related to the "Chemistry of Life". This is in line with Parette et al. [28] who claimed that PowerPoint animations allow presentation of instructional content in a clear and systematic format while students are engaged in the lessons.

Regarding downloaded 2D and 3D flash animations, in Cycle 1 they were used with their accompanying audio-narrations. One major difficulty faced by students was language barrier, which impeded on their comprehension of certain key words in the different animations. The mother tongue of the participant students is 'Creole' (French-based creole language); thus they lacked the appropriate phonetic skills that would have enabled them to understand the English-based audio narration. According to Wong [29], lack of phonetic skills can hinder listening ability. Nevertheless, it was found that when the animations were accompanied with written texts, together with the audio-narrations, students were better able to follow and understand the animations. Another important finding was that when the flash animations were played in a continuous manner, the participants encountered difficulties to retrieve information; the animations had to be viewed several times for students to develop conceptual understanding. This might be explained by the Cognitive Load Theory, which postulates that working memory can only accommodate a limited amount of information at a time; when this limit is exceeded, 'cognitive overload' results and processing (learning) is impaired [30]. Thus, when animations were viewed in a continuous manner, students could not grasp all information displayed at one go. These findings are also in line with Ayres et al. [31] who mentioned that students may be confused and overwhelmed throughout animation viewing if they are unable to properly process all incoming information before it disappears. Thus, in Cycles 2 and 3 of the present action research, two major amendments were made regarding use of flash animations – firstly, the animations were not played in a continuous fashion, but each animation was played with intermittent pauses during which students had to answer oral and/or written questions. Secondly, students were provided with written transcripts of the audio narrations. It was interesting to note that allowing students to view animations with intermittent pauses, during which they could answer questions and discuss with peers and teacher helped to enhance conceptual understanding better than viewing the animations continuously. Retrieval of information was also facilitated through use of animations with pauses, which allow for increased classroom interactions. Lowe [32] highlighted that it is important to break animations into chunks to help students integrate each piece of information. On the other hand, provision of written transcripts of the animations helped in minimizing students' language problems, thereby further facilitating retrieval of information. Most importantly, it was noted that use of flash animations also increased students' motivational and engagement levels; indeed, students were more motivated and engaged with use of the flash animations than with PowerPoint animations.

Our findings have also revealed that use of interactive online animations facilitated understanding of concepts like enzyme inhibition, competitive and non-competitive inhibition, as well as DNA structure. Most importantly, we found that students' participation, engagement and motivation were highest when the animations were interactive, allowing learners to interact with technology and with the concepts under study. Students' motivation and enjoyment of the interactive animations were evident during the lessons. They were found to appreciate animations with increased levels of interactivity; this impacted positively on their conceptual understanding. It was also observed that teacher-student interactions were greater in lessons involving use of interactive animations, as compared to non-interactive ones. Interactive animations are indeed reported to motivate students to invest more mental effort in learning, thereby enabling them to achieve higher performance [33]. Interactive animations were also found to allow participant students to take ownership of their learning thereby increasing their motivational level, in agreement with Deci and Ryan [34] who reported that increasing learners' sense of control can impact positively on their motivation.

It needs to be highlighted that during the first lesson involving use of online interactive animations, several students were slightly distracted by the fact that they were working online and had a tendency to open side pages of the internet. Nevertheless, as soon as the teacher reminded them to focus on their tasks, they became focused and engaged in the lessons. Interestingly, no such problems arose during subsequent lessons involving use of online interactive animations. Thus, our findings tend to support Hembrooke and Gay [35], who mentioned that when integrating technology in classroom teaching, it is essential to keep learners on task and to minimize electronic distractions.

Another problem that was noted in the first cycle of the action research was that students had a tendency to copy *verbatim* the short notes projected in the animations while answering questions in teacher-prepared worksheets. Reflecting on this problem leads to the insinuation that questions set in the worksheets were too straightforward and did not encourage thinking. Thus, in the later cycles, more challenging and higher-order questions were included in the worksheets in an attempt to foster higher order thinking skills amongst the students.

15.3.2 Students' Achievement Tests

After completing each cycle of the present action research, information was gathered about students' understanding of concepts taught by administering a Students' Achievement Test. Thus, three different tests were administered to participant

Table 15.1 Students' scores and percentage scores obtained in Tests 1, 2 and 3

Student's code name	Score out of 40 in Test 1 (%)	Score out of 40 in Test 2 (%)	Score out of 40 in Test 3 (%)
A1	28 (70 %)	29 (73 %)	26 (65 %)
A2	27 (68 %)	27 (68 %)	31 (78 %)
A3	19 (48 %)	16 (40 %)	20 (50 %)
A4	25 (63 %)	20 (50 %)	22 (55 %)
A5	30 (75 %)	34 (85 %)	35 (88 %)
A6	23 (58 %)	24 (60 %)	20 (50 %)
A7	25 (63 %)	25 (63 %)	28 (70 %)
A8	23 (58 %)	26 (65 %)	28 (70 %)
A9	27 (68 %)	28 (70 %)	30 (75 %)
A10	20 (50 %)	29 (73 %)	29 (73 %)
A11	19 (48 %)	23 (58 %)	29 (73 %)
A12	27 (68 %)	32 (80 %)	29 (73 %)
A13	26 (65 %)	22 (55 %)	23 (62 %)
A14	15 (38 %)	12 (30 %)	17 (42 %)

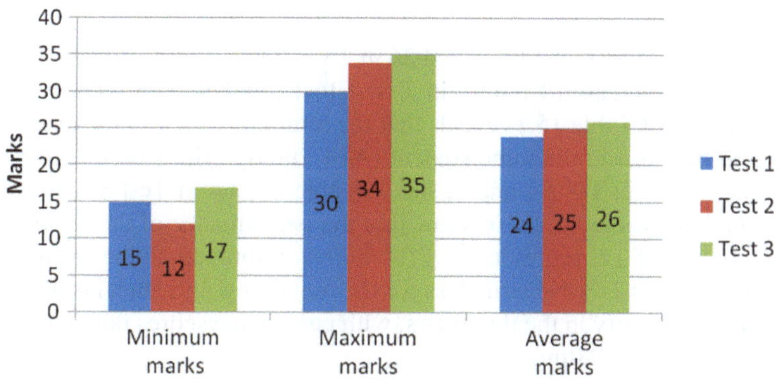

Fig. 15.1 Comparison of minimum, maximum and average marks scored by students in Tests 1, 2 and 3

students, namely Test 1, 2 and 3 after cycles 1, 2 and 3 respectively, to assess their conceptual understanding, and to obtain information about their performance after each cycle. The maximum score achievable for each test was 40.

The marks and percentage marks scored by students in the three tests are given in Table 15.1, while Fig. 15.1 gives a comparison of the minimum, maximum and average marks obtained by students in the three tests. (*Students have been given code names for ethical reasons*).

From Fig. 15.1, it can be seen that there has been a slight, but definite, improvement in students' performance from Test 1 to 2 and 3. The maximum mark rose from 30 in Test 1 to 34 and 35 in Tests 2 and 3 respectively, while the average

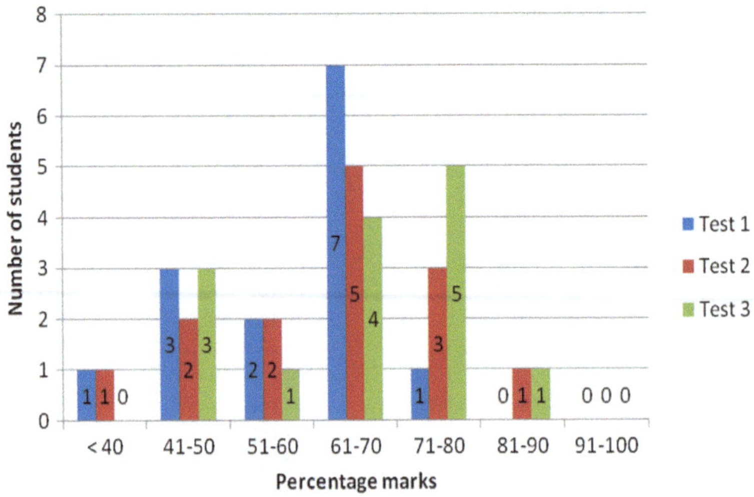

Fig. 15.2 Comparison of the number of students scoring in the different percentage ranges in the three achievement tests

marks rose from 24 in Test 1 to 25 and 26 in Tests 2 and 3 respectively. Though the minimum mark decreased from 15 in Test 1 to 12 in Test 2, it then rose to 17 in Test 3. Analysis of Table 15.1 reveals that the minimum mark in the three tests has been scored by one and the same student (**A14**); though she scored less than 40 % (Pass mark: 40 %) in Tests 1 and 2, she was able to pass in Test 3 with a score of 42 %. Data from Table 15.1 also show that the performance of the majority of the students (10 out of 14, representing 71.4 %) has improved as the action research moved from Cycles 1–3. Nevertheless for the remaining four students, their performance was satisfactory in the three tests, with percentage scores ranging from 50 % to 73 %, with no one failing.

In order to have a more detailed view about students' performance in the tests, a comparison of the number of students scoring in the different percentage ranges in the three achievement tests is given in Fig. 15.2.

It is noted that all students have obtained pass marks (≥40 %) in the three tests, except for the one student (**A14**) who failed (score < 40 %) in Tests 1 and 2, but managed to score a pass mark in Test 3. Both Figs. 15.1 and 15.2 point to an overall improvement as the research moved from Cycle 1 to Cycles 2 and 3. From Fig. 15.2, it is also noted that the number of students scoring percentage marks in the ranges 51–60 % and 61–70 % has decreased from Test 1 to 2 and 3, while there has been an increase in the number of students scoring in the range of 71–80 %.

In view of the above discussion and interpretation of data presented in Table 15.1 and Figs. 15.1 and 15.2, it is obvious that use of computer animations has impacted positively on students' conceptual understanding of "The Chemistry of Life". Students were able to develop understanding of the concepts involved and were able to perform well in the three tests. Further, the performance of most students

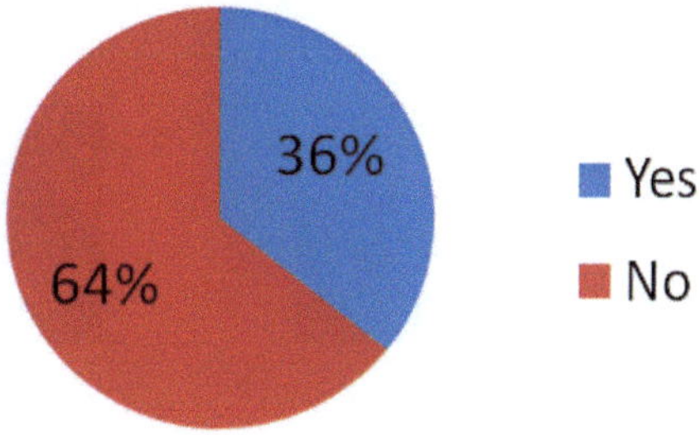

Fig. 15.3 Students' exposure to computer animations prior to this study

shows a definite improvement as we moved from cycle to cycle – this may be accounted for by two reasons. Firstly, students got more and more acquainted to use of computer animations as the research moved on and secondly, due to improvements and amendments made while implementing use of computer animations during the lessons. Thus, our findings tend to support results from other researchers who reported that computer animations can improve students' understanding of chemistry and biology concepts by facilitating understanding of chemical processes at the molecular level, dynamic phenomena and processes involving motion resulting in enhanced students' achievement and performance [17, 36–39].

15.3.3 Students' Questionnaire

After completing the three cycles of the action research, a questionnaire was administered to all participants to collect data regarding their attitudes and opinions on the use of computer animations in the teaching of the "Chemistry of Life". In this section, pertinent findings from the students' questionnaire are presented and discussed.

- **Students' exposure to computer animations prior to this study**
  When asked whether they have been exposed to use of computer animations during classroom teaching prior to this study, the majority of the students (64 %) answered negatively, while the remaining answered positively as shown in Fig. 15.3. On further probing, it was found that those who answered positively were those studying biology as a principal subject – their biology teacher made use of videos and animations to teach certain biological concepts using her personal laptop. However, it seemed that they were exposed only to teacher-controlled mode of presenting animations.
- **Students' responses to whether computer animations can help in better understanding of concepts**
  From Fig. 15.4, it is noted that most students (71.4 %) agreed or strongly agreed that computer animations helped them to better understand concepts, while the

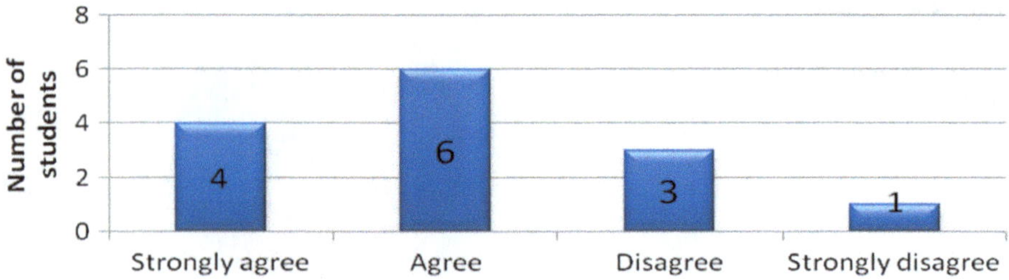

Fig. 15.4 Students' responses to whether computer animations can help in better understanding of concepts

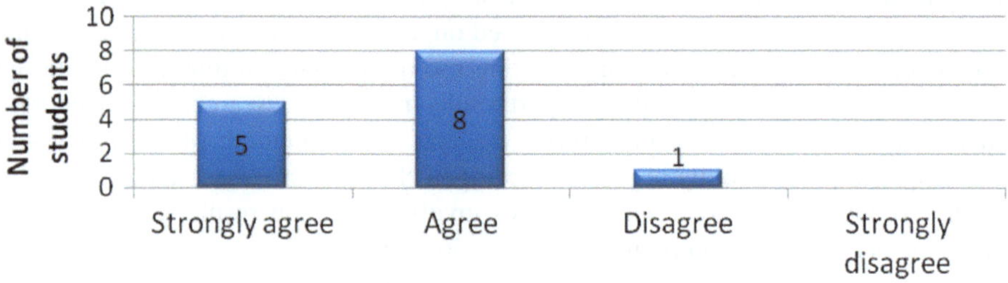

Fig. 15.5 Students' responses to whether computer animations can help in better retention of concepts

remaining 28.6 % answered negatively. Students who answered positively claimed that computer animations helped them to visualize concepts better allowing them to understand complex concepts such as protein structure, factors affecting enzyme activity, DNA replication and protein synthesis. Some even mentioned that use of computer animations helped them to understand concepts faster than in traditional classroom situation. These findings support results obtained through observation checklists and achievement tests, and are in line with findings from Starbek et al. [40] who reported that computer animations can enhance students' comprehension skills.

Students who answered negatively pointed out that they found it difficult to develop understanding of concepts from the animations. This might be due to the fact that they lacked spatial aptitudes, and preferred the traditional expository teaching method. Evidence suggests that the effectiveness of multimedia learning designs varies according to the learner's spatial aptitudes [41].

- **Students' responses to whether computer animations can help in better retention of concepts**

  Findings from Fig. 15.5 reveal that almost all students (92.9 %) agreed or strongly agreed that computer animations can help in retention of concepts.

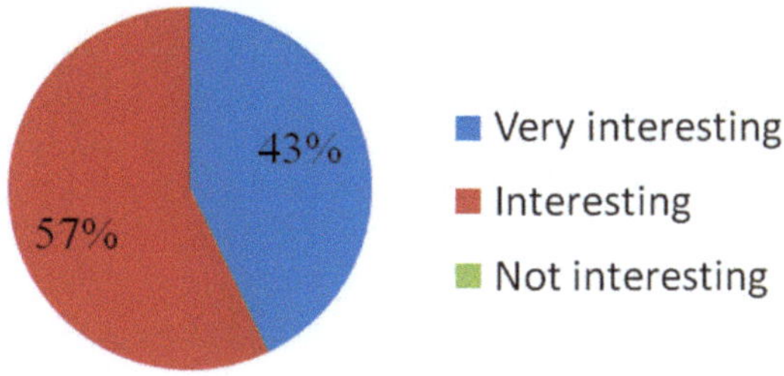

Fig. 15.6 Students' views regarding use of computer animations in the teaching of "Chemistry of Life"

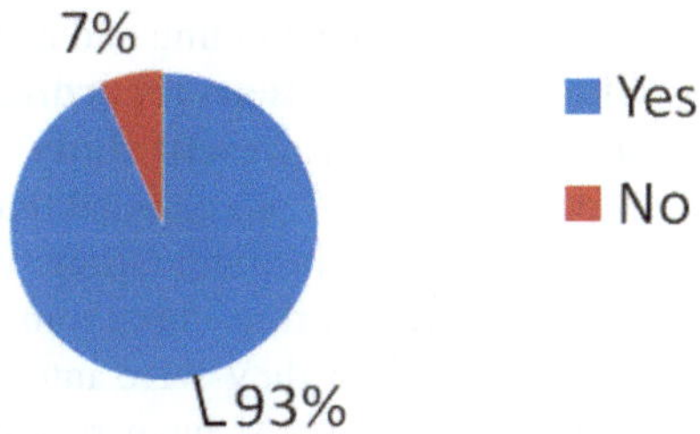

Fig. 15.7 Students' responses to whether use of computer animations has helped to motivate them

These students stated that they were able to recall about concepts taught through use of computer animations during the achievement tests. Our findings concur well with those of Rebetez et al. [42]; these authors asserted that computer animations do have a beneficial effect on students' retention. Only one student (7.1 %) claimed that computer animations did not help her to retain concepts; she mentioned that she had difficulty remembering facts about DNA, protein synthesis and ATP.

- **Students' views regarding use of computer animations in the teaching of "Chemistry of Life"**
  When asked about how they found use of computer animations in the teaching of the "Chemistry of Life", it is worth noting that all students unanimously stated that they found it very interesting (43 %) or interesting (57 %) as depicted in Fig. 15.6. Students said that computer animations allowed them to learn in a different way, and that they liked the 'colored images' and 'sound effect'. Indeed, adding graphics to scientific text has been reported to increase students' interest [43].
- **Students' responses to whether use of computer animations has helped to motivate them**
  From Fig. 15.7, it can be noted that the majority of students (13 out of 14, representing 93 %) claimed that use of computer animations in the teaching of "Chemistry of Life" has helped to motivate them. This finding not only confirms data collected through observation checklists, but is also in line with results from earlier research which has shown that use of computer animations and simulations does have beneficial effects on students' motivation [8, 44].

  Only one student (7 %) claimed that she was not motivated by use of computer animations and that she preferred teacher expository explanation and note-giving.

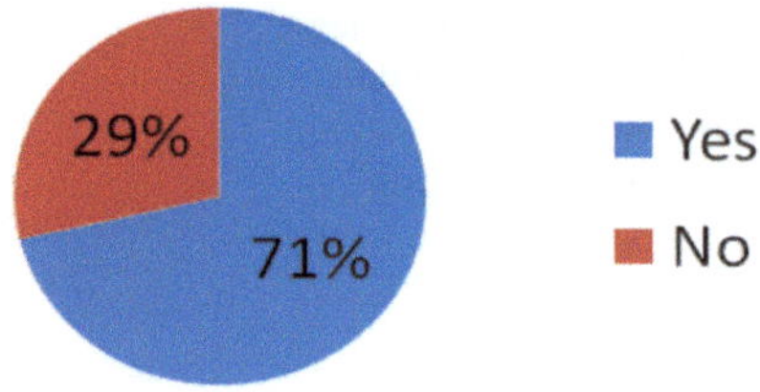

Fig. 15.8 Students' responses to whether computer animations can be used in the teaching and learning of other subjects

She was the same student who scored minimum marks in the three tests. This finding thus shows that lack of motivation may impact negatively on students' conceptual understanding and achievement.

- **Students' perceptions regarding the different types of computer animations used during the present study**
  Most students (57 %) claimed that they were more motivated during the lessons when the computer animations were interactive. This finding further confirms data gathered during classroom observation; students were found to be more motivated when they were interacting with the computer animations. Rotbain and co-workers [44] have recently reported that to use computer animations effectively in the teaching/learning process, it is important to make them interactive so as to increase students' involvement in their learning. Further, Deci and Ryan [34] pointed out that one way of enhancing learner's motivation is to increase his or her sense of control by providing choice or locus of control. Most students (57 %) also stated that use of interactive animations made the lessons more interesting than non-interactive ones; this further supports the finding that students were more motivated and more interested when using interactive animations during the lessons. These findings are in line with those of Lepper [45], who reported that students showed more enjoyment, better performance, and greater persistence at a variety of activities when dealing with animations having a higher degree of control and interactivity.

  Furthermore, most of the participant students (64 %) claimed that interactive animations were more effective in promoting thinking than non-interactive ones. These findings not only confirm data gathered through observation, where it was noted that students were more engaged in thinking while working with learner-controlled animations, but also concur well with results from the study by White and Bolker [46] who found that interactive animations promoted the active engagement and application of the learnt material.
- **Students' responses to whether computer animations can be used in the teaching and learning of other subjects**
  When asked whether computer animations can be used in the teaching and learning of other subjects, most of the participants (71 %) answered positively (Fig. 15.8), claiming that computer animations would help them understand certain concepts faster.

  Students who answered negatively to this question argued that use of computer animations is time consuming; this might affect syllabus completion or coverage.

15.4 Conclusions

The purpose of the present study was to enhance conceptual understanding of the "Chemistry of Life" at the CIE 'A'-Level by making use of computer animations during the lessons. Our findings have revealed that use of computer animations had a positive impact on the participant students, not only in terms of conceptual understanding, but also on their motivation, interest and engagement during the lessons. Findings from the different data-collection tools all pointed out to enhanced students' motivation and engagement during the lessons as well as improved understanding of the concepts under study. Thus, results from the present study lend further support to earlier claims that computer-based animations can deepen students' conceptual understanding and enhance their motivation, interest and engagement [40, 44, 47, 48].

In addition, our findings have also revealed that when designing computer animations for classroom teaching, it is important to include relevant cues and labeling. The latter were found to impact positively on students' conceptual understanding and motivational level as they facilitated retrieval of information from the animations; verbal cues are reported [25] to direct students' attention to key information thereby promoting conceptual understanding. We found that in the absence of appropriate cues and labeling, students encountered difficulties to follow the animations and to retrieve information that would help them to develop understanding of the concepts under study. On the other hand, embedded oral questions in teacher-prepared PowerPoint computer animations were found to promote students' thinking and teacher-students interactions, both of which contributed positively to students' engagement and understanding of concepts.

The use of computer animations in classroom teaching is more effective when students are allowed to view them with intermittent pauses rather than viewing the complete animation continuously. During the pauses, students can be engaged either in oral discussion with teacher and/or peers, or in answering guiding questions (in designed worksheets) that focus their attention on the concepts, and help them to retrieve relevant information. Such activities during the pauses were found to contribute positively to students' engagement and understanding. Furthermore, we would also like to emphasize that when designing worksheets to be used in conjunction with computer animations, teachers should integrate high-level cognitive questions that would help learners to construct understanding of the concepts addressed by the animations. This is in line with what Rotbain et al. [44] recommended based on a study on the use of computer animations in the teaching of high school molecular biology.

Findings from our study have also revealed that students' motivation and level of engagement during the lessons were highest when the animations were interactive. Learner-controlled interactive animations were found to be the most effective in enhancing students' motivation and engagement as compared to non-interactive ones. In addition, participant students have claimed that interactive animations were more motivating and more interesting, have helped in better retention of concepts, and were better to encourage thinking as compared to non-interactive ones.

We also report here that use of computer animations in classroom teaching can accommodate a variety of strategies and resources like cooperative learning, hands-on practical, use of analogies and specially designed worksheets amongst others. For instance, written questions (in worksheets) and oral questions used in conjunction with computer animations were found to encourage students' thinking, and to keep them engaged during the lessons, thereby facilitating retrieval of information and promoting conceptual understanding. When using practical activity in conjunction with computer animations, it was found that while practical work allowed students to develop conceptual understanding at the macroscopic level, computer animations facilitated conceptual understanding at the microscopic level [38, 49]. We also found that when using downloaded animations, language barrier can impede on students' comprehension of the accompanying oral narration or explanation. This barrier can be overcome by providing learners with a written transcript of the narrations, or by allowing learners to view the animations with teacher-recorded oral narration.

Based on experiences gained during the present study, we assert that prior planning and preparation by teacher allows proper implementation of lessons incorporating use of computer animations. All the animations to be used need to be carefully previewed by teacher and organized in labeled folders so that they are easily accessed during the lessons. The teacher also needs to ensure that the required software(s) are properly installed on the computers or laptops. Furthermore, relevant worksheets need to be prepared based on the animations for achieving the desired learning outcomes in line with the curriculum content. Prior planning also requires teacher to identify what resources and strategies to be used in conjunction with the computer animations. This would help to actively engage learners in retrieving relevant information from the computer animations for conceptual understanding.

In conclusion, our results clearly indicate that use of computer animations has impacted positively on students' motivation, interest, engagement and conceptual understanding of the "Chemistry of Life". In line with Rotbain et al. [44], the present study thus provides further support for the use of new technologies, such as computer animations, in classroom teaching.

Acknowledgements The authors sincerely acknowledge and thank all the sources of downloaded animations. We also thank all the students who participated in the study.

References

1. Cambridge International Examinations (1990–2006) Examiner report for advanced level chemistry. Cambridge, UK
2. Cambridge International Examinations (2007–2010) Examiner report for advanced level chemistry. Cambridge, UK
3. Bowen CW (1998) Item design considerations for computer-based testing of student learning in chemistry. J Chem Educ 75:1172–1175

4. Schönborn KJ, Anderson TR (2006) The importance of visual literacy in the education of biochemists. Biochem Mol Biol Educ 34:94–102
5. Malacinski GM, Zell PW (1996) Manipulating the "Invisible": learning molecular biology using inexpensive models. Am Biol Teach 58:428–432
6. Marbach-Ad G, Stavy R (2000) Students' cellular and molecular explanations of genetics phenomena. J Biol Educ 34:200–205
7. Bateman RC Jr, Booth D, Sirochman R, Richardson J, Richardson DC (2002) Teaching and assessing three-dimensional molecular literacy in undergraduate biochemistry. J Chem Educ 79:551–552
8. Abdoolatiff S, Narod FB (2009) Investigating the effectiveness of computer simulations in the teaching of "Atomic Structure and Bonding". In: Gupta-Bhowon M, Jhaumeer-Laulloo S, Li Kam Wah H, Ramasami P (eds) Chemistry education in the ICT age. Springer, Dordrecht, pp 85–100
9. Russel JW, Kozma RB, Jones T, Wykoff J, Marx N, Davis J (1997) Use of simultaneous, synchronized macroscopic, microscopic and symbolic representations to enhance the teaching and learning of chemical concepts. J Chem Educ 74:330–334
10. Lim CP (2007) Effective integration of ICT in Singapore schools: pedagogical and policy implications. Educ Technol Res Dev 55:83–116
11. Boodhena D (2008) Information and communication technology in the teaching and learning of chemical calculations at form IV level – an action research. PGCE dissertation, Mauritius Institute of Education, Réduit
12. Burke K, Greenbowe T, Windschitl M (1998) Developing and using conceptual computer animations for chemistry instruction. J Chem Educ 75:58–61
13. Rieber LP (1991) Animation, incidental learning, and continuing motivation. J Educ Psychol 83:318–328
14. O'Day DH (2007) The value of animations in biology teaching: a study of long-term memory retention. CBE Life Sci Educ 6:217–223
15. Williamson VM, Abraham MR (1995) The effects of computer animation on the particulate mental models of college chemistry students. J Res Sci Teach 32:521–534
16. Sanger MJ, Phelps AJ, Fienhold J (2000) Using a computer animation to improve students' conceptual understanding of a can-crushing demonstration. J Chem Educ 77:1517–1519
17. Kelly R, Jones L (2008) Investigating students' ability to transfer ideas learned from molecular animations of the dissolution process. J Chem Educ 85:303–309
18. Gardner H, Kornhaber M, Wake W (1996) Intelligence: multiple perspectives. Harcourt Brace, Fort Worth
19. Muñoz-Repiso AGV, Tejedor FJT (2009) Information and communication technologies in university teaching: implications in European higher education space. Int J Hum Sci 6:683–696
20. Selwyn N, Husen O (2010) The educational benefits of technological competence: an investigation of students' perceptions. Eval Res Educ 23:137–141
21. Ministry of Education and Human Resources (2006) Towards a quality curriculum. Strategy for reform. Republic of Mauritius
22. Ministry of Education, Culture and Human Resources (2009) Education and human resources strategy plan 2008–2020. Republic of Mauritius
23. Ministry of Education, Culture and Human Resources (2009) National curriculum framework: secondary. Republic of Mauritius
24. Cohen L, Manion L, Morrison K (2001) Research methods in education, 5th edn. Routledge Falmer, London
25. Mautone PD, Mayer RE (2001) Signaling as a cognitive guide in multimedia learning. J Educ Psychol 93:377–389
26. Grabe M, Grabe C (2007) Integrating technology for meaningful learning, 5th edn. Houghton Mifflin, Harcourt, Boston
27. Ainsworth SE (1999) The functions of multiple representations. Comput Educ 33:131–152

28. Parette HP Jr, Hourcade J, Blum C (2011) Using animation in microsoft powerpoint to enhance engagement and learning in young learners with developmental delay. http://works.bepress.com/jack_hourcade/10/. Accessed 15 June 2011
29. Wong R (1993) Pronunciation myths and facts. Engl Teach Forum 31:45–46
30. Sweller J (2004) Instructional design consequences of an analogy between evolution by natural selection and human cognitive architecture. Instr Sci 32:9–31
31. Ayres P, Marcus N, Chan C, Qian N (2009) Learning hand manipulative tasks: when instructional animations are superior to equivalent static representations. Comput Hum Behav 25:348–353
32. Lowe RK (2003) Animation and learning: selective processing of information in dynamic graphics. Learn Instr 13:157–176
33. Hegarty M, Kriz S, Cate C (2003) The roles of mental animations and external animations in understanding mechanical systems. Cognit Instr 21:325–360
34. Deci EL, Ryan RM (1985) Intrinsic motivation and self-determination in human behavior. Plenum Press, New York
35. Hembrooke H, Gay G (2003) The laptop and the lecture: the effects of multitasking in learning environments. J Comput High Educ 15:46–64
36. Stith B (2004) Use of animation in teaching cell biology. Cell Biol Educ 3:181–188
37. Falvo D (2008) Animations and simulations for teaching and learning molecular chemistry. Int J Technol Teach Learn 4:68–77
38. Velazquez-Marcano A, Williamson V, Ashkenazi G (2004) The use of video demonstrations and particulate animation in general chemistry. J Sci Educ Technol 13:315–323
39. Marbach-Ad G, Rotbain Y, Stavy R (2008) Using computer animation and illustration activities to improve high school students' achievement in molecular genetics. J Res Sci Teach 45:273–292
40. Starbek P, Erjavec M, Peklaj C (2010) Teaching genetics with multimedia results in better acquisition of knowledge and improvement in comprehension. J Comput Assist Learn 26:214–224
41. Garg A, Norman GR, Spero L, Maheshwari P (1999) Do virtual computer models hinder anatomy learning? Acad Med 74:S87–S89
42. Rebetez C, Betrancourt M, Sangin M, Dillenbourg P (2010) Learning from animation enabled by collaboration. Instr Sci 38:471–485
43. Harp SE, Mayer RE (1997) The role of interest in learning from scientific text and illustrations: on the distinction between emotional interest and cognitive interest. J Educ Psychol 89:92–102
44. Rotbain Y, Marbach-Ad G, Stavy R (2008) Using a computer animation to teach high school molecular biology. J Sci Educ Technol 17:49–58
45. Lepper MR (1988) Motivational considerations in the study of instruction. Cognit Instr 5:289–309
46. White BT, Bolker ED (2008) Multimedia in biochemistry and molecular biology education. Interactive computer simulations of genetics, biochemistry, and molecular biology. Biochem Mol Biol Educ 36:77–84
47. Adegoke B (2010) Integrating animations, narratives and textual information for improving physics learning. J Res Educ Psychol 8:725–748
48. Rosen Y (2009) The effects of an animation-based on-line learning environment on transfer of knowledge and on motivation for science and technology learning. J Educ Comput Res 40:451–467
49. Johnstone AH (2000) Teaching of chemistry – logical or psychological? Chem Educ Res Pract 1:9–15

Others

- http://www.youtube.com/watch?v=lijQ3a8yUYQ
- http://www.youtube.com/watch?v=eUS6CEn4GSA&feature=related
- http://www.youtube.com/watch?v=wM2LWCTWlrE
- https://mywebspace.wisc.edu/jonovic/web/proteins.html
- http://www.wiley.com/legacy/college/boyer/0470003790/animations/enzyme_binding/enzyme_binding.htm
- http://www.wiley.com/legacy/college/boyer/0470003790/animations/enzyme_inhibition/enzyme_inhibition.htm
- http://www.youtube.com/watch?v=ZGHkHMoyC5Ihttp://www.youtube.com/watch?v=ZGHkHMoyC5I
- http://learn.genetics.utah.edu/content/begin/dna/builddna/
- http://dnaftb.org/20/animation.html
- http://www.wiley.com/legacy/college/boyer/0470003790/animations/replication/replication.htm
- http://www.wiley.com/legacy/college/boyer/0470003790/animations/translation/translation.htm
- http://www-class.unl.edu/biochem/gp2/m_biology/animation/gene/gene_a2.html
- http://www-class.unl.edu/biochem/gp2/m_biology/animation/gene/gene_a3.html
- http://learn.genetics.utah.edu/content/begin/dna/transcribe/
- http://www.yteach.co.uk/index.php/resources/deoxyribonucleic_acid_DNA_gene_population_species_race_phenotype_adaptation_inheritance_recombination_mutations_crossing_over_environmental_variation_t_page_8.html
- http://www.yteach.co.uk/page.php/resources/view_all?id=malaria_mutagens_mutation_phenotype_point_mutation_gene_mutation_proliferation_sickle_cell_disease_spontaneous_mutations_substitution_unscheduled_DNA_synthesis_UDS_t_page_10&from=search
- http://www.youtube.com/watch?v=9UpwV1tdxcs
- http://www.linkstudio.info/portfolio/animation05.htm

Chapter 16
$NaBH_4$-Mediated Complete Reduction of the α,β-Unsaturated Ketone Units of Chalcones in the Synthesis of Flavans

Ishmael B. Masesane and Ofentse Mazimba

Abstract In this chapter, the $NaBH_4$-mediated reduction of both the carbon-carbon double bond and the carbonyl group of chalcones and subsequent acid mediated cyclisation of the resulting saturated alcohols to flavans is described. The chalcones were readily prepared from commercially available acetophenone and salicylaldehyde derivatives. Through this sequence of reactions, 3′,4′-dimethoxyflavan and 3′,4′,8-trimethoxyflavan were prepared in good overall yields.

16.1 Introduction

Borohydrides form part of the chemist's classic arsenal of reducing agents employed in organic synthesis. Sodium and potassium borohydrides in methanol or ethanol for example are used for the reduction of ketones and aldehydes [1–6]. The use of borohydrides in the reductions of ketones and aldehydes is classically thought to be a chemoselective reaction. Isolated double bonds are normally inert to borohydride reaction conditions. However, a few reports have appeared in the literature outlining the $NaBH_4$-mediated reduction of both the carbon-carbon and carbon-oxygen double bonds of α,β-unsaturated carbonyl compounds to give saturated alcohols [7–9]. In this paper we report the synthesis of 3′,4′-dimethoxyflavan and 3′,4′,8-trimethoxyflavan involving the non-chemoselective $NaBH_4$ mediated reduction of the corresponding chalcones to give saturated alcohols which were cyclized in acid to give the flavans.

Flavans are a class of flavonoids distinguished by the benzopyran core and are widely distributed in the plant kingdom. It is of interest to develop an efficient general method for the synthesis of flavans because many of them exhibit significant biological and pharmacological activity [10–14].

I.B. Masesane (✉) • O. Mazimba
Department of Chemistry, University of Botswana, Private Bag, 00704 Gaborone, Botswana
e-mail: masesane@mopipi.ub.bw; Ofentse.Mazimba@mopipi.ub.bw

M. Gupta Bhowon et al. (eds.), *Chemistry: The Key to our Sustainable Future*,
DOI 10.1007/978-94-007-7389-9_16,

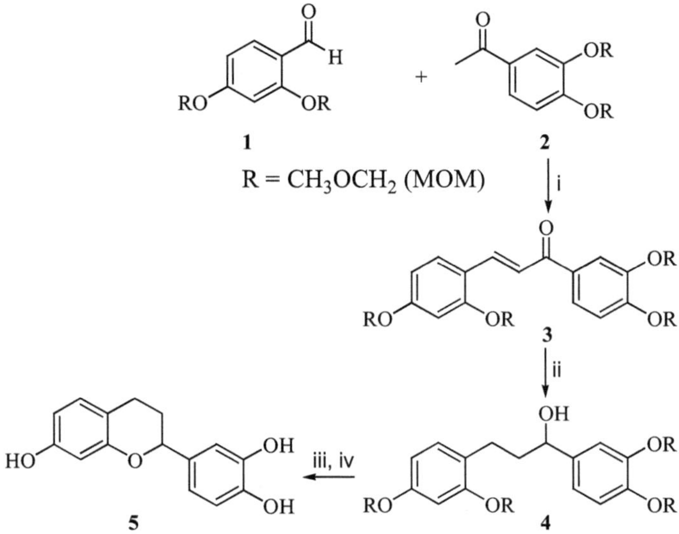

Scheme 16.1 Reagents and conditions: (i) KOH, CH_3OH, 85 %. (ii) H_2/Pd, EtOH, 99 %. (iii) BF_3. Et_2O, 1,4-dioxane. (iv) deprotection, 90 % from **4**

Prior to our work discussed in this chapter, Xue and co-workers [8] described the reaction of salicylaldehyde derivative **1** and acetophenone derivative **2** in the synthesis of the natural occurring flavan **5** (Scheme 16.1). To begin, a solution of **1** and **2** in the presence of KOH was stirred at room temperature to give chalcone **3**. To set the stage for the cyclisation reaction, the *trans* carbon-carbon double bond of **3** must either be isomerized to the *cis* form or completely reduced. In this case, chalcone **3** was treated with H_2 in the presence of a catalytic amount of Pd to give intermediate **4** in 99 % yield. It is instructive to draw attention to the fact that both the carbon-carbon and carbon-oxygen double bond of **3** were reduced by H_2/Pd, a reagent usually used for carbon-carbon double bond reduction. To complete the synthesis, Lewis acid mediated cyclisation of intermediate **4** and acidic de-protection of the MOM-protected hydroxyl groups delivered the desired naturally flavan **5** in good yield.

Initial work of our group [9] involved reaction of salicylaldehyde **6** with methoxyacetophenone derivatives of type **7** to give chalcones of type **8**, followed by $NaBH_4$-mediated reduction of chalcones **8** and final cyclisation of the corresponding intermediates to give flavans of type **9** (Scheme 16.2). During this study, we noted that using *m*-methoxyacetophenone **7a** resulted in the lowest yields in all the three reactions involved in the synthesis of the flavans. In an effort to understand these low yields, we decided to investigate the effect of using 3,4-dimethoxyacetophenone instead of *m*-methoxyacetophenone, on the percentage yields of the reactions.

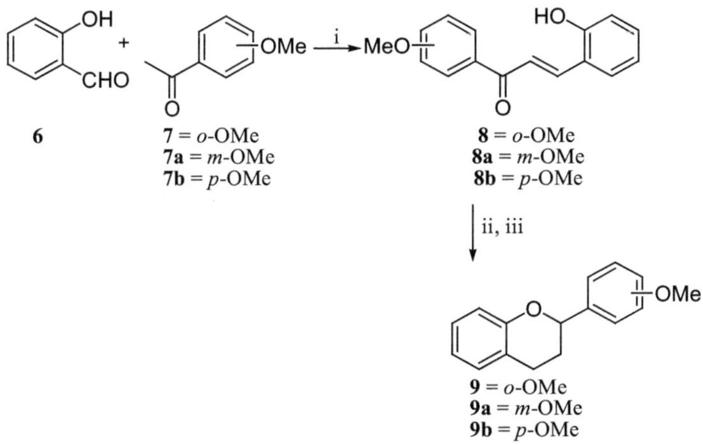

Scheme 16.2 Reagents and conditions: (i) 40 % $NaOH_{(aq)}$, ethanol, 60°C, 2 h, 83 % [**8**], 66 % [**8a**], 78 % [**8b**]. (ii) $NaBH_4$, MeOH, 25°C, 10 min. (iii) AcOH, reflux, 30 min, 70 % [**9**], 38 % [**9a**], 78 % [**9b**]

16.2 Results and Discussion

Our synthesis of 3′,4′-dimethoxyflavan commenced with the commercially available 3,4-dimethoxyacetophenone **10** and salicylaldehyde **6** which were dissolved in ethanol and allowed to react in the presence of NaOH to give chalcone **11** in 79 % yield. Subsequent treatment of a solution of chalcone **11** in methanol with $NaBH_4$ led to the reduction of both the carbon-carbon double bond and the carbonyl group of the chalcone to give alcohol **12** in 85 % yield. Lastly, treatment of alcohol **12** with acetic acid at reflux for 30 min gave 3′,4′-dimethoxyflavan **13** in 82 % yield (Scheme 16.3).

For the preparation of 3′,4′,8-trimethoxyflavan, a solution of acetophenone derivative **10** and salicylaldehyde derivative **6a** in ethanol and aqueous NaOH was refluxed for 2 h and then neutralized with HCl to give chalcone **11a** in 85 % yield. Non-chemoselective $NaBH_4$ mediated reduction of the alkene and carbonyl group of chalcone **11a** afforded alcohol **12a** in 91 % yield. Subsequent cyclization of **12a** in acid gave the desired flavan **13a** in 88 % yield (Scheme 16.3).

It is instructive at this juncture to draw attention to the fact that the use of 3,4-dimethoxyacetophenone gave significantly higher percentage yields for the three reactions involved in the synthesis of the flavans than when *m*-methoxyacetophenone was used. However it is not clear yet to what this difference should be attributed, so we offer no rationalization at this point.

The $NaBH_4$-mediated reduction of the conjugated carbon-carbon double bond is thought to proceed by conjugate addition of a hydride from the borohydride ion to

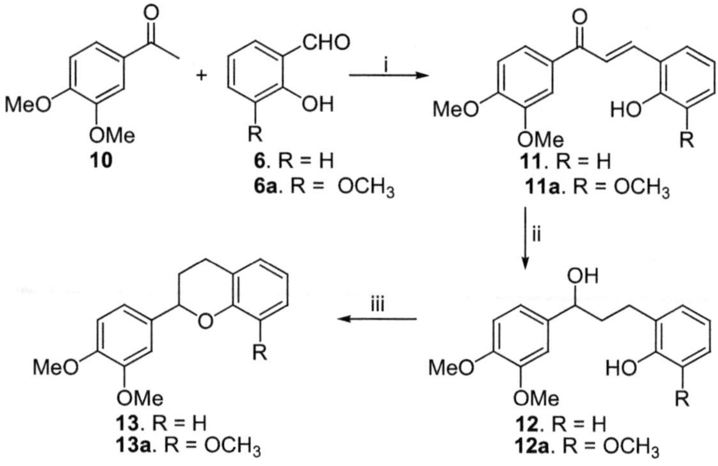

Scheme 16.3 Reagents and conditions: (i) 40 % $NaOH_{(aq)}$, ethanol, reflux, 2 h, 79 %(**11**), 85 % (**11a**). (ii) $NaBH_4$, MeOH, 25°C, 10 min, 85 %(**12**), 91 %(**12a**). (iii) AcOH, reflux, 30 min, 82 % (**13**), 88 %(**13a**)

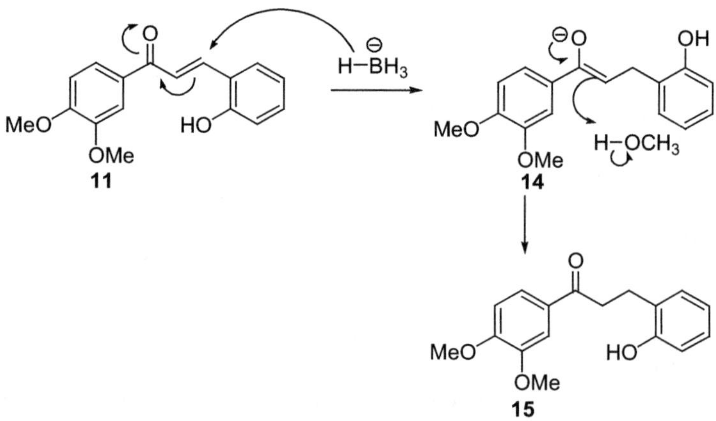

Scheme 16.4 Possible mechanism of the $NaBH_4$ mediated reduction of the conjugated C–C double bond

the β-carbon atom of the chalcone **11** to give enolate **14**. Protonation of enolate **14** affords saturated ketone **15** (Scheme 16.4) which is then reduced by $NaBH_4$ to give the observed alcohol. It is instructive to note that while we used $NaBH_4$ in the non-chemoselective reduction of α,β-unsaturated carbonyl system, Xue and co-workers [8] reported the reduction of related compounds using H_2/Pd.

16.3 Methodology

16.3.1 General

All chemicals were purchased from Aldrich and were used without further purification. Column chromatography was performed using Merck column-silica gel 60 particle size 0.040–0.063 mm. Infrared spectra were measured on a Perkin-Elmer spectrum100 spectrophotometer fitted with a universal attenuated total reflectance sampling device. ^{1}H NMR, ^{13}C NMR and 2D NMR spectra were acquired on a 300 MHz Bruker Avance DPX spectrometer using standard pulse sequences and referenced to the residual solvent signal. Mass spectra were acquired using a Waters GCT Premier mass spectrometer.

16.3.2 Experimental Procedure

To a solution of acetophenone (40 mmol) and salicylaldehyde (40 mmol) in EtOH (50 ml) was added aq. NaOH (40 %, 10 ml) dropwise and the reaction was refluxed for 2 h. The solution/suspension was poured into cold water and the reaction mixture was neutralized with 2 M HCl until the solution was acidic. The yellow precipitate was collected, washed with water and recrystallized from EtOH to yield the clean chalcone. Sodium borohydride powder (5 eq, 33.5 mmol) was slowly added to a stirred methanolic solution of the chalcone (6.7 mmol) at 25 °C. The resulting suspension was stirred for 10 min and after cooling the reaction mixture was quenched with 2 M HCl. The organic layer was separated and the aqueous phase extracted with EtOAc (3 × 15 ml). The combined organic phase was dried with $MgSO_4$ and concentrated to obtain the saturated alcohol. Glacial acetic acid (10 ml) was added to the saturated alcohol (4.4 mmol). The solution was heated at reflux for 30 min, then cold water (15 ml) followed by saturated $NaHCO_3$ (30 ml) were added, and the solution was extracted with EtOAc (2 × 20 ml). The combined organic layer was washed with water and brine and dried over anhydrous $MgSO_4$. The solvent was then evaporated under reduced pressure and the residue was purified by silica gel column chromatography eluting with petroleum ether-EtOAc (v/v, 4:1) to give the flavan.

16.3.3 Spectral Data of Flavans

Flavan **13**; brown gum, υ_{max} (neat): 2962, 2937, 1611, 1519, 1445 cm^{-1}; δ_H (300 MHz, acetone-d_6): 2.09 (1H, *m*, H-3a), 2.20 (1H, *m*, H-3b), 2.77 (1H, *m*, H-4a), 2.98 (1H, *m*, H-4b), 3.82 (3H, *s*, MeO-3′), 3.83 (3H, *s*, MeO-4′), 5.02 (1H, *dd*, $J = 2.4$, 10.5 Hz, H-2), 6.82 (1H, *t*, $J = 7.5$ Hz, H-5′), 6.84 (1H, *t*, $J = 1.6$ Hz,

H-2′), 6.97 (1H, *br s*, H-6′), 6.99 (1H, *br d*, J = 1.8 Hz, H-7), 7.08 (1H, *t*, J = 1.8 Hz, H-8), 7.09 (1H, *t*, J = 1.8 Hz, H-6), 7.10 (1H, *br s*, H-5); δ_C (75 MHz, acetone-d_6): 24.8 (C-4), 29.5 (C-3), 55.2 (MeO-3′), 55.3 (MeO-4′), 77.4 (C-2), 110.1 (C-2′), 111.6 (C-8), 116.5 (C-5′), 118.3 (C-6), 120.0 (C-6′), 122.0 (C-4a), 127.0 (C-7), 129.5 (C-5), 134.6 (C-1′), 149.1 (C-4′), 149.4 (C-3′), 155.3 (C-8a); HRMS (EI) found M^+, 270.1258 $C_{17}H_{18}O_3$ requires 270.1256.

Flavan **13a**; brown gum, υ_{max} (neat): 2937, 2844, 1611, 1519, 1445 cm^{-1}; δ_H (300 MHz, acetone-d_6): 2.08 (1H, *m*, H-3a), 2.16 (1H, *m*, H-3b), 2.78 (1H, *m*, H-4a), 2.97 (1H, *m*, H-4b), 3.77 (6H, *s*, MeO-3′ and 4′), 3.82 (3H, *s*, MeO-8), 5.00 (1H, *dd*, J = 3.3, 10.2 Hz, H-2), 6.68 (1H, *dd*, J = 2.1 7.2 Hz, H-6), 6.77 (1H, *dd*, J = 2.7, 7.8 Hz, H-6′), 6.79 (1H, *dd*, J = 2.1, 7.2 Hz, H-7), 6.97 (1H, *d*, J = 7.8 Hz, H-5′), 7.00 (1H, *dd*, J = 2.1, 7.2 Hz, H-5), 7.10 (1H, *d*, J = 2.7 Hz, H-2′); δ_C (75 MHz, acetone-d_6): 19.6 (C-4), 24.8 (C-3), 55.22 (MeO-4′), 55.23 (MeO-3′), 55.4 (MeO-8), 77.3 (C-2), 109.9 (C-7), 110.3 (C-2′), 111.7 (C-5′), 118.4 (C-5), 119.5 (C-6′), 121.2 (C-6), 122.6 (C-4a), 134.6 (C-1′), 145.0 (C-8a), 148.8 (C-4′), 149.0 (C-3′), 149.4 (C-8); HRMS (EI) found M^+, 300.1363. $C_{18}H_{20}O_4$ requires 300.1362.

16.4 Conclusions

The use of $NaBH_4$ in the reduction of both carbon-carbon double bond and keto-group of chalcones has been described. This reduction reaction has been shown to be central in the synthesis of flavans from chalcones. The use of chalcones for the synthesis of flavans is also discussed.

Acknowledgements We thank the Royal Society of Chemistry for a research grant, the University of Botswana for a studentship (to OM), Mr D. Mosimaneathebe for MS experiments and Mr Marape for NMR experiments. We also thank Dr Patrick G. Steel of the University of Durham, UK for donating chromatography columns to our laboratory.

References

1. Nieminen TEA, Hase TA (1987) Selective reduction of ketones with sodium borohydride-acetic acid. Tetrahedron Lett 28:4725–4728
2. Cho BT, Kang SK, Kim MS, Ryu SR, An DK (2006) Solvent-free reduction of aldehydes and ketones using solid-acid activated sodium borohydride. Tetrahedron 62:8164–8168
3. Brown HC, Wheeler OH, Ichikawa K (1957) Chemical effects of steric strains: XIII. Kinetics of the reaction of sodium borohydride with carbonyl groups: a convenient tool for investigating the reactivities of aldehydes and ketones. Tetrahedron 1:214–220
4. Kim J, De Castro KA, Lim M, Rhee H (2010) Reduction of aromatic and aliphatic keto esters using sodium borohydride/MeOH at room temperature: a thorough investigation. Tetrahedron 66:3995–4001

5. Periasamy M, Thirumalaikumar M (2000) Methods of enhancement of reactivity and selectivity of sodium borohydride for applications in organic synthesis. J Organomet Chem 609:137–151
6. Bowden K, Hardy M (1966) The reduction of substituted acetophenones by sodium borohydride. Tetrahedron 22:1169–1174
7. Fortunato JM, Ganem B (1976) Lithium and potassium trialkylborohydrides. Reagents for direct reduction of alpha, beta-unsaturated carbonyl compounds to synthetically versatile enolate anions. J Org Chem 41:2194–2200
8. Xue JJ, Zhang XS, Liang XZ, Li Y (2003) The first total synthesis of (±)-3′,7-dihydroxy-4′-methoxyflavan. Chinese Chem Lett 14:443–444
9. Mazimba O, Masesane IB, Majinda RR (2011) An efficient synthesis of flavans from salicylaldehyde and acetophenone derivatives. Tetrahedron Lett 52:6716–6718
10. Aron PM, Kennedy JA (2008) Flavan-3-ols: nature, occurrence and biological activity. Mol Nutr Food Res 52:79–104
11. Li J, Jiang Y (2007) Litchi flavonoids: isolation, identification and biological activity. Molecules 12:745–758
12. Heim KE, Tagliaferro AR, Dennis J, Bobilya DJ (2002) Flavonoid antioxidants: chemistry, metabolism and structure-activity relationships. J Nutr Biochem 13:572–584
13. Waffo-Téguo P, Hawthorne ME, Cuendet M, Mérillon JJ, Kinghorn AD, Pezzuto JM, Mehta RG (2001) Potential cancer-chemopreventive activities of wine stilbenoids and flavans extracted from grape (*Vitis vinifera*) cell cultures. Nutr Cancer 40:173–179
14. Cushnie TPT, Lamb AJ (2005) Antimicrobial activity of flavonoids. Int J Antimicrob Agents 26:343–356

Chapter 17
Workshop on Unlocking the Potential for Low-Cost Teaching in Third World Countries

Jared C. Ogunde, Antony J. Rest, and Raymond G. Wallace

Abstract The sciences, i.e. Biology, Chemistry and Physics, together with Mathematics are regarded as *difficult* by students, especially in third World Countries, with the consequences that the number of students taking them was relatively small, courses are expensive to stage, students get low grades, and the number of students embarking on careers in these subjects is small whereas the needs for indigenous scientists, engineers, technicians and medical practitioners are huge. A Workshop was organised at ICPAC 2012 to highlight and address the problems and to promote "*hands-on*" Chemistry and "*live*" Chemistry via multimedia resources. In order that these resources can be available to both urban and rural students, the participants were introduced to the Chemistry Aid solar classrooms whereby rural students could gain access to IT and modern teaching and learning via low cost solar equipment which could be used both for schools and in their communities. Other modern technology was also demonstrated which could enable on-line resources to be customised for science curricula. To enable students to realise that Chemistry is all around them, low

J.C. Ogunde (✉)
Scientific Advisory and Information Network (SAIN) and Chemistry Aid Kenya, Makongeni, Fatima Court, 3rd Fl. Suite 19, Marcus Garvey Rd. off Argwings Khodhek Rd., P.O. Box 120–00510, Nairobi, Kenya
e-mail: ogundecj@yahoo.com

A.J. Rest
Chemistry Video Consortium, Educational Techniques Group Trust of the Royal Society of Chemistry (UK) and Chemistry Aid, University of Southampton, University Road, Southampton SO17 1BJ, UK
e-mail: ajr@soton.ac.uk; www.Chemistry-aid.com; www.etgt.co.uk

R.G. Wallace
Educational Techniques Group Trust of the Royal Society of Chemistry (UK), Chemistry Aid, and School of Science and Technology, Nottingham Trent University, Clifton Lane, Nottingham NG11 8NS, UK
e-mail: ray.wallace@ntu.ac.uk; dr_ray_wallace@hotmail.com; www.Chemistry-aid.com; www.etgt.co.uk

M. Gupta Bhowon et al. (eds.), *Chemistry: The Key to our Sustainable Future*,
DOI 10.1007/978-94-007-7389-9_17,

cost resources and experiments using chemicals available in kitchen and supermarkets were demonstrated in an interactive session. The practicalities of making the new resources available via grants and sponsorship were part of a "brainstorming" session. Feedback from participants was very positive about solar classrooms and "hands-on" Kitchen and Supermarket Chemistry using plastic equipment.

17.1 Introduction

A principal goal for Governments of Developing Countries is to develop an indigenous body of scientists, engineers, industrialists, medical practitioners and technicians.

Some problems with this goal are that:

a. Science subjects, e.g. Biology, Chemistry, Mathematics and Physics, are regarded as difficult by students;
b. Traditional science subjects, e.g. Biology, Chemistry, Mathematics and Physics, are expensive to teach in schools, colleges and universities;
c. Students in rural villages have no electricity and water with which to support the teaching of science subjects;
d. The grades achieved by students studying Biology, Chemistry, Physics and Mathematics at all stages are very much lower than those for Humanities and the number of students achieving A grades in science subjects is low (Fig. 17.1).

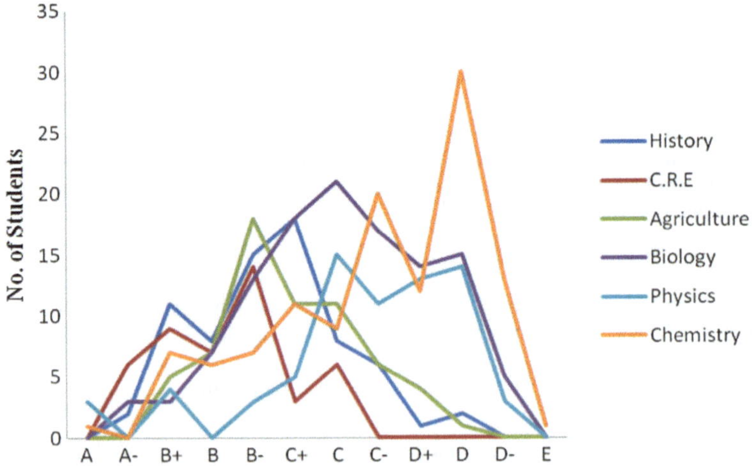

Fig. 17.1 Subject grade distribution in chemistry, physics, biology, agriculture, C.R.E (Christian Religious Education) and history for the Kenya Certificate of Secondary Education in a rural school in Western Kenya for 2010

17.1.1 How Can These Problems Be Overcome?

The aim of the Workshop was to share experiences and innovations for supporting the teaching and learning of Chemistry in Developing Countries, especially for students in schools in rural regions and their communities. Pilot studies have been carried out by the Scientific Advisory and Information Network of Kenya (SAIN), the Chemistry Video Consortium (CVC) and Chemistry Aid in Kenya, Uganda and Zambia (2003–2012). These pilot activities provided the basis for activities and discussions in the workshop.

17.2 Background

In 2003 the Scientific Advisory and Information Network of Kenya (SAIN) contacted the Chemistry Video Consortium (CVC), a project team based at the University of Southampton, developing multimedia resources for teaching Chemistry in the UK, to ask for assistance in supporting Chemistry teaching in Kenya. In collaboration with the Education Department of the Royal Society of Chemistry, the CVC forwarded books and multimedia resources and the Chemistry Aid project was born [1].

In subsequent years the CVC and SAIN explored a number of ways to enable students in Africa and Developing Countries to access modern aspects of teaching and learning Chemistry, Physics and Biology through "hands-on" pilot schemes (see below).

Microscale Chemistry. Microscale Chemistry [2] has the potential to drastically reduce the costs of equipment and chemicals needed for practical Chemistry and to enable experiments which normally require fume cupboards to be carried out on the open bench (Fig. 17.2). Pilot evaluation was carried out in Kenya in 2004 and seemed promising. Further research revealed, however, that the manufacture of Microscale equipment in East Africa would not be feasible and so this approach was abandoned.

Transfer of Equipment from UK. Many companies and institutions in the UK offer redundant equipment for recycling in the UK and overseas. This recycling is primarily carried out by a charity called LabAid (www.labaid.org). For example,

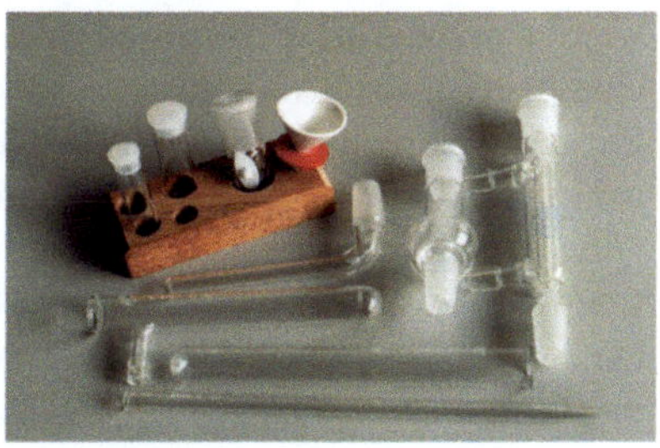

Fig. 17.2 Microscale equipment; taken from CD ROM number 11 of "Practical Laboratory Chemistry" produced by CVC

Fig. 17.3 Science bus used for chemistry road shows in Kenya

a consignment of equipment was sent to the Londiani Township Secondary School in Kenya for use in neighbouring schools and colleges in 2010. This was a very successful exercise and other "pump priming" shipments could be envisaged. It would not be feasible, however, for LabAid to service the needs of all Developing Countries.

Multimedia Resources. The CVC carried out a pilot Road Show in Kenya, Uganda and Zambia in 2005 to evaluate the potential for using multimedia resources to provide "*live*" Chemistry experiments and demonstrations of practical Chemistry experiments [3, 4] (Fig. 17.3). It was found that while schools and colleges in towns and cities had mains electricity supplies to run multimedia equipment (the video equipment was of low basic standard, e.g. VHS video tape), in the rural villages there was no mains electricity and a petrol generator was needed to show the videos. A further Road Show was carried out in 2009. In this Road Show it was found that CDROMs could now be played in towns and cities and that most urban students had mobile phones. For rural villages, however, there was still no mains electricity and the prospect of such a fundamental energy supply in the near, medium and long term future remains poor.

Use of Plastic Equipment. Chemistry experiments traditionally use glass apparatus. In Developing Countries glassware is costly, easily broken and difficult to replace. This raises problems for schools and colleges for students sitting practical examinations as part of national examinations. It was demonstrated during Road Shows that inexpensive plastic equipment, readily obtainable from market stores and supermarkets, can be used as a substitute for glassware in routine Chemistry practical work. Additionally, plastic trays/sheets and Ziploc plastic bags can be used in place of traditional laboratory benches (Figs. 17.4, 17.5 and 17.6). In the case of the former, small drops of reagents can be used, thus keeping down costs (Johnson JM, personal communication, johnson.j.m@btinternet.com) and plasticware can be safely disposed of. In the case of the latter, kitchen chemicals (see below), local minerals, and local fruits, flowers and vegetables can be used. Experiments using gases can be devised safely and easily and in tiny quantities using plastic Petri dishes [5]. Improvements in practical Chemistry facilities and "*hands-on*" experiments have resulted in dramatic improvements in motivation and Chemistry examination grades, e.g. from D to B, in a pilot project in rural schools in the Kabale district of Uganda (Johnson JM, personal communication, johnson.j.m@btinternet.com).

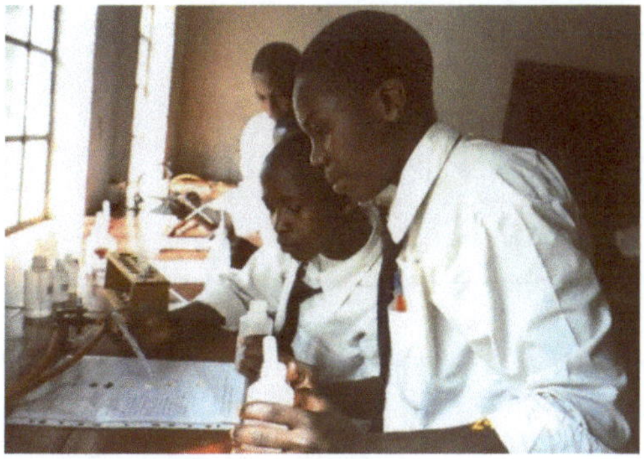

Fig. 17.4 Students in Uganda using plastic sheets for chemical reactions

Fig. 17.5 Students in a chemistry club in the UK using ziploc plastic bags

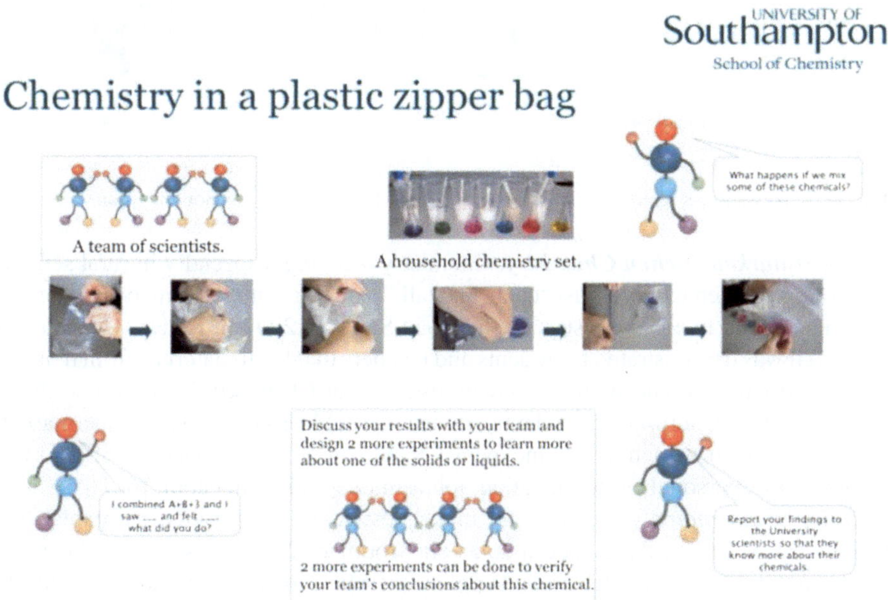

Fig. 17.6 Interactive open day demonstrations at the University of Southampton

News release on the Science Daily

December 21, 2011 | Author admin

ScienceDaily®

Your source for the latest research news

This week the Solar Powered Classroom sprung up on the 'Science Daily' site.

Click here to view the webpage.

Posted in Uncategorized | No Comments »

Solar Energy Generator Press Release

December 16, 2011 | Author admin

Southampton University Media Centre has just sent out a Press Release on the Solar Energy Generator, the press release reads:

A major difficulty in teaching science subjects in Developing Countries, especially in rural schools, is that students are rarely able to get "hands-on" experience of experiments. In the case of Chemistry this is partly because of a lack of equipment, chemicals and facilities but mainly because of a lack of electricity and running water.

Keith Wilkinson, a teacher at the International School at Lusaka, Zambia, and Professor Antony Rest of Southampton University have devised a solar powered solution based on a Dell M109S low powered data/video projector (47 watt) and low cost solar energy panels. Most current data/video projectors require 200-300 watt and cannot be economically sustained for many minutes.

A paper on the solar powered classroom has been presented and demonstrated at the 5th International Conference on ICT for Development, Education and Training at Lusaka in May 2010. (Conference abstract)

Contacts: krwilkinson@live.com and ajr@soton.ac.uk

Fig. 17.7 News/press releases from science daily and the University of Southampton

Supermarket/Kitchen Chemistry. Chemistry teaching, especially in rural schools and colleges, often involves just "chalk and talk" teaching with learning by rote and no experiences of "live" Chemistry. In the Road Show in 2007 in Kenya, Uganda and Zambia it was demonstrated to students and teachers that local materials from markets and supermarkets could be used to demonstrate that Chemistry is all around them. This was often a surprise to students and teachers. Materials for such supermarket/kitchen Chemistry teaching are inexpensive and can be easily disposed of after use [Johnson JM, personal communication, johnson.j.m@btinternet.com; [6, 7]].

Solar Classrooms. In order to make resources universally available, the Chemistry Aid team developed a solar classroom in 2010 [8] so that students in towns and cities, where electricity supplies are often intermittent, and particularly for students in rural villages, with no mains electricity, can have access to modern teaching and learning and to IT facilities (Figs. 17.7, 17.8 and 17.9;

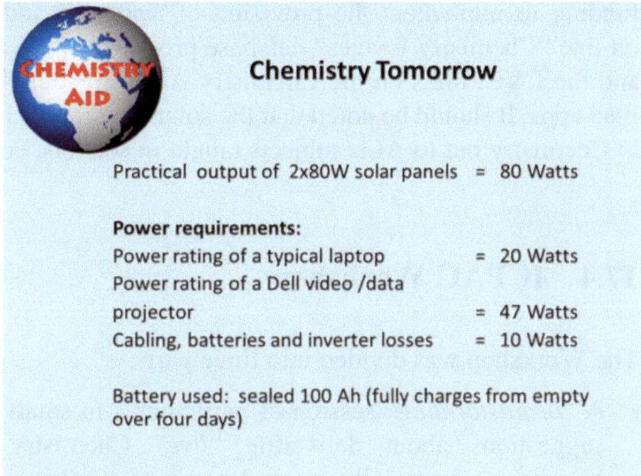

Fig. 17.8 Power balance between solar panels and solar classroom equipment

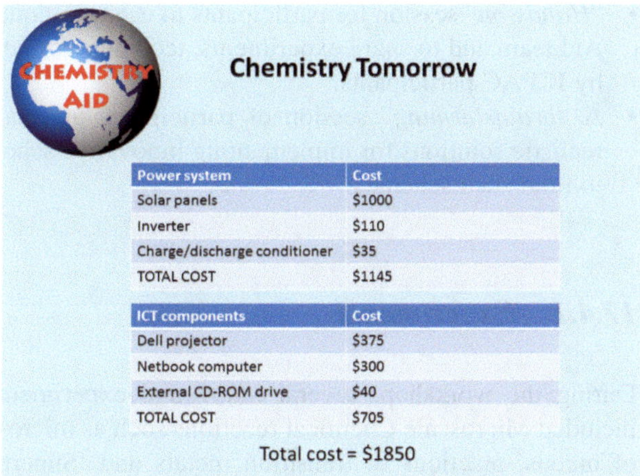

Fig. 17.9 Typical costs of solar classroom equipment

www.Chemistry-aid.com/news). In addition to the solar classrooms being used in the daytime in schools, such facilities can be used in the evenings and weekends for local communities, e.g. for training, information about agriculture and health, and for social events. Request for information about solar classrooms have been received from many countries, e.g. Burma, Cameroon, Haiti, India, Malawi, Mauritius, and Thailand.

17.3 The Future

A basic principle of the work of Chemistry Aid is that resources should be available to ALL students. It was not until the breakthrough with solar classrooms that this principle was achieved. The next steps in various countries will be to generate funding to implement the provision of resources and facilities, e.g. through the on-line "Chemistry Images" database provided by the Royal Society of Chemistry and the CVC, links on the Chemistry Aid web site, FACEBOOK, YouTube, and iPad apps. It should be noted that the solar classroom concept is applicable not only to Chemistry but to ALL subjects taught in schools, colleges and universities.

17.4 ICPAC Workshop

The Workshop was divided into three parts:

- A "*brainstorming*" session of participants in small groups to provide practical suggestions about delivering "*live*" Chemistry and practical Chemistry experiments especially for students in rural schools and colleges where there is no electricity and running water.
- "*Hands-on*" session for participants to use techniques brought by the Chemistry Aid team and to share experiments, techniques, materials and expertise brought by ICPAC participants.
- A "*brainstorming*" session of participants in small groups to come up with realistic solutions for implementing innovative schemes and applying for grants to fund such schemes.

17.4.1 Experiments

During the workshop, several innovative experiments were carried out. These included Microscale Chemical reactions such as micro-scale displacement reaction of metals, reactions of transition metals and 'Supermarket Chemical' reactions (also referred to as Kitchen Chemistry).

Microscale Chemistry Experiments: Displacement Reaction of Metals

In these experiments Microscale reactions between metals and metal salt solutions were carried out. More reactive metals react with less reactive metals in solution.

Worksheet

Plain photocopy papers were provided to the workshop participants. They were required to draw a worksheet with five columns and five rows as shown below (Fig. 17.10).

Fig. 17.10 Worksheet 1

	$CuSO_4$	$Mg(NO_3)_2$	$ZnCl_2$	$Fe(NO_3)_3$
Copper turnings				
Mg ribbon				
Zn granules				
Iron nail NOT steel (or Fe filings)				

Metals were delineated in columns as shown above and the aqueous solutions of the metals, $CuSO_4$, $Mg(NO_3)_2$, $ZnCl_2$ and $Fe(NO_3)_3$ appear in rows (Fig. 17.10). The Worksheets were placed in A4 plastic wallets.

Procedure

The metals, copper, magnesium, zinc and nail were placed in the corresponding column on the Worksheet. Two drops of copper sulphate solution were then added to each metal in the $CuSO_4$ column. The procedure was then repeated for $Mg(NO_3)_2$, $ZnCl_2$ and $Fe(NO_3)_3$ solutions. A few minutes were allowed for reactions to take place.

Small Scale Chemistry Experiments: Typical Reactions of Transition Metals

Requirements

The chemicals required for these experiments are 0.1 M solutions of the transition metal cations: Cr^{3+}, Mn^{2+}, Fe^{2+}, Fe^{3+}, Co^{2+}, Ni^{2+}, Cu^{2+}, and Zn^{2+}. In addition 2.0 M solutions of NaOH and NH_3 are required.

Plain photocopy papers were provided to the Workshop participants. The participants were then asked to make a worksheet as shown below (Fig. 17.11).

Procedure

The Worksheet was then placed in an A4 plastic wallet (Fig. 17.12). A drop of each cation solution was then placed in each box. In the box with the heading 'one drop', a drop of NaOH solution was added. In the other box the NaOH solution was added in excess until some change was observed. This was repeated with NH_3 solution. The participants were asked to observe if any precipitate was formed and whether the precipitate changed when excess of NaOH or NH_3 solutions is added in excess.

Cation	1 drop NaOH	Excess NaOH	One drop NH_3	Excess NH_3
Cr^{3+}				
Mn^{2+}				
Fe^{2+}				
Fe^{3+}				
Co^{2+}				
Ni^{2+}				
Cu^{2+}				
Zn^{2+}				

Fig. 17.11 Worksheet 2

Fig. 17.12 Picture showing results of the experiments

Chemical Reactions in Supermarket Ziploc Bags Experiments

The materials used in these experiments are readily available in supermarkets, are safe and can be easily disposed. In these experiments Small Scale reactions can be used to demonstrate acid/base reactions as well as exothermic and endothermic reactions.

Requirements

In order to carry out these experiments, several reagents were required such as calcium chloride powder, sodium hydrogen carbonate, and citric powder among other chemical reagents. Also required is the red cabbage juice which was made by cutting the red cabbage into small pieces and then boiling in water until the water was dark purple. The purple cabbage juice was used as an acid–base indicator.

Procedure for Exothermic Reaction

A teaspoon of calcium chloride powder was placed into a Ziploc plastic bag and a teaspoon of sodium hydrogen carbonate was also added to the same Ziploc plastic bag. About 15 cm^3 of the purple cabbage juice was placed in a mini plastic cup. The mini plastic cup with the cabbage juice was then placed inside the Ziploc bag. Air was then squeezed out. The purple cabbage juice was then emptied into the Ziploc plastic bag (Fig. 17.5). Observations of what happened were made.

Procedure for Endothermic Reaction

A teaspoon of citric acid powder was placed into a Ziploc bag. In the same bag a teaspoon of baking soda (sodium hydrogen carbonate) powder was added. About 15 cm^3 of purple cabbage juice was placed in a mini plastic cup and then carefully transferred into the bag. Air was then squeezed out and the liquid in the cup emptied into the bag. Observations of what happened were made (Fig. 17.6).

17.4.2 Brainstorming Sessions

The ideas from the Brainstorming sessions were recorded on "*flip charts*":

- *Ways Ahead*

  Use of solar energy to provide electricity for rural schools and colleges so as to enable students to use IT resources and teachers to incorporate multimedia courseware in their teaching, e.g. to provide "live" Chemistry experiments (see www.Chemistry-aid.com/news).

  Use of mobile classrooms, e.g. via Chemistry Bus Road Shows (See Fig. 17.3) to take innovative ideas to students, teachers, parents and educational administrators so as to effect changes in teaching and learning practices.

  Involve students in more "*hands-on*" experiments to increase interest and motivation in Chemistry and other science subjects, e.g. Biology (Institute of Biology, UK) and Physics (Institute of Physics, UK) [9].

  Provide students with examples of multimedia resources on industrial Chemistry via "*Industrial Chemistry for Schools and Colleges*" [10] so that students, teachers and parents can relate Chemistry in schools and colleges to the Chemistry in industrial processes which generate wealth for the nation.

  The use of plastic equipment *instead* of glassware, which is easily broken, is costly to replace, and often takes a long time to replace, enables students to have more "hands-on" experiences of Chemistry.

  In addition to support for Chemistry, the Chemistry Bus on its Road Shows could include presentations from other sources, e.g. Water Aid [11] and sanitation [12]. The latter is important because girls in rural areas from the age of 11 can lose up to 5 days per month of schooling because of menstruation.

- *Networking Ideas*
  The Internet and e-mail can be used to provide help and support networks on a national and international basis [13]. For example, there is a Chemistry Teachers network in the United Kingdom in which teachers ask for help and share ideas by submitting them to:

  CHEMISTRY-TEACHERS@YAHOOGROUPS.CO.UK

  The Chemistry Aid web site (www.Chemistry-aid.com) could be used to share news and innovations about Chemistry teaching.

  Networking sites can also serve as a database for lists of e-mails of teachers and lecturers.
- *Use of Modern Technology*
  The Workshop participants discussed the use of modern technology, e.g. mobile phones, YouTube, FACEBOOK, iPad to download and transfer resources.

  Rather than each school, college, and university downloading its own resources, it will almost certainly be best to create a "*National Centre*" which can download from on-line sources, e.g. the Royal Society of Chemistry Learn-Chemistry site, and re-master them according to the needs of a national curriculum and to produce packages of resources in the format and language appropriate to that country.
- *Funding*
  The Workshop participants discussed the use of national and international funding in order to progress the implementation of innovative schemes, e.g. via grants, sponsorship and discounting arrangements from manufacturers for the cost of equipment.

17.5 Feedback from Workshop Questionnaires

Positive feedback has been received primarily concerning the use of solar classrooms and supermarket/kitchen Chemistry.

17.6 Conclusions

It was concluded from the "flip chart" comments that:

- Many opportunities already exist for students and teachers to obtain improved and more interesting resources but these would be enhanced, especially for rural regions, by the use of solar classroom technology to support teaching and learning of Chemistry
- National and international resources exist for funding from various funding bodies and institutions but that such resources need to be made more available to Developing Countries

- It is a vital requirement in harnessing both the above conclusions that good communication structures are implemented and disseminated via Networking amongst teachers and administrators
- In the limited time available in the Workshop, there could only be a "taster" of the types of materials, experiments, technology, techniques and funding that are available but that, through the Bibliography (see References) and the Internet, participants could share and network resources and deliver improved Chemistry teaching and learning in schools, colleges and universities in Developing Countries.

Acknowledgements The authors would like to acknowledge support from Chemistry Aid colleagues, the Royal Society of Chemistry, the Educational Techniques Group Trust (a sub-group of the Educational Division of the Royal Society of Chemistry), the Chemistry Video Consortium (CVC), the University of Southampton, Nottingham Trent University, Chemistry Aid and Chemistry Aid colleagues, and the Scientific Advisory and Information Network (SAIN) of Kenya. The authors would also like to thank the Organising Committee of ICPAC for the opportunity to present the Workshop at ICPAC 2012 and for the supply of chemicals and equipment used in the Workshop.

References

1. Osborne CJ, Rest AJ (2004) Chemistry aid. Royal Society of Chemistry News, June issue, p 7
2. Skinner J (1998) Microscale chemistry. Royal Society of Chemistry, UK. www.amazon.co.uk/Microscale-Chemistry-Experiments-John-Skinner/dp/1870343492 and www.Microscale.org/about.asp
3. Rest AJ (1998) Practical laboratory chemistry. Chemistry Video Consortium and Viewtech Educational Media, see www.Chemistry-aid.com/products and www.viewtech.co.uk
4. Lister T, Rest AJ (2000) Practical chemistry for schools and colleges. Chemistry Video Consortium and Royal Society of Chemistry, see www.Chemistry-aid.com/products
5. Skinner J (1998) Microscale chemistry. Royal Society of Chemistry, London, experiments 24–30
6. Selco JI, Bruno M, Chan S (2012) Students doing chemistry: a hand-on experience for K-12. J Chem Educ 89:206–210
7. Lister T, Blumenthal H (2005) Kitchen chemistry. Royal Society of Chemistry, UK, see www.rsc.org/learn- Chemistry/resource/listing?searchtext = Kitchen + Chemistry
8. Jadoloh A, Johnson J, Ogunde J, K'osura H, Rest T, Wilkinson K (2010) Using solar generators to support IT and "hands-on" science in developing countries. In 5th international conference on ICT for development, education and training in Africa, Lusaka, Zambia, paper D5, p 373. www.Chemistry-aid.com/news
9. Brattan D, Burdett N, Curry P, Jadoloh A, Johnson J, Abu Bakar M, Ogunde J, Rest T, Selco J, Sleightholme G, Wallace R, Wilkinson K (2009) High-tech and low-tech solutions to providing "hands-on" experiences of chemistry. Paper S301_003, 42nd IUPAC conference, Glasgow
10. Lister T (2007) Alchemy? Chemistry and industrial processes for schools and colleges. Royal Society of Chemistry, see www.docstoc.com/330355/FOR-SCHOOLS-AND-COLLEGES
11. Water Aid – Clean water and sanitation for Africa, Asia and South America. www.wateraid.org

12. Global Finnish Dry Toilet project. www.drytoilet.org/intro.html. Accessed Oct 2013
13. Elecanal-Guavais AM, Bin Hitam R, Rest A (2010) Supporting chemistry in South East Asia via chemistry aid. Asian Pacific Education Research Conference, Kuala Lumpur. Accessed Oct 2013

Chapter 18
Percolation Studies of Single- and Multi-Walled Carbon Nanotubes/Poly (methyl methacrylate) Nanocomposites

Riyadh M. Mungur and Soonil D.D.V. Rughooputh

Abstract Using an ultrasonication method, the dispersion of single-walled carbon nanotubes (SWCNTs) and multi-walled carbon nanotubes (MWCNTs) in a polymer matrix, poly(methyl methacrylate) (PMMA), were used to produce SWCNT/PMMA and MWCNT/PMMA composites. Composites were made using purified SWCNTs and MWCNTs with different SWCNT and MWCNT loadings (ranging from 0.2 % to greater than 2.0 % by weight) and viscous Plexiglas® (PMMA) solutions. DC electrical conductivity measurements were performed on high quality samples, attaining conductivities of the order of 10^3 S/m for high loading. The electrical percolation threshold (φ_c) and the critical exponent, t, for SWCNT/PMMA and MWCNT/PMMA samples have been obtained. Our measured values are consistent and within the range reported in the literature. The t values obtained by our method are consistent with a quasi-two dimensional network for MWCNT nanocomposites and three-dimensional network for SWCNT nanocomposites.

18.1 Introduction

The concept of nanotechnology originated with the American physicist Richard P. Feynman in a talk to the American Physical Society in December 1959, entitled "*There's Plenty of Room at the Bottom*" [1]. Nanotechnology can, in one way, be defined as the ability to engineer and explore new attributes through controlling characteristics at a relatively very small scale – at the atomic level [1].

In the last decade, much attention has been given to the use of carbon nanotubes (CNTs) as filler in conductive nanocomposites in order to harness their exceptional electrical properties [2]. By now, more than 200 publications have been reported on the electrical percolation threshold of CNT embedded in different polymer

R.M. Mungur • S.D.D.V. Rughooputh (✉)
Department of Physics, Faculty of Science, University of Mauritius, Réduit, Mauritius
e-mail: mungur.riyadh281@gmail.com; rughooputh.sddv@gmail.com

M. Gupta Bhowon et al. (eds.), *Chemistry: The Key to our Sustainable Future*,
DOI 10.1007/978-94-007-7389-9_18,

systems [3]. Due to the much higher electrical conductivity and aspect ratio of CNTs, outstanding electrical properties can be achieved with CNT nanocomposites containing far less filler material than usual fillers like short carbon fibers or carbon flakes. Potential applications of nanocomposites as functional materials include organic field emitting displays, photovoltaic cells, highly sensitive strain sensors, and, because of their capability of dissipating electrostatic charges, as shielding devices from electromagnetic radiation. Significant research efforts have been directed towards the synthesis and characterization of CNT – polymer nanocomposites to further improve and explore their functional and mechanical properties with the aim of identifying their potential capabilities in several different applications [4].

The aim of writing this paper is threefold. Firstly, there is still a lot of controversy in the experimental determination of critical exponent values for SWCNT and MWCNT composites as reported in the literature. Secondly, we wanted to situate our experimental values using a specific procedure of preparing the high quality films of CNT embedded in a polymer (i.e. PMMA) which have not been reported yet, and, thirdly, to gain additional insight in the theory and in the design aspects of functional nanocomposites.

18.2 Theory

Percolation theory deals with the effects of varying, in a random system, the richness of interconnections present [5]. The intellectual appeal of the percolation model resides in its almost game-like mathematical aspects and the fact that it provides a well-defined and intuitively satisfying model for spatially random processes [5]. Many authors have reported that the electrical behavior of polymer nanocomposites filled with CNTs turns out to be very complex, especially in the narrow region around the percolation threshold, which seems to strongly depend on the processing conditions. The single most seductive aspect of the percolation model is the presence of a sharp phase transition at which long-range connectivity suddenly appears [5].

The well-known scaling law describing the statistical percolation behaviour in the vicinity of the percolation limit, which refers to a situation where randomly distributed conducting elements form percolating paths [6]:

$$\sigma = \sigma_0(\varphi - \varphi_c)^t \tag{18.1}$$

where σ_0 is a scaling factor that may be comparable to the effective conductivity of the filler, φ is the CNT weight fraction, and φ_c is the percolation critical concentration. The parameter t is a critical exponent that governs the scaling law in the vicinity of percolation and has been associated to the dimensionality of the system [4].

18.3 Experimental

18.3.1 Materials

In Table 18.1, we give the physical properties of:

(a) Single-walled carbon nanotubes (SWCNTs) 8,653 M/MJ supplied from Carbolex Inc.® and,
(b) Multi-walled carbon nanotubes (MWCNTs) M7110328 supplied from Timesnano Company® (China).

PMMA Plexiglas® (density: 1.19 g/cm^3) was used in pellet form.

Analytical chloroform ($CHCl_3$) ReagentPlus® (Sigma-Aldrich Co. LLC®) was used as solvent to dissolve PMMA and to disperse the carbon nanotubes.

Silver Conductive Paint (Kemo L100™) highly conductive approximately 0.02–0.1 Ω/cm^2 was supplied from Kemo-Electronic GmbH Leher Landstraße 20 D-27607 Langen®.

18.3.2 Sample Preparation

The method used in fabricating thin films of CNT-reinforced PMMA composites is sonicated solution mixing and casting. All glassware was carefully washed with distilled water followed by ethanol for removal of possible contaminants. Different masses of PMMA pellets are weighed and left to dissolve in chloroform for 24 h and kept in vials. Different masses of CNT powder are weighed and stored in separate vials. The CNT powder is poured into chloroform and underwent an ultrasonication process for 30 min. Then, this ultrasonicated solution of CNT is then added to the viscous solution of PMMA and further ultrasonicated for 30 min. The resulting solution is then used to cast a thin film on microscope slides kept at a temperature of 50 °C. At this temperature, high quality homogeneous films of the nanocomposites were obtained free from bubbles.

Table 18.1 Physical properties of CNTs

	CNT	
Properties	SWCNT	MWCNT
Outer diameter	0.7–1.4 nm	30–50 nm
Length	0.5–3 μm	10–20 μm
Purity	>90 wt%	>95 wt%
Density	1.80 g/cm^3	1.66 g/cm^3
Ash	<1.2 wt%	<1.5 wt%
Specific surface area	>60 m^2/g	>60 m^2/g
Electrical conductivity	>10^2 S/cm	>10^2 S/cm

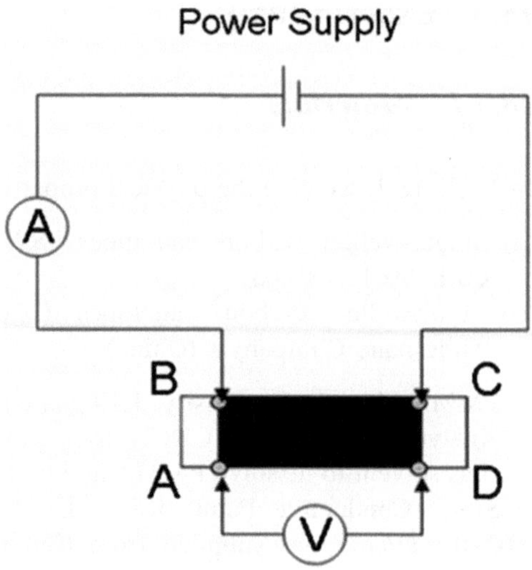

Fig. 18.1 Schematic circuit for four-probe electrical conductivity measurement

18.3.3 DC Conductivity Measurements

After the application of silver conductive paint acting as electrical contacts on the four corners of the thin film CNT-reinforced PMMA composites, four-point probe dc conductivity measurements were conducted on the rectangular thin films using the Van der Pauw technique [7] as shown in the circuit in Fig. 18.1. Through a record of a series of ammeter and voltmeter readings conducted at room temperature and pressure, the critical parameters φ_c, σ_0 and t were determined via graphical illustrations (see Sect. 18.4).

18.3.4 Size Distributions

Dilute suspensions (of SWCNT and MWCNT), of the order of 1.0 % by volume as per the requirements of DLS Particle Size Analyzer (90 Plus Particle Size Analyzer, Brookhaven Instruments Corporation™, USA) were prepared and the latter was used to characterize the particle size and distribution of SWCNT and MWCNT, separately. Analyses were performed at 25 °C with angle of detection of 90°.

Figures 18.2 and 18.3 show respectively the histograms of SWCNT and MWCNT agglomerate sizes measured by particle size analyzer. In this work, size is considered as the 'effective diameter of CNT agglomerates' and is given as the corresponding mean value ± one standard error after dispersion in chloroform by using the method described as follows: *Ultrasonication of CNTs in addition to PMMA in* $CHCl_3$ *for 1 h.*

For SWCNT and MWCNT, a mean value of 125.4 ± 10.3 nm (polydispersity of 0.203) and 114.7 ± 6.3 nm (polydispersity of 0.161) was obtained respectively.

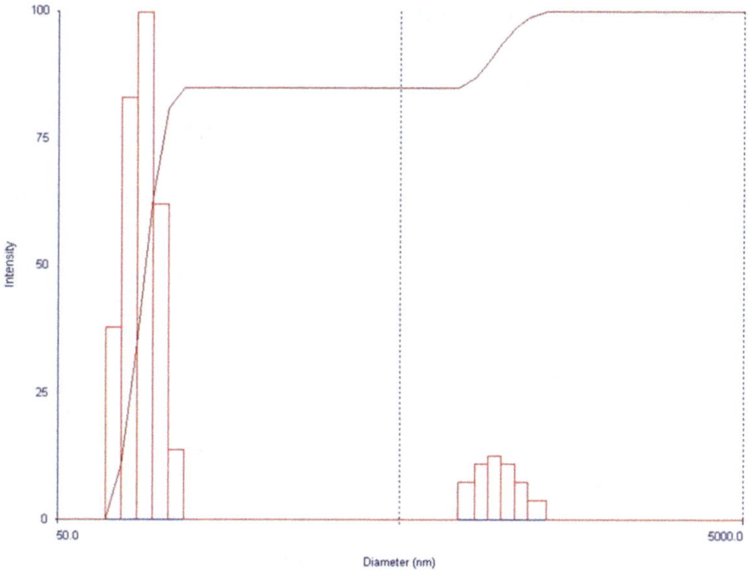

Fig. 18.2 Size distribution of SWCNT

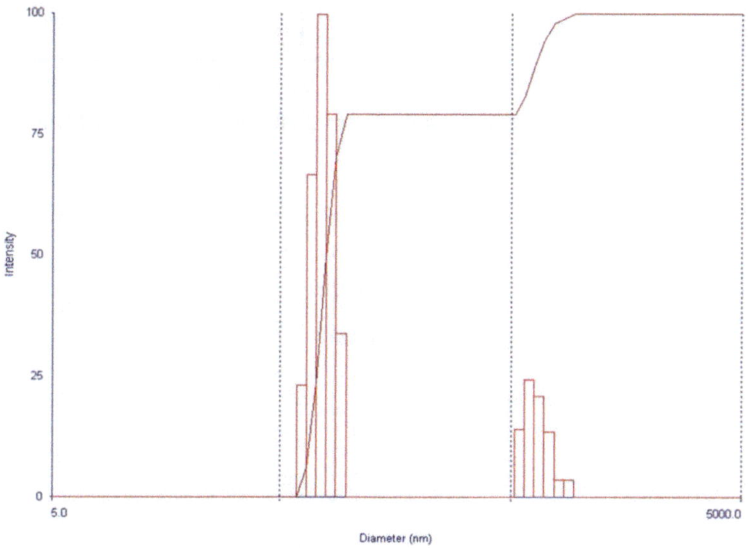

Fig. 18.3 Size distribution of MWCNT

18.4 Results and Discussion

The variation of direct current conductivity of the:

- SWCNT/PMMA composite films with varying SWCNT concentration is shown in Fig. 18.4 (top).
- MWCNT/PMMA composite films with varying MWCNT concentration is shown in Fig. 18.4 (bottom).

The overall behavior of conductivity variation can be explained in terms of percolation theory. At low concentration of CNT, proper percolation paths are not

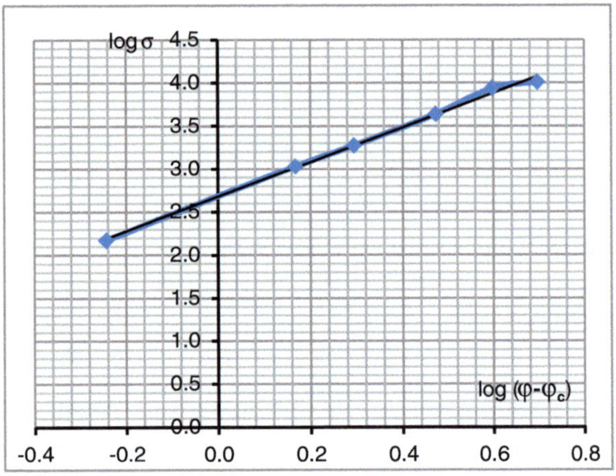

(For φ_c= 0.03; t = 2.004 and log σ_0= 2.6814; R^2 = 0.9964)

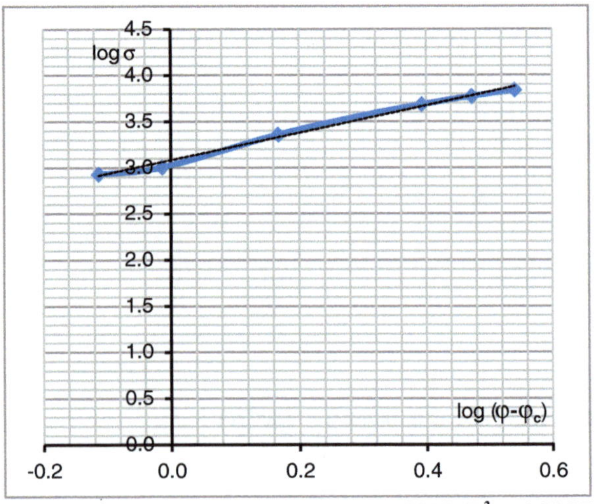

(For φ_c = 0.03; t = 1.469 and log σ_0= 3.0838; R^2 = 0.9895)

Fig. 18.4 Graph of log σ against log (φ−φ$_c$) (*top*: SWCNT/PMMA composites; *bottom*: MWCNT/PMMA composites)

set up by the carbon nanotube network. As the concentration of the CNT in the composite increases, the percolation paths via conducting carbon nanotubes are set up whence the nanotubes form a conducting network in the polymer matrix. From the electrical conductivity plots of SWCNT/PMMA and MWCNT/PMMA composites, the percolation threshold, φ_c, the critical exponent and the scaling factor can be extracted, especially with the help of a log-log plot of conductivity against $(\varphi-\varphi_c)$ as shown in Fig. 18.4. The latter provides as slope, the t value and the y-intercept, $\log \sigma_0$.

Reported ranges for SWCNT composites in the literature are: $\varphi_c \sim 0.17$–1.3 and $t \sim 1.3$–2.3 and, for MWCNT composites, $\varphi_c \sim 0.003$–0.65 and $t \sim 1.8$–2.3. A few researchers have reported values of t of ~ 4.9 and even higher than 7 [8, 9]. Moreover, Stauffer [6] showed that the critical exponent t depends on the dimensionality of the system, with values of $t \sim 1.3$ (or slightly higher) representing a two-dimensional network and $t \sim 2$ (or slightly higher) a three-dimensional one. A lower value of t is generally ascribed to a percolation network consistent with a large number of dead arms and with a network that deviates from a classical random network. The value of percolation threshold seems to depend on the type of nanotubes used (such as single walled, double walled or multi walled carbon nanotubes) and on the bundle size of the nanotubes actually taking part in the conduction.

Whilst the value of φ_c is traditionally taken to be the onset of the sharp transition read off from the linear plot, this practice is not correct and will be the subject of a detailed study in a following publication. We believe that this has led to the reporting of many erroneous values of the critical exponent, scaling factor and the percolation threshold in the literature; hence, the controversial nature of the physical interpretation of the critical exponent t for polymer nanocomposites filled with CNTs [4].

The electrical percolation threshold (φ_c) and the critical exponent for SWCNT/PMMA and MWCNT/PMMA samples measured in this work are as follows: For SWCNT/PMMA: $\varphi_c \sim 0.03$ % by weight, $t \sim 2.00$ and $\sigma_0 \sim 480$ S/m. For MWCNT/PMMA: $\varphi_c \sim 0.03$ % by weight, $t \sim 1.47$ and $\sigma_0 \sim 1{,}213$ S/m. Based on the dimensional dependence [4], the t values obtained by our method are consistent with a 3D network for SWCNT nanocomposites and a quasi-2D network for MWCNT nanocomposites.

18.5 Conclusions

Conductivities of systems of SWCNT/PMMA and MWCNT/PMMA nanocomposites increase due to the introduction of conducting paths in the polymer matrix. Percolation theory and scaling laws are applicable to CNT-based composites, with enhanced maximum conductivity achievable at relatively low filler concentrations will, no doubt, be important for many applications. The concentration of nanotubes required for the unique region from insulator to conductor transition referred to as percolation threshold concentration or percolation limit, the critical exponent and the scaling factor have been deduced experimentally and are consistent to those of published [4–9]. Based on the dimensional dependence, the critical exponents are consistent with a polymer matrix system adopting a 3D/quasi2D conducting network for SWCNT/MWCNT nanocomposites, respectively. We can therefore conclude that from our results and those reported in the literature that the variation in the values obtained for electrical percolation threshold, φ_c and critical parameters, t and σ_0 depend on the polymer matrix, the processing technology, and the type of carbon nanotube utilised. We also believe that many authors have reported wrong values by adopting traditional methods of percolation threshold analysis.

Acknowledgements The authors thank the University of Mauritius for providing the facilities to carry out this work.

References

1. Ramsden J (2009) Essentials of nanotechnology. Bookboon ISBN: 978-87-7681-418-2 1 edition; 126 pp
2. Grossiord N, Kivit PJJ, Loos J, Meuldijk J, Kyrylyuk AV, van der School P, Koning CE (2008) On the influence of the processing conditions on the performance of electrically conductive carbon nanotube/polymer nanocomposites. Polymer 49:2866–2872
3. Bauhofer W, Kovacs JS (2009) A review and analysis of electrical percolation in carbon nanotube polymer composites. Compos Sci Technol 69:1486–1498
4. Li J, Ma PC, Chow WS, To CK, Tang BZ, Kim J-K (2007) Correlations between percolation threshold, dispersion state, and aspect ratio of carbon nanotubes. Adv Funct Mater 17:3207–3215
5. Zallen R (1998) The physics of amorphous solids. Wiley Classics Library, Virginia Tech, Blacksburg
6. Stauffer G (1985) Introduction to percolation theory. Taylor & Francis, London
7. Daghero D (2002) Resistivity measurements: the conventional and van der Pauw techniques. PhD thesis, National Institute for Physics of the Matter, Torino
8. Aguilar JO, Bautista-Quijano JR, Avilés F (2010) Influence of carbon nanotube clustering on the electrical conductivity of polymer composite films. Express Polym Lett 4:292–299
9. Spitalsky Z, Tasis D, Papagelis K, Galiotis C (2010) Carbon nanotube–polymer composites: chemistry, processing, mechanical and electrical properties. Prog Polym Sci 35:357–401

Chapter 19
Chemistry Aid: How Innovative Solutions to Chemistry Education Are Making a Difference

Jared C. Ogunde, Aggrey Omolo, and Antony J. Rest

Abstract In Kenya, as in many developing countries, the government has placed great importance on science education. This importance on science education is difficult to realize, however, because of the challenges facing science education, especially Chemistry education. These challenges include a lack of well equipped laboratories, the high cost of chemicals and equipment, a lack of mains electricity, lack of mains water, and students' lack of laboratory gear such as goggles and lab coats. Many students continue to perform dismally in Chemistry as well as other sciences. In addressing this problem, the Scientific Advisory and Information Network (SAIN) Kenya and the Chemistry Video Consortium (CVC) UK have initiated a program, the Chemistry Aid initiative that has been working on devising and promoting the use of innovative solutions to problems facing Chemistry education in Kenya and other developing countries.

19.1 Introduction

Science education in Kenya is regarded as important to the economic development of the nation. There are however numerous challenges that are experienced in the teaching and learning of sciences in Kenya. This situation is not unique to Kenya alone. Many developing countries also face similar challenges in the teaching and

J.C. Ogunde (✉) • A. Omolo
Scientific Advisory and Information Network (SAIN) and Chemistry Aid Kenya, Makongeni, Fatima Court, 3rd Fl. Suite 19, Marcus Garvey Rd. off Argwings Khodhek Rd., P.O. Box 120–00510, Nairobi, Kenya
e-mail: ogundecj@yahoo.com; omolo1973@yahoo.com

A.J. Rest
Chemistry Video Consortium, Educational Techniques Group Trust of the Royal Society of Chemistry (UK) and Chemistry Aid, University of Southampton, University Road, 82 High Street, West End, SO17 1BJ Southampton, UK
e-mail: ajr@soton.ac.uk; www.Chemistry-aid.com; www.etgt.co.uk

M. Gupta Bhowon et al. (eds.), *Chemistry: The Key to our Sustainable Future*,
DOI 10.1007/978-94-007-7389-9_19,

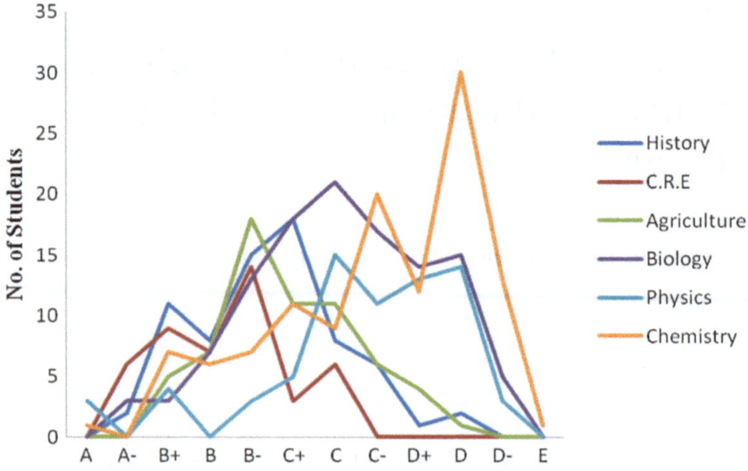

Fig. 19.1 Subject grade distribution in chemistry, physics, biology, agriculture, C.R.E (Christian Religious Education) and history – Kenya Certificate of Secondary Education (KCSE) 2010

learning of sciences especially Chemistry. Chemistry has therefore become unpopular among many students in developing countries. Lack of properly equipped laboratories and high cost of chemicals are some of the challenges that many developing countries are facing in Chemistry education. Also, in many rural schools there is lack of electricity and running tap water. In addition to the above challenges, students find Chemistry to be very abstract. In developing countries the students also lack important gear such as lab coats and goggles [1].

Due to the numerous challenges facing the teaching and learning of Chemistry (as well as other sciences and mathematics), the grades achieved by students in national examinations are lower for these subjects compared to humanity subjects as portrayed by Fig. 19.1.

In order to tackle the challenges faced in Chemistry education, the Scientific Advisory and Information Network (SAIN) a non-profit making, non-governmental organization (NGO) in Kenya has partnered with the Chemistry Video Consortium (CVC) of the University of Southampton in UK to develop and offer innovative solutions to the challenges facing the teaching and learning of Chemistry in developing countries. The two organizations have been working under the banner of the Chemistry Aid initiative (see http://www.chemistry-aid.com) to promote and encourage the use of innovative solutions.

19.2 Innovative Approaches to Chemistry Education

Our innovative approaches to the challenges facing Chemistry education are geared towards making Chemistry 'lively' and therefore very interesting to both teachers and students. These approaches are as follows:

(a) Promoting and encouraging the use of ICT multimedia resources (audio-visual resources) in the teaching and learning of Chemistry;
(b) Promoting and encouraging the use of readily available chemicals and materials that can be used to perform Chemistry experiments;
(c) Use of sporting activities in encouraging Chemistry education;
(d) Devising a simple solar power generator that can be used in schools that do not have mains power connection.

19.2.1 Promoting and Encouraging the Use of ICT Multimedia Resources

For a long time in many Kenyan schools, Chemistry has been taught as a "theoretical" subject making the students to have a feeling that Chemistry is an abstract subject. The lack of properly equipped laboratories has meant that students taking Chemistry have not been able to have a "hands on" experience on Chemistry. In order to overcome this challenge the Chemistry Video Consortium (CVC) has provided ICT multimedia resources, mainly Chemistry CDRoms and DVDs to the SAIN for distribution to schools in order to enhance the teaching and learning of Chemistry in Kenya (Fig. 19.2). These materials have been developed by the Royal Society of Chemistry (RSC), UK and the CVC [2]. Also, the RSC has made Chemistry educational materials available for download in developing countries [3].

The ICT multimedia resources have been found to have several benefits as follows:

(a) They have been found to be very motivating to the students. Through the use of these materials students have been encouraged to learn Chemistry;
(b) Since these resources mainly involved the use of computers students have been motivated to develop ICT skills even when they are out of class or even school;
(c) The use of these resources by students has encouraged independent learning as well as active learning;
(d) The students have been able to see practical experiments that they are not able to perform due to lack of laboratory equipment. The students are therefore able to see "live" Chemistry via the ICT multimedia materials;
(e) The ICT multimedia materials have instructional consistency.

19.2.2 Promoting and Encouraging the Use of Readily Available Chemicals and Materials

Readily available chemicals and materials can be used in doing Chemistry experiments. The Chemistry Aid initiative has been promoting and encouraging the use of materials such as:

"Chemistry Aid" helps to better African chemistry

COPIES OF A CD-rom entitled *Practical Chemistry for Schools and Colleges* have been despatched to Kenya in an education programme that is expected to be extended to other African countries. The initiative has been dubbed "Chemistry Aid".

It follows a request from the Scientific Advisory and Information Network in Nairobi, Kenya for help with teaching practical chemistry. Many schools in Kenya rely on visits to better equipped schools for practical science lessons and examinations. In 2003, the average mark of the 198,747 students that sat the Kenya Certificate of Secondary Education (KCSE) chemistry exam was only 26.85% – the lowest mean score in the three science subjects.

Dr Colin Osborne, the RSC's education manager for schools and colleges, said: "Chemistry is probably one of the most difficult subjects to teach in developing countries because students do not have access to chemicals and equipment and teachers cannot demonstrate chemical reactions. Yet such countries need indigenous chemists as part of their development plans."

The CD-roms were jointly produced by the Chemistry Video Consortium (CVC), which is based at the University of Southampton, and the RSC. They show laboratory experiments and stress practical techniques. They can be used by teachers to present topics to students and by students for individual study. The RSC said the initiative could be expanded through the British Council and the Commonwealth.

A CD-rom on practical chemistry is helping Kenyan students to improve their grades

The encouraging response could open up possibilities for providing other resources for teaching chemistry, as the CVC/RSC has a database called "Chemistry Images", which includes VHS tapes, laser disks and CD-roms for teaching, learning and training. Dr Osborne said the scheme "could even be extended to developing countries worldwide".

i Offers of help, support, donations, and sponsorship should be sent to: Dr Colin Osborne, Education Department, Burlington House (address on page 16), tel +44 (0)207 437 8656, or Dr Tony Rest, School of Chemistry, University of Southampton, Southampton SO17 1BJ, tel +44 (0)2380 593595. Colin Osborne can also provide further information

Regional meetings come to a conclusion

THE 2004 SERIES of regional meetings has now been completed. Over 200 members attended the nine meetings, which were held in London, Dublin, Huddersfield, Leicester, Stirling, Liverpool, Bristol, Aberystwyth and Cambridge.

There were two main topics for discussion. The first subject, which created a good deal of debate, was the proposed changes to the RSC structure of governance. Though the changes were supported by a large majority of members, there were some concerns voiced over how to ensure that the new bodies would be representative of the RSC as a whole.

The second item on the agenda was the RSC's Campaign for the Chemical Sciences. This presentation provided members with an overview of how the RSC was evolving into a more proactive organisation and how it intended to become an advocate for the chemical sciences.

i If you would like more information on regional meetings, contact David Lindores at Thomas Graham House (address on page 16), tel +44 (0)1223 422264, email lindoresd@rsc.org. Summaries of the meetings are available at www.rsc.org/lap/rsccom/regions/regindex.htm

Knovel takes the strain out of chemical sciences' research

ALMOST 700 SCIENTIFIC databanks and electronic books are now available to RSC members through the Knovel web-based search engine.

The chemical sciences' collection includes: *Lange's Handbook of Chemistry*, *Endocrine Disrupting Chemicals* and *Yaw's Handbook of Thermodynamic and Physical Properties of Chemical Compounds*.

Nazma Masud, information officer with the RSC's Library & Information Centre, said: "Imagine having to consult these works individually. The simple interface on this immense resource allows the construction of search strategies in an intuitive manner to search the entire collection at once or in groups of one of 16 subject areas. The types of searches can range from using basic keywords to specifying ranges of numeric property values such as vapour pressure or dissociation constant in appropriate units."

For example, an advanced search for alcohols with a melting point of 50C to 100C and boiling point of 200C to 300C retrieved 54 matches in seven books, including Knovel Critical Tables. Results include the name, links to synonyms, structure, melting point and boiling point, the ranges of which hyperlink to new windows that give full bibliographic details of the data source. In addition, the results can be copied and pasted into Word or Excel using an export feature, which automatically cites the work.

i For further information, contact Martin Cadman or Nazma Masud at library@rsc.org. To start searching, go to www.rsc.org/lic/knovel_library.htm

WWW.RSC.ORG

RSC NEWS | JUNE 2004 7

Fig. 19.2 June 2004 RSC news article

(a) Use of plastic Ziploc bags in doing Chemistry experiments [4].
(b) Use of kitchen/supermarket chemicals such as bi-carbonate, vinegar, lime juice in doing Chemistry experiments (Fig. 19.3).
(c) Use of plastic document wallets in doing 'micro-scale' Chemistry experiments.

Fig. 19.3 Students during a 'hands-on' supermarket chemical reactions workshop

One advantage of the use of readily available resources is that it has been found to cut on the cost of doing "hands on" experiments. Also the materials are environmentally friendly hence are easy to dispose after use. Since these experiments focus on everyday life, and could be done outside the laboratory/classroom, they were of great interest to the students. The plastic Ziploc bags and plastic document wallets can be used in place of the laboratory bench.

19.2.3 Use of Sporting Activities in Encouraging Chemistry Education

Sports are a great youth attraction all over the world. The SAIN and the CVC under the Chemistry Aid initiative have launched the Sports 4 Science program aimed at using sports in enthusing young people in science (Fig. 19.4). The program uses sports to promote science education among the youth. The SAIN and CVC acknowledge that it is possible to associate science with youth sporting activities. One of the activities in our 'Sports 4 Science' program is football tournament for the teens. During the tournaments the teens are motivated to perform well in science subjects by giving them talks on how science is related to football. We have informed the teens that the sporting gear such as the uniforms, shoes, gloves and the soccer ball are made from materials that are products of chemical processes. As a result, the teens have been able to associate their favourite sporting activity with Chemistry. Some scholars have argued that youth participation in sporting activities is essential for their education [5]. Sporting activities can therefore be used to influence young learners' attitude in Chemistry.

Fig. 19.4 Teens during a Sports 4 Science tournament organized by the SAIN and CVC

Fig. 19.5 Solar power generator kit: hand held projector, phone, laptop and solar panel

19.2.4 Solar Power Program

A major challenge facing many rural schools in developing countries is lack of mains electricity. In order that schools benefit from the use of ICT multimedia resources, they need electricity. A way of overcoming the challenge of lack of electricity is by the use of portable power generators. However, portable power generators have several disadvantages and are associated with several safety hazards. Also, in developing countries, producing power via a petrol powered generator is expensive due to the cost of petrol fuel and maintenance. Solar power is an alternative to petrol driven generators, so under the Chemistry Aid initiative a simple Solar Power Generator has been devised for use in solar powered classroom. The solar powered generators enable students to make use of the ICT multimedia materials, hence experience the "hands-on" experiences of chemical experiments. The generators can be connected to portable computers such as laptop and data projectors [6] (Fig. 19.5).

19.3 Conclusions

In several developing countries, the teaching and learning of Chemistry as well as other Sciences and Mathematics remain a big challenge. Most schools especially in the rural areas lack important facilities such as laboratories and equipment. Also chemicals are expensive and most schools cannot afford to offer practical laboratory sessions. This is in addition to lack of mains electricity. In order to overcome these challenges, there is need of developing countries to embrace innovative solutions. The Chemistry Aid Initiative of the SAIN and CVC is a model that can be used by developing countries in applying innovative solutions to the challenges facing teaching and learning of Chemistry as well as other Sciences.

Acknowledgements We, the authors, would like to acknowledge the support of the Chemistry Video Consortium (CVC) especially from Prof. Antony Rest who is the project coordinator, the Royal Society of Chemistry (RSC), the support of our colleagues in the Scientific Advisory and Information Network (SAIN), as well as the entire Chemistry Aid team.

References

1. Elecanal-Guavais AM, Bin Hitam R, Rest AJ (2010) Supporting chemistry in South East Asia via chemistry aid. Asian Pacific education research conference, Kuala Lumpur
2. Osborne C, Rest AJ (2004) Chemistry aid. Royal Society of Chemistry (RSC) News article, p 7
3. "Chemnet" Database, Royal Society of Chemistry. www.rsc.org/learn-Chemistry/resource/res00000929/communities-rsc-chemnet
4. Selco J (2008) Chemical reactions in Ziploc bags. http://ectn-assoc.cpe.fr/news/letter/2008/092_200804_Ref2_Africa_Report_JIS.pdf
5. Hartmann D (2008) High school sports participation and educational attainment: recognising, assessing and utilizing relationship. Report to LA84 Foundation
6. Jadoloh A, Johnson J, Ogunde J, K'osura H, Rest T, Wilkinson K (2010) Using solar generators to support IT and "hands-on" science in developing countries. Fifth international conference on ICT for development, education and training in Africa, Lusaka, p 373. www.chemistry-aid.com/news, http://www.Chemistry-aid.com/news

Chapter 20
Synthesis and Characterization of Some New Metal Complexes of Condensation Reaction Products of 3-Amino-1,2,4-Triazole with Isatin, N-Acetylisatin and Bis (2,3-Dioxoindolin-1-yl)Mercury(II)

Ahlam J. Abdulghani and Zainab Z. Ahmed

Abstract The condensation reaction of 3-amino-1,2,4-triazole (amitrol, 3-AT) with N-acetylisatin caused the opening of the indole ring and led to the formation of the new ligand (Z)-2-(1H-1,2,4-triazol-3-ylimino-2-(2-acetamido-phenyl)-N-(1H-1,2,4-triazol-3-yl) acetamide (**1**), while the reaction with isatin and bis (2,3-dioxoindolin-1-yl)mercury(II), (Hg(isatin)$_2$) caused no ring opening and the Schiff base (Z)-3-(1H-1,2,4-triazol-3-ylimino) indolin-2-one (**2**), and the new mercury complex 2-(E)-3-(1H-1,2,4-triazol-3-ylimino)-2-oxoindolin-1-yl)((Z)-3-(1H-1,2,4-triazol-3-ylimino)-2-oxoindolin-1-l)mercury(II) (**3**) were formed. New complexes of Co(II), Ni(II), Cu(II), Pd(II) and Zn(II) ions with the two ligands (**1**) and (**2**) were prepared. The studied products were characterized by elemental analysis, NMR, IR, mass spectra and thermal analysis. In all the complexes the metal was coordinated to the nitrogen atoms of the triazole ring and to the azomethine group.

20.1 Introduction

The interest in the structural properties of N-acetylisatin derivatives was primarily attributed to their cytotoxic activity, and a variety of pharmacological activities such as bactericides, antitumor and antiviral [1]. Some Schiff bases of N-acetylisatin have shown fibrinolytic, muscle relaxant, antiallergic, immunosuppressant, hypotensive respiratory depression and antidiuretic effects [2]. The chemistry of 1,2,4-triazole derivatives have attracted the attention of chemists owing to their bactericidal, fungicidal and pharmacological activities in addition to their use as antiulcer agents [3]. Due to the presence of more than one donor atom in the triazole ring, Schiff base derivatives of 3-AT were considered powerful chelating agents

A.J. Abdulghani (✉) • Z.Z. Ahmed
Department of Chemistry, College of Science, University of Baghdad, Jaderiya, Baghdad, Iraq
e-mail: ahlamjameel@scbaghdad.edu.iq; zainabzahid489@yahoo.com

M. Gupta Bhowon et al. (eds.), *Chemistry: The Key to our Sustainable Future*,
DOI 10.1007/978-94-007-7389-9_20,

giving rise to bi- and polynuclear complexes [4, 5]. Different isatin derivatives incorporating heterocyclic rings like thiazole, thiadiazole, benzothiazole, and imidazole moieties and their corresponding cobalt(II), copper(II), nickel(II) and zinc(II) complexes were reported [2]. The isatin derivatives showed good antibacterial, antiviral, antifungal and anticancer activities against different strains, and most of them had their activity enhanced on complexation with the metal ions. The reaction of N-acetylisatin with some alcohols, amines, pyrrolidine and water yielded products resulting from nucleophilic attack at the C-2 carbonyl that leads to heterocyclic ring cleavage.

In this work we study the stability of N_1-acetyl and Hg(II) substituted isatin ring toward nucleophilic attack of 3-amino-1,2,4-triazole (3-AT). We also study the coordination behavior of the N-acetyl condensation product (**1**) with some divalent metal ions and compared it with that of the Schiff base (Z)-3-(1H-1,2,4-triazol-3-ylimino) indolin-2-one (**2**), which has been prepared previously by us from the condensation reaction of unsubstituted isatin and 3-AT [5].

20.2 Methodology

20.2.1 General

All chemicals were purchased either from Fluka or BDH Chemicals and were used without further purification, with the only exception of ethanol which was distilled prior to use. N-acetylisatin was prepared according to methods reported in the literature [6–8]. Purity of products was ascertained by using TLC techniques using a mixture of acetone: chloroform (1:1, 1:2 and 2:1 v/v) and acetone, chloroform, and ethanol individually as eluents.

The melting points of all samples were determined using a Gallenkamp M.F.D. 600-010f melting point apparatus. Infrared spectra were measured on a Shimadzu FTIR 8400S spectrometer in the range of 4,000–200 cm^{-1} using KBr or CsI pellets. Electronic spectra of compounds were recorded on a Shimadzu 1650 UV-visible recording spectrophotometer (200–1,100 nm). Elemental analysis of samples was determined on a Euro vector EA 3000A. The metal contents were determined by atomic absorption technique using a Shimadzu AA 6200 flame atomic absorption spectrophotometer. The mass spectra were recorded on a Shimadzu Qp 5050A. ^{1}H NMR and ^{13}C NMR spectra were obtained from a Bruker Ultra Shield 300 MHz NMR spectrometer. Thermal analysis by TG, DTG and DTA technique was recorded on a STA 409 PC Lux thermal analyzer under a nitrogen atmosphere at 25–1,000 °C. The molar conductivity of complexes in DMF and DMSO (10^{-3} M) at room temperature was measured on a Radiometer Copenhagen CDM 83 conductivity meter. Magnetic moments (μ_{eff}) were measured at room temperature by using a Bruker Magnet B.M. 6 and a magnetic susceptibility balance Model MSB-MK-1.

20.2.2 Synthesis of Compounds (1)–(3)

1-(Z)-2-(1H-1,2,4-triazol-3-ylimino-2-(2-acetamido-phenyl)-N-(1H-1,2,4-triazol-3-yl) acetamide (**1**): To a stirred solution of N-acetylisatin (1.115 g, 5.9 mmol) in dry ethanol (30 ml) heated to 70 °C was added a solution of 3-AT (0.5 g, 5.9 mmol) in ethanol followed by six drops of glacial acetic acid. The mixture was heated under reflux for 4 h with continuous stirring. The color of solution changed from red orange to yellow. The resultant solution was left overnight at room temperature and a yellow precipitate was formed. The product was filtered, washed with hot ethanol followed by ether and vacuum dried. Yield 65.5 %; m.p. 164 °C; IR (ν, cm^{-1}): 3255 (N-H), 1720 (C=O, acetyl), 1689 (C=O, amide), 1650 (C=N, azomethine), 1609 (C=N, 3-AT), 1527 (N-C=N, 3AT), 1275 (N-N=C, 3-AT); MS [EI]: $C_{16}H_{19}N_9O_3$ (M), m/z (I%): 189 [M-$C_6H_{10}N_7O$] (16), 162 [(M-$C_6H_{10}N_7O$-HCN] (90), 146 [(M-$C_6H_{10}N_7O$-$COCH_3$] (100), 135 [(M-$C_6H_{10}N_7O$-COCN] (11), 120 [(M-$C_6H_{10}N_7O$-$COCNCH_3$] (80), 111 [(M-$C_6H_{10}N_7O$-C_2NH_2] (40), 84 [M-$C_{14}H_8N_5O_2$] (81), 77 [(M-$C_6H_{10}N_7O$-$C_3H_2N_2O_3$] (5), 66 [M-$C_6H_{10}N_7O$-COCN-CH_3-COCN] (69), 43 [M-$C_{14}H_{13}O_2N_9$] (90); Elemental analysis (%) calculated for $C_{16}H_{19}N_9O_3$: C 50.5, H 3.2, N 32.7 %, found: C 49.9, H 3.7, N 32.2; ^1H NMR δ (ppm, DMSO):13.9 (2H, N-H triazole ring), 10.5–10.9 (2H, N-H amide), 7.2–8.1 (4H, Ph), 8.3–8.4 (2H, olefinic protons), 1.1–1.2 (3H, CH_3 of EtOH), 1.9–2.9 (3H, $COCH_3$), 3.3 (2H, CH_2CH_3); ^{13}C NMR δ (ppm): 169.4 (2C, CO amide), 162.6 (1C, C=N), 144 (4C, C=N of triazole ring), 103–124 (4C, Ph (C-1,C-2)), 132–135 (2C, Ph(C-3,C-6)), 24–25 (3C, aliphatic carbon atoms); Electronic spectra (DMF): λ_{max} (nm) 242 sh, 321 and 400, ν_{max} (cm^{-1}) 41322, 31152 and 25000, ε_{max} 6693 and 247660 ($\pi \rightarrow \pi^*$) and 2231 ($n \rightarrow \pi^*$); DMSO: λ_{max} (nm) 274, 340, ν_{max} (cm^{-1}) 36496 and 29411, ε_{max} 10127 and 2310 ($\pi \rightarrow \pi^*$).

2-(Z)-3-(1H-1,2,4-triazol-3-ylimino) indolin-2-one (**2**): This ligand has been prepared as in our previous work [5a] with some modification. To a stirred solution of isatin (0.1 g, 0.68 mmol) in boiling super dry ethanol (10 ml) was added a solution of 3-AT (0.05 g, 0.68 mmol) dissolved in the same solvent (2 ml), followed by addition of six drops of glacial acetic acid. The mixture was heated under reflux for 14 h with continuous stirring. An orange yellow precipitate was formed. The product was filtered, washed with hot ethanol followed by ether and vacuum dried. Yield 58.3 %; m.p. 258–260 °C; IR (ν, cm^{-1}): 3124 (NH isatin ring), 1728 (C=O), 1645 (C=N), 3184, 1612, 1530, 1270 (NH, C=N, N-C=N, N-N=C respectively, triazole ring); MS [EI]: $C_{10}H_7N_5O$ (M), m/z (I%): 213.2 $[M]^+$ (81), 185 $[M\text{-}CO]^+$ (22), 170 [M-OCNH] (7), 158 [(M-CO)-HCN) (34), 144 [M-$C_2H_2N_3$] (17), 130 [(M-CO-HCN)-N_2] (41), 116 [M-OC_2N_4H] (14), 103 [(M-CO-HCN-N_2)-HCN] (100), 90 [(M-OC_2N_4H)-CN] (25); Elemental analysis (%) calculated for $C_{10}H_7N_5O$: C 56.30, H 3.28, N 32.86, found C 56.80, H 3.37, N 32.66; ^1H NMR δ (ppm, DMSO): 14.4–14.8 (d, 1H, triazole NH), 10.95 (1H, isatin NH), 6.7–8.4 (4H, aromatic), 8.7 (s, 1H, triazol olefinic C-H); ^{13}C NMR δ (ppm): 164.1–164.6 (1C, carbonyl of isatin ring), 155 (1C, azomethine carbon), 152 (1C, triazole ring at 3), 151 (1C, N-C=N, triazole), 148–99.9 (6C, aromatic carbons).

3-(E)-3-(1H-1,2,4-triazol-3-ylimino)-2-oxoindolin-1-yl)-(Z)-3-(1H-1,2,4-triazol-3-ylimino)-2-oxoindolin-1-yl)mercury(II) (**3**): To a stirred solution of Hg(isatin)$_2$ [9] (0.1 g, 2 mmol) in DMF (2.85 ml) containing four drops of glacial acetic acid was added a solution of 3-AT (0.0406 g, 4 mmol) in DMF (1.1 ml). The mixture was heated under reflux for 30 min. A white precipitate was formed within 5 min. The mixture was left to cool to room temperature. The product was filtered, washed with DMF, ethanol, ether and vacuum dried. Yield 52 %; m.p. 202 °C.

20.2.3 Synthesis of Metal Complexes of (1) and (2)

To a stirred solution of each ligand in hot ethanol was added an ethanolic solution of the metal(II) chloride (except Zn(II) ion as acetate and Pd(II) ion as dichloro (benzonitrile)palladium(II) [$PdCl_2(PhCN)_2$] [10] in mole ratios of 2:1 and 1:1 of (**1**):M and (**2**):M respectively. The mixture was heated under reflux for 1 h. Precipitation of complexes took place within 5–30 min. The products were filtered hot, washed thoroughly with hot ethanol, followed by ether and vacuum dried. The zinc (II) complex of (**1**), **C**$_4$, was prepared by adopting the same procedure but DMF was used instead of absolute ethanol and the reaction was refluxed for 4 h. The product was filtered, washed with DMF, hot ethanol followed by ether and vacuum dried.

20.3 Results and Discussion

20.3.1 Synthesis

The synthesis route of the two ligands (**1**) and (**2**) and compound (**3**) are illustrated in Schemes 20.1 and 20.2 respectively. The physical properties and elemental analyses results of the prepared compounds are described in Table 20.1. The ligands and their metal complexes were soluble in DMF and DMSO, sparingly soluble in ethanol, methanol, chloroform and dichloromethane but insoluble in less polar solvents.

20.3.2 NMR Spectra

The ^{1}H and ^{13}C NMR spectra of (**1**) (Fig. 20.1a, b) confirmed the presence of amide NH protons and amide carbonyls at C(3) of triazole ring and at C(2) of phenyl group moiety. The ^{1}H NMR spectrum (Fig. 20.1a) exhibited a signal at δ 10.5–10.9 ppm which was attributed to chemical shifts of amide protons [11, 12]. Chemical shifts assigned to (N-H) protons of triazole rings of (**1**) and (**2**) (Fig. 20.1a, c) were observed at δ 13.9 and 14.5 ppm respectively [13–15].

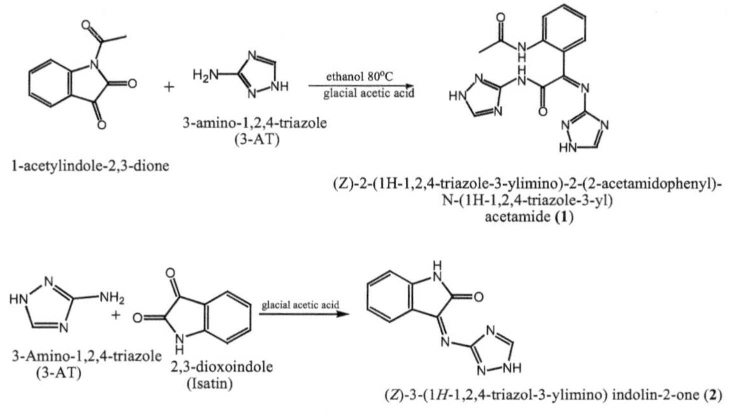

Scheme 20.1 Synthetic routes to (**1**) and (**2**)

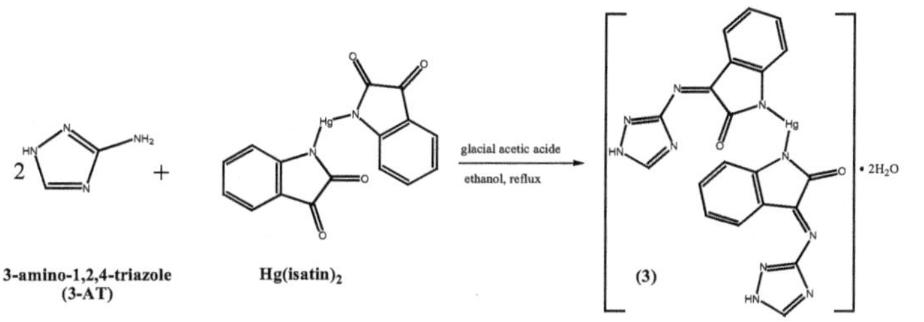

Scheme 20.2 Synthetic route to (**3**)

The ^{1}H NMR spectrum of (**2**) in DMSO (Fig. 20.1c) exhibited chemical shifts at δ 8.70 and 10.95 ppm assigned to olefinic (CH) proton of triazole ring [12] and isatin (N-H) proton [2, 16] respectively.

The ^{13}C NMR spectra of (**1**) and (**2**) (Fig. 20.1b, d) exhibited peaks assigned to carbon atoms of azomethine groups at δ 162.6 and 155.0 ppm respectively [12, 17–19] which confirms the condensation of the amine with C-3 carbonyl of isatin ring. The signals observed at δ 169.4 and 164.0 ppm in the two spectra were attributed to the carbons of amide carbonyl [13] and isatin ring carbonyl [12] respectively. The peaks appearing at δ 103–135 ppm and 99.9–148 ppm [12] were assigned to the aromatic carbons of the two ligands respectively [12, 20].

20.3.3 Mass Spectra

The mass spectra of the two ligands (Figs. 20.2 and 20.3 respectively) were obtained using electron impact mode of fragmentation at 70 eV [12, 21, 22]. The peak corresponding to $[M]^{+\cdot}$ was not observed in the spectrum of (**1**). Other

Table 20.1 Physical properties and analytical data for the prepared compounds

Cpd	Color	M.p. (°C)	Yield	% Found (calc.) C	H	N	M	Cl
(1)	Yellow	164	65.5	49.90 (50.50)	3.70 (3.20)	32.20 (32.70)	–	–
C_1 Co(II)	Blue	>300	22.0	26.34 (26.88)	3.10 (2.80)	20.63 (20.61)	18.20 (18.70)	22.45 (22.72)
C_2 Ni(II)	Light green	246	33.2	23.54 (23.78)	3.40 (3.50)	17.57 (17.83)	16.33 (16.61)	20.50 (20.10)
C_3 Cu(II)	Green	243	21.4	23.65 (23.45)	3.10 (3.48)	17.95 (17.58)	17.62 (17.78)	19.44 (19.82)
C_4 Zn(II)	Pale yellow	>300	72.9	39.52 (40.00)	3.20 (3.88)	23.11 (23.32)	11.90 (12.09)	–
C_5 Pd(II)	Light brown	>300	58.3	23.50 (23.02)	2.50 (2.33)	17.80 (17.26)	29.50 (29.16)	19.86 (19.46)
(2)	Orange yellow	260	58.3	56.80 (56.30)	3.37 (3.28)	32.66 (32.86)	–	–
C_6 Co(II)	Green	>300	34.0	32.15 (31.70)	2.90 (2.90)	18.83 (18.49)	15.60 (15.40)	19.10 (18.76)
C_7 Ni(II)	Yellow-brown	>300	17.0	32.50 (33.20)	2.55 (2.96)	19.60 (19.40)	16.11 (16.27)	19.96 (19.68)
C_8 Cu(II)	Green	200	50.6	31.95 (31.27)	2.66 (2.86)	18.64 (18.26)	16.55 (16.59)	18.71 (18.50)
(3)	Off-white	202	52.0	35.90 (36.30)	2.85 (2.40)	21.35 (21.19)	30.60 (30.36)	–

characteristic peaks were observed at m/z values 189.1, 162, 146, 135, 120, 111, 84, 77, 66 and 43. The mass spectrum of compound **(2)** exhibited a peak assigned to $[M]^{+\cdot}$ m/z 213.4 which was located at 214 in our previous work [5a]. Structural assignments of the main fragments of **(1)** and **(2)** are described in Schemes 20.3 and 20.4.

20.3.4 IR Spectra

The important IR vibrations and tentative assignments of compounds **(1)**, **(2)** and their metal complexes are given in Table 20.2. The infrared spectra of the two ligands exhibited the absence of bands assigned to stretching vibrations of both C=O groups at position (C-3) of isatin ring and the NH_2 group of 3-AT, and the appearance of a new band related to azomethine group [11, 12, 16, 23, 24] which confirms the formation of Schiff base ligands. The spectrum of compound **(1)** exhibited two bands at 1720 and 1689 cm^{-1} attributed to acetyl and (C-2) amide carbonyls respectively [12, 16, 24]. A new sharp strong band appeared at 3255 cm^{-1} and was attributed to N-H amide stretching vibration caused by ring

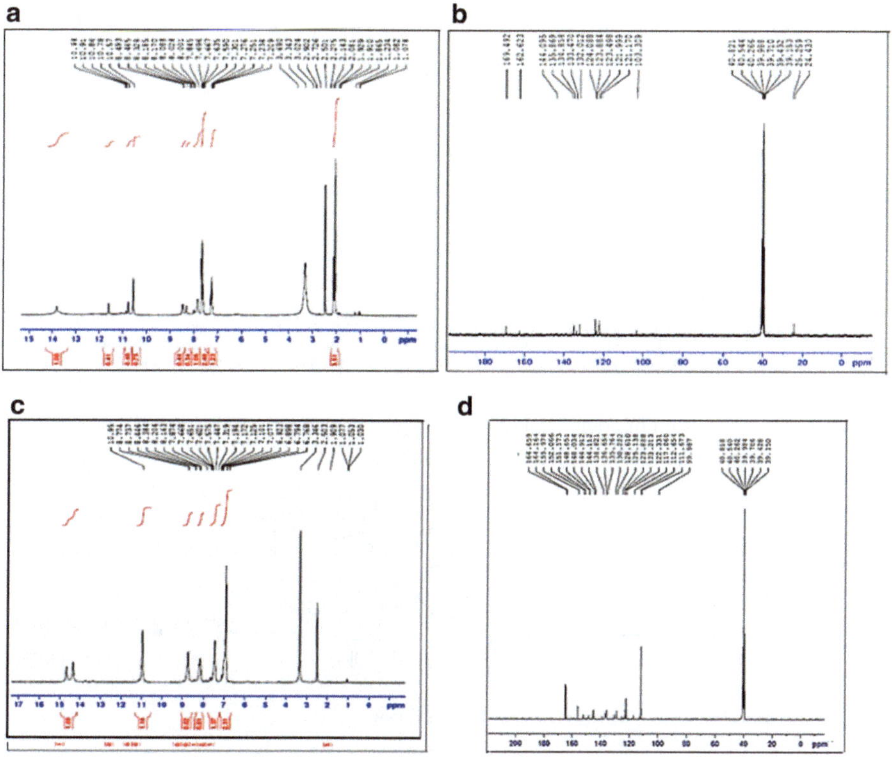

Fig. 20.1 ^{1}H and ^{13}C NMR spectra of compounds (**1**) (**a**, **b**) and (**2**) (**c**, **d**)

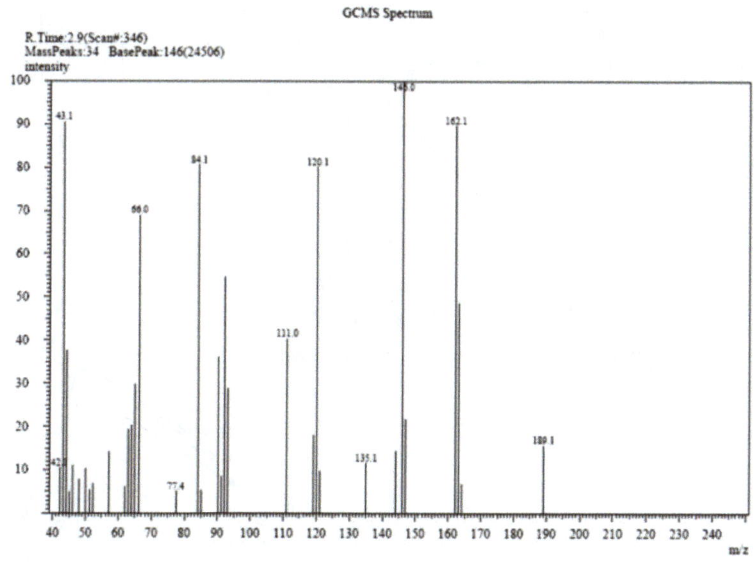

Fig. 20.2 Mass spectrum of compound (**1**)

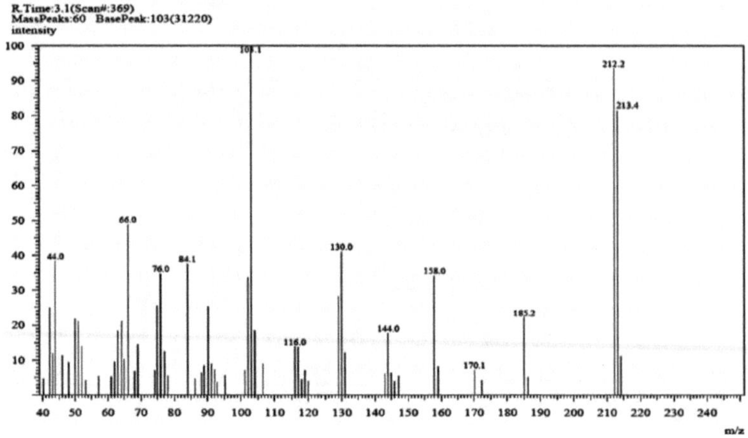

Fig. 20.3 Mass spectrum of compound (**2**)

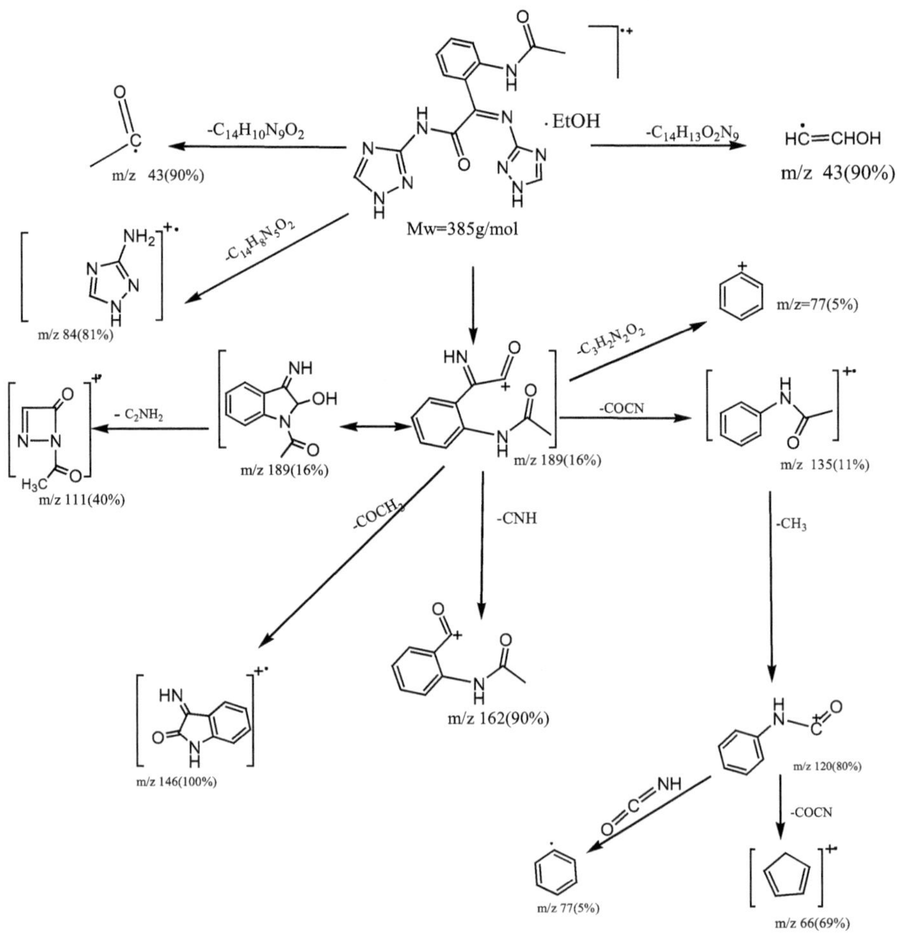

Scheme 20.3 Fragmentation pathways of compound (**1**)

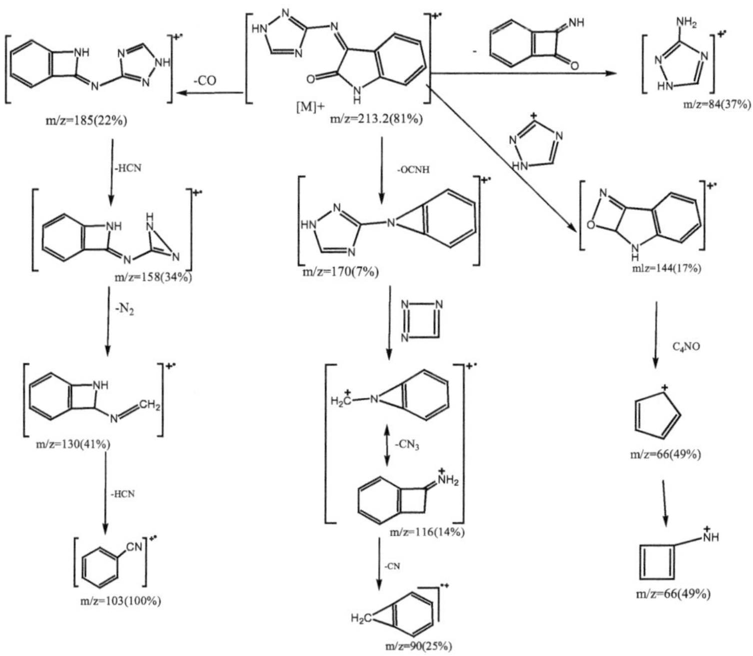

Scheme 20.4 Fragmentation pathways of compound **(2)**

opening of N-acetylisatin [2, 12, 25]. The band observed at 3409 cm^{-1} was attributed to OH stretching vibration of lattice ethanol [12]. Vibrational modes of C(2) carbonyls were observed unchanged in the IR spectra of $Hg(isatin)_2$ and compound **(3)**. This indicates that the condensation of 3-AT with N-acetylisatin led to ring opening while reaction with isatin and $Hg(isatin)_2$ occurred without ring opening (Scheme 20.2).

The spectra of the metal complexes exhibited shifts in the azomethine and both $\nu_{C=N}$ and $\nu_{N\text{-}C=N}$ vibrations of triazole ring to lower vibrations indicating bidentate nature of the two ligands [5, 6, 13, 26, 27]. The spectra of complexes of **(2)**, $\mathbf{C_6}$-$\mathbf{C_8}$ exhibited also shifts of bands assigned to C(2)-carbonyl group of isatin ring to higher frequencies which rules out the coordination of metal ions [5, 16, 23, 27]. The spectrum of $Hg(isatin)_2$ exhibited the absence of N-H stretching vibration of isatin ring with no significant changes in the vibrational modes of the two carbonyls inferring bonding of nitrogens only to the metal ion [28]. The strong sharp bands appearing at 1500–1481 cm^{-1} were attributed to aromatic C=C vibrations [11, 12]. The spectrum of **(3)** exhibited the absence of C(3) carbonyl and the appearance of four new bands attributed to stretching vibrations of azomethine group [16, 24] and C=N, N-C=N, and N-N=C vibrations of the triazole ring [5, 13] (Table 20.2). The band observed at 3201 cm^{-1} was assigned to N-H stretching vibration of triazole

Table 20.2 IR vibrations (cm^{-1}) of compounds **(1)** and **(2)** and the metal complexes

	N-H amide		C=O		Triazole						
Cpd	(triazole)	Amide C=O acetyl	C-2	Schiff C=N	C=N	N-C=N	N-N=C	υ_{H2O} Lattice (coordinated)	M-O	M-N	M-Cl
(1)	3255	1720	1689	1650	1609	1527	1275	–	–	–	–
	(3200)										
C_1	3325	1735	1643	1580	1550	1520	1275	3410	–	393	308
Co(II)	(3150)										
C_2	3260	1735	1643	1589	1535	1500	1296	3340	455	385	301
Ni(II)	(3160)							(3480, 850, 760)			
C_3	3263	1750	1650	1580	1558	1460	1296	3500	520	480	293
Cu(II)	(3180)							(3394, 894, 736)			
C_4	3224	1740	1620	1596	1560	1527	1226	(3394–3340, 871, 750)	439	493	–
Zn(II)	(3190)								578		
C_5	3260	1740	1650	1620	1550	1450	1218	3394	516	462	316[a]
Pd(II)	(3143)										270[b]
(2)	(3184)	–	1728	1645	1612	1530	1270	–	–	–	–
C_6	(3167)	–	1735	1635	1542	1504	1211	(3409, 880, 760)	455	375	301
Co(II)											
C_7	(3153)	–	1735	1635	1550	1465	1275	(3440, 850, 763)	455	354	285
Ni(II)											
C_8	(3155)	–	1740	1620	1550	1480	1310	(3348, 871, 763)	520	478	293
Cu(II)											
$Hg(Is)_2$		1735	1685	–	–	–	–	3460–3500	–	447	–
(3)	(3201)	–	1682	1650	1580	1434	1257	3340	–	460	–

[a]Terminal
[b]Bridging

ring [5, 13]. Characteristic bands of coordinated water $\nu_{(OH)}$, $\rho_{r(H_2O)}$ and $\rho_{w(H_2O)}$ vibrations) were observed in the spectra of **C$_2$-C$_4$** and **C$_6$-C$_8$** [11, 27] whereas vibrational modes of lattice water were observed in the spectra of **C$_1$**, **C$_2$**, **C$_3$** and **C$_5$** [26, 27, 29], Hg(isatin)$_2$ [24] and **(3)**. Bands assigned to M-N, M-O and M-Cl stretching modes [18, 27, 30, 31] appeared at lower frequencies. Vibrational modes of monodentate and ionic acetate of the zinc complex, **C$_4$**, were observed at 1334 and 1404 cm^{-1} respectively [27].

20.3.5 Electronic Spectra, Magnetic Moments and Conductivity Measurements

The spectrum of compound **(1)** in DMF exhibited a sharp single band at 31,152 cm^{-1} with a shoulder at 41,322 cm^{-1}. In DMSO, the spectrum exhibited a sharp single band at 36,496 cm^{-1} with one shoulder at 29,411 cm^{-1}. The spectrum of compound **(2)** in DMF exhibited a multiplet band at 37,174, 29,674 and 23,530 cm^{-1}. All these bands were attributed to $\pi \rightarrow \pi^*$ transitions [12, 32, 33]. The spectra of metal complexes are described in Table 20.3. All complexes exhibited shifts and changes in the profiles of ligand bands with the appearance of bands in the visible and near IR regions assigned to charge transfer bands and ligand field d-d transitions. The spectrum of the Co(II) complex of **(1)**, **C$_1$**, in DMF exhibited two new bands assigned to the spin-allowed transitions $^4A_2 \rightarrow {}^4T_1$ (ν_2) and to $^4A_2 \rightarrow {}^4T_1(P)$ (ν_3) related to the tetrahedral geometry of Co(II) complexes [29, 32, 33]. By applying band ratio $\nu_3/\nu_2 = 1.59$ on the Tanabe-Sugano diagram, the values of Dq/$\bar{B}$, $\bar{B}$, 10Dq, β and ν_1 assigned to $^4A_2 \rightarrow {}^4T_2(F)$ were obtained. The magnetic moment of the complex (4.9 B.M.) indicates a binuclear structure with spin orbital contribution [33, 34]. The spectrum of the Ni(II) complex, **C$_2$**, in DMF exhibited three new bands. The first two low intensity bands were attributed to the transitions $^3T_1 \rightarrow {}^3A_2$ (ν_2) and $^3T_1 \rightarrow {}^3T_1(P)$ (ν_3) respectively and the third high intensity band was assigned to LMCT transitions of tetrahedral geometry [32, 33]. By applying the band ratio ν_3/ν_2 (1.6) on the Tanabe-Sugano diagram, the energy of ν_1 assigned to the transition $^3T_1 \rightarrow {}^3T_2$ was calculated [32–34]. The magnetic moment (3.1 B.M.) is in agreement with tetrahedral nickel(II) complexes [33, 34]. The spectrum of the Cu(II) complex, **C$_3$**, in DMF exhibited a new low intensity band which appeared as a doublet with two maxima at 15,000 and 13,888 cm^{-1}. The average value (14,444 cm^{-1}) was attributed to $^2E_g \rightarrow {}^2T_{2g}$ transition of a tetragonally distorted Cu(II) octahedral complex [22]. The magnetic moment of the complex (1.83 B.M.) refers to binuclear Cu(II) complexes [33, 34]. The spectrum of the zinc(II) complex, **C$_4$**, in DMSO showed one additional band at 30,770 cm^{-1} which was assigned to a ligand to metal charge transfer (LMCT) transition

Table 20.3 Electronic spectral parameters, magnetic moments and molar conductivity of metal complexes

Cpd	Band position	Assignment	$Dq/\overline{B}$	$\overline{B}/cm^{-1}$ (β)	μ_{eff}/B.M. (geometry)	$10Dq/cm^{-1}$	Molar conductivity/$Smol^{-1}cm^2$ DMF (DMSO)
C_1, Co(II)	4507 ν_1 (calc.)	$^4A_2 \rightarrow {}^4T_2$	1.1	452	4.9	4,972	29.5
	10193 ν_2	$^4A_2 \rightarrow {}^4T_1(F)$		(0.783)	(T_d)		Non-electrolyte
	16277 ν_3	$^4A_2 \rightarrow {}^4T_1(P)$					
	28057, 35874	$\pi \rightarrow \pi^*$					
C_2, Ni(II)	4614 ν_1 (calc.)	$^3T_1 \rightarrow {}^3T_2$	0.765	839	3.1	4,933	153.2
	10131 ν_2	$^3T_1 \rightarrow {}^3A_2$		(0.814)	(T_d)		1:2
	16369 ν_3	$^3T_1 \rightarrow {}^3T_1$					
	21153	LMCT					
	37037, 38314	$\pi \rightarrow \pi^*$					
C_3, Cu(II)	14444 (av.)	$^2E_g \rightarrow {}^2T_{2g}$	–	–	1.83	–	23
	30864, 36900, 38314	$\pi \rightarrow \pi^*$			(O_h)		Non-electrolyte
C_4, Zn(II)	30770	LMCT	–	–	Diamag.	–	(67)
	35336, 41382	$\pi \rightarrow \pi^*$			(O_h)		1:1
C_5, Pd(II)	13994	$^1A_{1g} \rightarrow {}^1A_{2g}$	–	–	Diamag.	–	170
	18085	$^1A_{1g} \rightarrow {}^1B_{1g}$			(sq.)		1:2
	37037, 28676	$\pi \rightarrow \pi^*$					
C_6, Co(II)	5327 ν_1(calc.)	$^4T_{1g} \rightarrow {}^4T_{2g}$	0.826	710.6	4.5	5,864	30.3
	11886 ν_2	$^4T_{1g} \rightarrow {}^4A_{2g}$		(0.773)	(O_h)		Non-electrolyte
	15988 ν_3	$^4T_{1g} \rightarrow {}^4T_{1g}$					
	23357, 39024, 41493	$\pi \rightarrow \pi^*$					
C_7, Ni(II)	11728 ν_1	$^3A_{2g} \rightarrow {}^3T_{2g}$	1.61	738 (0.717)	2.8	11,818	33.8
	18450 ν_2	$^3A_{2g} \rightarrow {}^3T_{1g}$ (F)			(O_h)		Non-electrolyte
	26800 ν_3 (calc.)	$^3A_{2g} \rightarrow {}^3T_{1g}(P)$					
	25510, 36231, 41841	$\pi \rightarrow \pi^*$					
C_8, Cu(II)	13939	$^2E_g \rightarrow {}^2T_{2g}$	–	–	1.72	–	26.7
	24264, 37313	$\pi \rightarrow \pi^*$			(O_h)		Non-electrolyte
(**3**), Hg(II)	28818	LMCT			Diamag.		Non-electrolyte
	38610	$\pi \rightarrow \pi^*$					

characteristic of d^{10} metal complexes [32, 33]. The spectrum of the diamagnetic Pd (II) complex, $\mathbf{C_5}$, in DMF exhibited one additional band at 18,085 cm^{-1} attributed to the transition $^1A_{1g} \rightarrow {}^1B_{1g}$ of square planar geometry [35]. The bands observed in the spectra of the Co(II), Ni(II) and Cu(II) complexes of **(2)**, $\mathbf{C_6}$-$\mathbf{C_8}$ and their magnetic moment values are indicative of octahedral geometries [33, 34]. The energies of ν_1 for the Co(II) complex, $\mathbf{C_6}$, and ν_3 for the Ni(II) complex, $\mathbf{C_7}$, assigned to the transitions $^4T_{1g} \rightarrow {}^4T_{2g}$ and $^3A_{2g}(F) \rightarrow {}^3T_{1g}(P)$ respectively were calculated by applying the band ratios ν_3/ν_2 (1.345) and ν_2/ν_1 (1.573) respectively on appropriate Tanabe-Sugano diagrams. The magnetic moments of the Co(II) and Cu (II) complexes (4.5 and 1.72 B.M. respectively) are in agreement with mononuclear complexes [33, 34, 36], while that of the Ni(II) complex (2.8 B.M.) refers to a binuclear complex [33, 34]. The spectra of $Hg(isatin)_2$ and compound **(3)** in DMSO exhibited high intensity bands assigned to $\pi \rightarrow \pi^*$ transitions at 31,447 and 36,900, and 38,610 cm^{-1} respectively, and lower intensity bands assigned to LMCT transitions at 22,075 and and 28,818 cm^{-1} respectively [32]. The values of the nephelauxetic ratio β for the cobalt(II) and Ni(II) complexes indicate the presence of covalent character in the bond between the ligands and metal ions [32, 33]. Results of conductivity in DMF or DMSO indicate the electrolytic nature of $\mathbf{C_2}$, $\mathbf{C_4}$ and $\mathbf{C_5}$ with ionic ratios of 1:2, 1:1 and 1:2 respectively and non-electrolytic nature of the other complexes [37].

From the above data together with elemental analyses and IR spectra, the suggested structures of complexes are shown in Scheme 20.5.

20.3.6 Thermal Analysis of Metal Complexes

Thermographs of selected complexes are shown in Fig. 20.4.

Water molecules in the crystal lattice of complexes were lost at the temperature range of 50–150 °C as was observed in the TG curves [38]. At the early stages the decline in DTA curves refers to endothermic reactions [11, 38, 39], while at temperatures higher than 350 °C the progress of weight loss was combined with continuous rise of DTA curve. This is attributed to further decomposition of metal complexes leading to fragments that undergo exothermic reactions [38]. At the final stages of the thermal decomposition of metal complexes of **(1)**, $\mathbf{C_1}$, $\mathbf{C_2}$, $\mathbf{C_3}$ and $\mathbf{C_5}$, the end products correspond to the formation of $CoCl_2$ + CoO, 2NiO, CuO + $CuCl_2$ and $PdCl_2$ + PdO respectively as final residues [38], while 95 % of the Hg(II) complex **(3)** has been sublimed. This confirms the difference in the bonding nature between the four transition metal complexes and the Hg (II) complex [38].

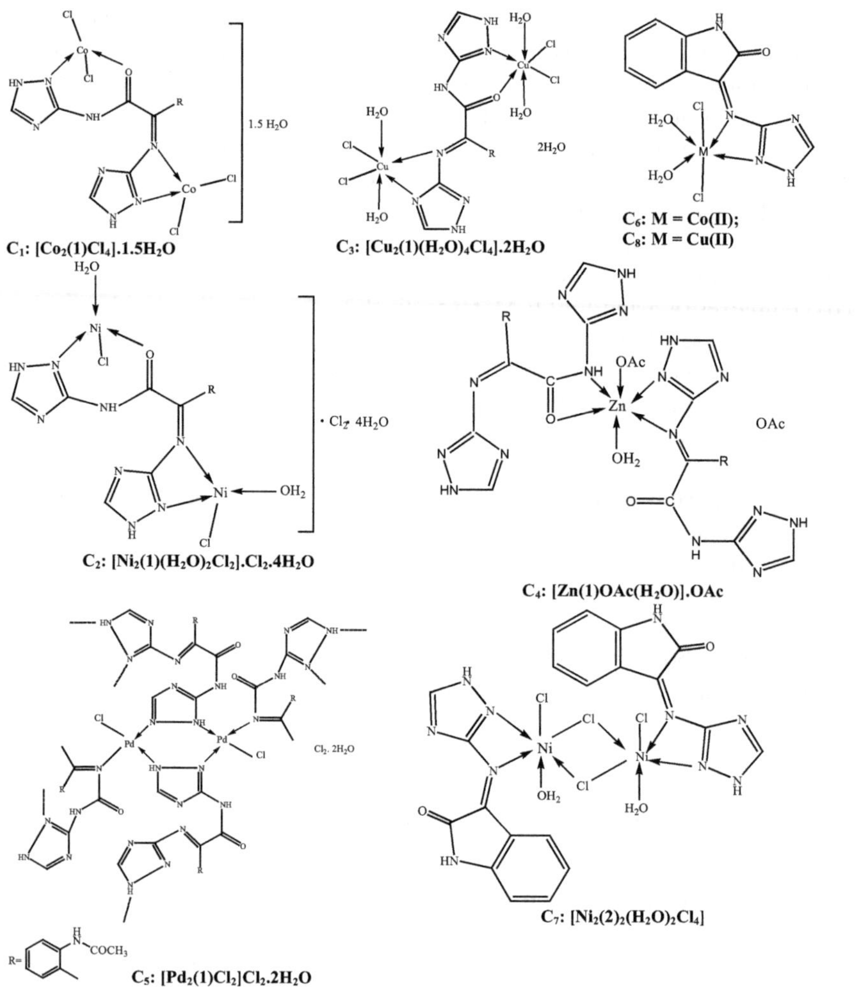

Scheme 20.5 Proposed structures of the prepared metal complexes

20.4 Conclusions

The condensation reaction of 3-AT with isatin and Hg(isatin)$_2$ resulted in the formation of the azomethine linkage at C(3) of isatin ring without cleavage of the isatin ring while, like other reported amines, reaction of 3-AT with N-acetylisatin led to ring opening and consequently to different coordination behavior of **(1)** compared with **(2)**. In all the metal complexes the present ligands showed bidentate behavior in which at least one of the three nitrogens of triazole ring is participating in coordination with the metal ions. The thermal analysis of the selected metal complexes of compound **(1)** proved their binuclear structure.

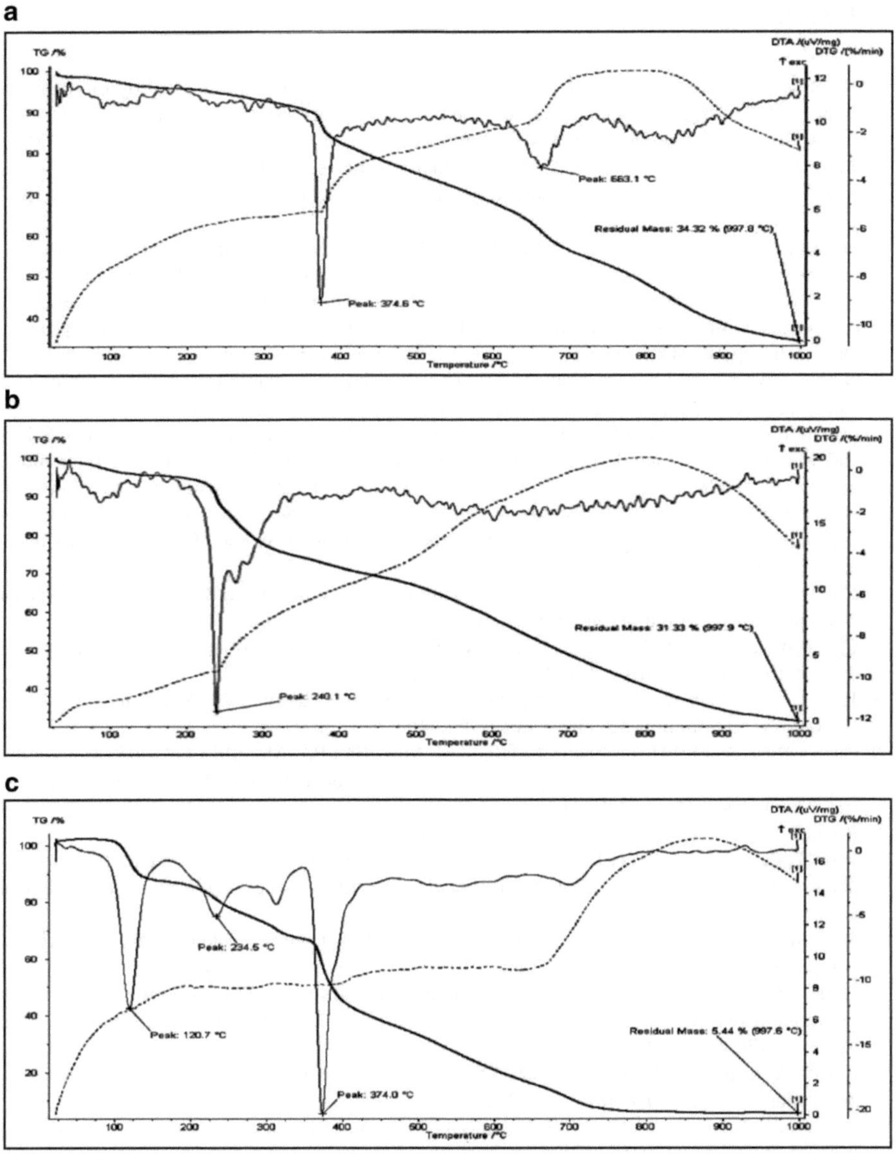

Fig. 20.4 TG, DTG and DTA curves of (**a**) $[(Co_2(\mathbf{1})Cl_4)]1.5H_2O$ (**C_1**); (**b**) $[Cu_2(\mathbf{1})Cl_4(H_2O)_4].2H_2O$ (**C_3**); (**c**) $C_{20}H_{12}N_{10}OHg.2H_2O$ (**3**)

References

1. Natmov P, Anastasova F (2001) Experimental and theoretical vibrational study of isatin, its 5-(NO_2, F, Cl, Br, I, CH_3) analogues and the isatinato anion. Spectrochim Acta A 57:469–481
2. (a) Da Silva JFM, Garden S, Pinto AC (2001) The chemistry of isatins: a review from 1975 to 1999. J Braz Chem Soc 12:1–105; (b) Cerchiaro G, Ferreira AM (2006) Copper complexes with oxindole-derivatives as potential pharmacological agents. J Braz Chem Soc 17:1473–1485
3. (a) Shaban MAE, Nasr AZ, Taha MAM (1990) Synthesis of condensed 1,2,4-triazolo-heterocycles. J Islamic Acad Sci 3:174–179; (b) Abdulghani AJ, Sahan SM (2012) Synthesis, structure and characterization of new amic acid derivatives of 3-amino-1,2,4-triazole and their complexes with some metal ions. In: Chemistry for sustainable development, Springer, Dordrecht, pp 103–120
4. Naylor AW (1964) Complexation of 3-amino-1,2,4-triazole in plant metabolism. J Agric Food Chem 121:21–25
5. (a) Abdulghani AJ, Alabidy NM (2011) Synthesis and biological activity study of new Schiff and Mannich base ligands derived from isatin and 3-amino-1,2,4-triazole and their metal complexes. J Chem Chem Eng 5:61–72; (b) Sahan SM (2002) Synthesis, structure and characterization of new Schiff bases and amic acid derivatives of 3-amino-1,2,4-triazole and their complexes with some metal ions. MSc thesis, College of Science Department of Chemistry, University of Baghdad
6. Chung YM, Im YJ, Kim JN (2002) Baylis-Hillman reaction of isatin derevatives: isatins as a new entry for the Baylis-Hillman reaction. Bull Korean Chem Soc 23:1651–1654
7. Cheah WC, Black DS, Gash WK, Kumar N (2008) Synthesis of anti-bacterial peptidomimetics derived from N-acylisatins. Tetrahedron 49:2965–2968
8. Jacobs TL, Winstein S, Linden GB, Robson JH, Levy EF, Seymour D (1955) 2-Hydroxycinchoninic acid. Org Synth 3:456–457
9. Garcia-Raso A, Fiol JJ, Molins E, Calafat AM, Marzilliet PA, Marzilli LG (1995) Metallation of isatin(2,3-indolinedione). X-ray structure and solution behavior of bis(isatinato)mercury (II). Metal Based Drugs 2:81–90
10. Rochow EG (1960) Inorg Synth 4:218. McGraw Hill, New York
11. Rosu T, Pasculescu S, Laza V, Chifiriuc C, Cernat R (2006) Copper(II) complexes with ligands derived from 4-amino-2,3-dimethyl-1-phenyl-3-pyrazolin-5-one: synthesis and biological activity. Molecules 11:904–914
12. Silverstein RM, Webster FX, Kiemle DJ (2005) Spectrometric identification of organic compounds, 7th edn. Wiley, New York
13. Gudasi KB, Marvalli PB, Goudar TR (2005) Thermokinetic and spectral studies of Nb (V) complexes with 3-substituted 4-amino-5-mercapto-1,2,4-triazole Schiff bases. J Serb Chem Soc 70:643–650
14. Cansiz A, Koparr M, Demirda A (2004) Synthesis of some new 4,5-substituted-1,2,4-triazole-3-thiolderivatives. Molecules 9:204–212
15. Doerg DR, Nicmezura WP (1989) Suicide inactivation of lactoperoxidase by 3-amino-1,2,4-triazole. Res Toxicol 2:100–103
16. Sridhar SK, Ramesh A (2001) Synthesis and pharmacological activities of hydrazones, Schiff and Mannich bases of isatin derivatives. Biol Pharm Bull 24:1149–1152
17. Bergman J, Engqvist R, Stalhandske C, Walberg H (2003) Studies of the reactions between indole-2,3-diones (isatins) and 2-amino benzylamine. Tetrahedron 59:1033–1048
18. Agarwal RK, Singh L, Sharma DK (2006) Synthesis, spectral and biological properties of copper(II) complexes of thiosemicarbazones of Schiff bases derived from 4-aminoantipyrine and aromatic aldehydes. Bioinorg Chem Appl 59509:1–10 doi:10.1155/BCA/2006/59509
19. El-Saied FA, Ayad MI, Issa RM, Aly SA (2000) 4-Azomalononitrile antipyrine complexes of some first row transition metals. Pol J Chem 74:919–926
20. Pandeya SN, Raja AS (2002) Synthesis of isatin semicarbazones as novel anticonvulsant role of hydrogen bonding. J Pharm Pharm Sci 5:266–271

21. El-Ajaily MM, El-Ferjani RM, Maihub AA (2007) Preparation and physical investigation of complexes derived from 4-dimethyl-aminobenzaldehyde and 4-aminoantipyrene Schiff base with Ni(II), Cu(II), Rh(III), and Pt(IV) ions. Jordan J Chem 2:287–296
22. Maquestiau A, Beugnies D, Flammang R, Freiermuth B, Wentrup C (1990) Flash-vacuum pyrolysis of 1-acylbenzotriazole: direct observation of cyclopenta-2,4-dienylidenemethaneimines by tandem mass spectrometry and low-temperature infrared spectrometry. Org Mass Spectrom 25:197–203
23. Kriza A, Parnau C (2001) Complexes of tin(IV) with some bidentate Schiff bases derived from 1-H-indol-2,3-dione. Acta Chimica Slov 48:445–452
24. Dyer JR (1965) Applications of absorption spectroscopy of organic compounds. Prentice-Hall, Englewood Cliffs
25. Plostnieks J (1969) Oxindole derivatives. US Patent 3,428,649
26. Pandeya SN, Sriram D, Nath D, DeClercq E (1999) Synthesis, antibacterial, antifungal and anti-HIV activity of Schiff and Mannich bases of isatin with N-[6-chlorobenzothiazol-2-yl] thiosemicarbazide. Indian J Pharm Sci 61:358–361
27. Nakamoto K (1997) Infrared and Raman spectra of inorganic and coordination compounds, 5th edn. Wiley, New York
28. Al-Seif FA, Khalil MMH (2007) Reactions and characterization of group 6 and 8 metal carbonyl complexes of 5-methylisatin. J Saudi Chem Soc 11:69–276
29. Abdulghani AJ, Khaleel AMN (2009) Synthesis and characterization of new Schiff bases and amides derived from N(1) substituted isatin with dithiooxamide and their Co(II), Ni(II), Pd (II) and Pt(IV) complexes. Bioinorg Chem Appl. doi:10.1155/2009/413175
30. Abd El-Wahab ZH, Mashaly MM, Faheim AA (2005) Synthesis and characterization of cobalt (II), cerium(III), and dioxouranium(VI) complexes of 2,3-dimethyl-1-phenyl-4-salicylidene-3-pyrazoline-5-one mixed ligand complexes, pyrolytic products and biological activities. Chem Pap 59:25–36
31. Malawska B (2005) New anticonvulsant agents. Curr Top Med Chem 5:69–85
32. (a) Sutton D (1968) Electronic spectra of transition metal complexes, Mc Graw-Hill, New York; (b) Kermagoret A, Braunstein P (2008) Synthesis of nickel complexes with bidentate N, O-type ligands and application in the catalytic oligomerization of ethylene. J Chem Soc Dalton Trans 12:1564–1573. doi:10.1039/b716111e
33. (a) Lever ABP (1968) Inorganic electronic spectroscopy. Elsevier, London; (b) Solomon E, Lever A (1999) Inorganic electronic structure and spectroscopy. Wiley, New York; (c) Figgis B (1966) Introduction to ligand fields. Inter-Science, Wiley, New York
34. Bruger K (1967) Coordination chemistry: experimental methods. Butterworth, London
35. Chandra S, Jain D, Sharma AK, Sharma P (2009) Coordination modes of a Schiff base pentadentate derivative of 3-aminoantipyrine with cobalt(II), nickel(II) and copper metal ions: synthesis, spectroscopic and antimicrobial studies. Molecules 14:174–190
36. Sayyed M, Mokle S, Bukhar M, Mankar A, Surwase S, Bhusare S, Vibhute Y (2006) Synthesis of some new 2,3-diaryl-1,3-thiazolidin-4-one as antibacterial agents. Arkivok ii:187–192
37. Geary WJ (1971) The use of conductivity measurements in organic solvents for the characterization of coordination compounds. Coord Chem Rev 7:81–122
38. (a) Duval C (1963) Inorganic thermogravimetric analysis, 2nd edn. Elsevier, London; (b) Barbooti MM, Al-Sammerrai DA, Al-Ansari RM, Al-Badri HT, Roomaya AF (1984) Thermoanalytical investigations on heterocyclic organic compounds. Part IV. Thermal decomposition of 1-benzyl-1,2,3-triazolino-4,5-(N-phenyl) dicarboxiimides. Thermochim Acta 19:139–147; (c) Jasim HH (1985) The Rmo analytical measurements for binary systems consisting of non-refractory metallic oxides and peroxo salts. MSc thesis, College of Science Department of Chemistry, University of Baghdad
39. Ercage A, Yildirim SO, Akkurt M, Ozgur MU, Heinemann FW (2006) Novel isatin-Schiff base Cu(II) and Ni(II) complexes. X-ray crystal structure of bis-(4-hexylphenyl imino)-1H-indole-2(3H)-one]dichloro copper(II) complex. Chinese Chem Lett 17:243–246

Chapter 21
Propericiazine as a Reagent for the Spectrophotometric Determination of Osmium

Thimme A. Gowda

Abstract Propericiazine (PPC) is proposed as a new sensitive and selective reagent for the spectrophotometric determination of microgram amounts of osmium. PPC forms an orange-red colored species with osmium(VIII) or osmium (VI) instantaneously at room temperature in 6 M hydrochloric acid medium. The orange-red species exhibits maximum absorbance at 511 nm. Beer's law is valid over the concentration range 0.1–8.0 ppm for osmium(VIII) and 0.20–10 ppm for osmium(VI). Sandell's sensitivity of the reaction is 9.4 ng/cm^2 for osmium(VIII) and 13.5 ng/cm^2 for osmium(VI). The effects of time, temperature, acidity, order of addition of reagents, reagent concentration and diverse ions are investigated. The application of the proposed method in the determination of osmium content in synthetic syserkite mineral has been explored.

21.1 Introduction

The use of several chromogenic reagents in the spectrophotometric determination of osmium has been reviewed by several workers [1–5]. Most of the methods reported suffer from the disadvantages of one kind or the other. The various experimental conditions, such as reaction time, temperature, acidity, reagent concentration and certain foreign ions, that affect the color development and hence the analytical method require rigid control over them. Although 1,5-diphenylcarbohydrazide [6] is a sensitive reagent for osmium, it has many disadvantages. The color development is dependent on temperature, time of standing in aqueous phase, the acid concentration, the period between the extraction of the complex into chloroform and subsequent absorbance measurement and even the order of addition of reactants.

T.A. Gowda (✉)
Haranahalli Ramaswamy Institute of Higher Education,
Salagame Road, 573 201 Hassan, Karnataka, India
e-mail: ankapuratgowda@rediffmail.com

M. Gupta Bhowon et al. (eds.), *Chemistry: The Key to our Sustainable Future*, DOI 10.1007/978-94-007-7389-9_21,

In addition, the associated ions of platinum metals and base metals interfere. For this reason, despite the lack of sensitivity, thiourea [7] remains the reagent of choice for osmium. 2-Amino-8-naphthol-3,6-disulphonic acid [8] has been claimed to be a good reagent for osmium. Although there is no interference by other platinum metals, the method lacks sensitivity.

Propericiazine (PPC), 2-cyano-10-[3-(4-hydroxy-1-piperidyl)propyl]phenothiazine, is a tranquilizing agent which was used for the spectrophotometric determination of ruthenium(III) [9] and palladium(II) [10]. In the present work, PPC has been proposed as a reagent for the spectrophotometric determination of osmium in solution. The proposed method offers the advantages of simplicity, rapidity, good sensitivity and reasonable selectivity. The application of the method in the determination of osmium content in the industrial samples such as syserkite mineral has also been investigated.

21.2 Methodology

21.2.1 Materials

Osmium(VIII) solution: A stock solution of osmium(VIII) was prepared by dissolving 1 g sample of osmium tetroxide (Johnson Matthey Chemicals Ltd, London) in about 100 ml of 0.2 M sodium hydroxide in a glass-stoppered flask as described by Ayres and Wells [11]. The orange-red solution was transferred to a 1-l volumetric flask and diluted with doubly distilled water. The solution was standardised by the method of Klobbie as follows: To 25 ml of the solution were added 15 ml of 6 M sulphuric acid and 2 g of potassium iodide. The liberated iodine was titrated against 0.01 N sodium thiosulphate. The difficulty in determining the endpoint due to the green color of the reduced osmium was circumvented by adding 5 ml of benzene and titrating with standard sodium thiosulphate until a colorless layer of benzene was obtained. The sample solutions were prepared by diluting the stock solution in order to give convenient strengths.

Osmium(VI) solution: Osmium(VI) solution was prepared by dissolving a known amount of potassium osmate (Johnson Matthey Chemicals Ltd, London) in a known volume of 1 M hydrochloric acid.

Osmium(IV) solution: Osmium(IV) solution was prepared by dissolving 0.54 g of potassium chloroosmate (Johnson Matthey Chemicals Ltd, London) in a known volume of 1 M hydrochloric acid and stored in a refrigerator.

PPC solution: A 0.2 % solution of PPC was prepared in doubly distilled water and stored in an amber bottle in a refrigerator.

Diverse ions: Aqueous solutions of diverse ions of suitable concentrations were prepared using analytical grade reagents.

Beckman Model DB spectrophotometer with matched 1 cm thick silica cells (Beckman Instruments, Inc., Fullerton, USA) was used for absorbance measurements.

21.2.2 Standard Procedure for the Determination of Osmium(VIII) or Osmium(VI)

To an aliquot of the sample solution containing 2.5–200 μg of osmium(VIII) or 5–250 μg of osmium(VI), were added 15 ml of 10 M hydrochloric acid and 3 ml of 0.2 % PPC solution and diluted to 25 ml with doubly distilled water. The solution was shaken well and the absorbance was measured at 511 nm against a reagent blank prepared in the same way. The amount of osmium(VIII) or osmium (VI) was calculated using the standard calibration curve.

21.3 Results and Discussion

PPC is readily oxidized to the orange-red colored species by osmium(VIII) or osmium(VI) at room temperature (25 ± 2 °C) in hydrochloric, sulphuric or phosphoric acid medium (Fig. 21.1). The sensitivity and stability of the orange-red species, which is believed to be a free radical [12–14], depend on the nature and concentration of the acid employed. The sensitivity in the three-acid media is in the order HCl ≅ H_3PO_4 ≅ H_2SO_4. The color development in acetic acid medium is very slow. The orange-red free radical formed in sulphuric acid medium is stable for only 10 min. Nitric acid cannot be used as it oxidizes PPC to an orange-red intermediate. Maximum color intensity is achieved in 4.5–8.0 M hydrochloric acid or 4.0–6.0 M phosphoric acid medium. The maximum color intensity is not observed below 4.5 M hydrochloric acid. The reagent undergoes further oxidation at acidities higher than 8 M hydrochloric acid or 6.0 M phosphoric acid. The stability of the red species in 6 M hydrochloric acid and phosphoric acid is 2 and 1 h respectively. Hydrochloric acid medium (6 M) was selected for further studies because of its longer stability and less interference of foreign ions. Osmium(IV) did not produce any colored species with PPC. The structures of PPC and the orange-red species are shown in Fig. 21.1.

21.3.1 Absorption Spectra

The absorption spectra of the orange-red colored oxidation product of PPC, osmium (VIII), osmium(VI) and PPC in 6 M hydrochloric acid medium are shown in Fig. 21.2.

The orange red-colored species exhibits maximum absorbance at 510–512 nm and the reagent, osmium(VIII) and osmium(VI) do not absorb near this wavelength range thus promoting excellent analytical conditions. Further studies were made at 511 nm.

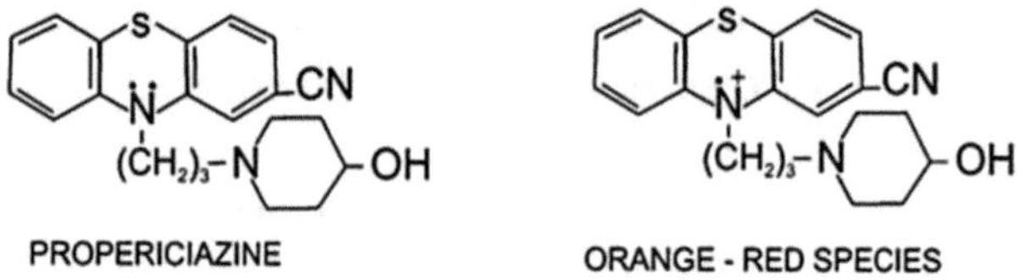

Fig. 21.1 Structures of propericiazine and *orange-red* species

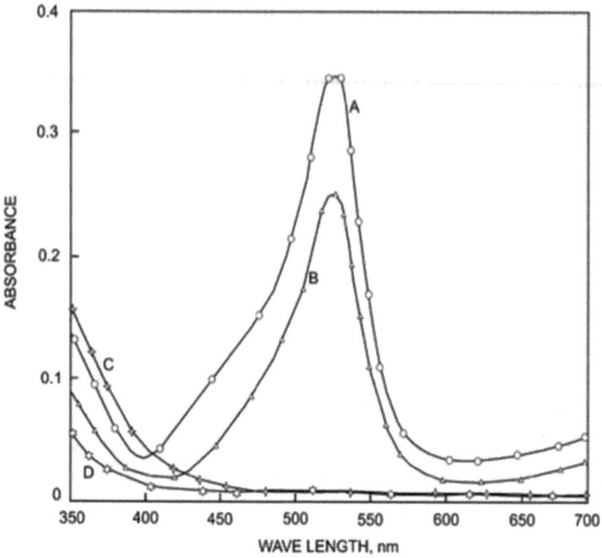

Fig. 21.2 Absorption spectra of solutions of *red* species, reagent blank and osmium(VIII)/osmium(VI) in hydrochloric acid medium. *A*: Solution of *orange red* species of PPC with osmium(VIII) versus reagent blank (concentration of osmium(VIII) = 5 μg/ml). *B*: Solution of *orange red* species of PPC with osmium(VI) versus reagent blank (concentration of osmium (VI) = 5 μg/ml). *C*: Reagent PPC blank versus water. *D*: Solution of osmium(VIII)/osmium (VI) versus water (concentration of osmium(VIII)/osmium(VI) = 5 μg/ml)

21.3.2 Effect of Reagent Concentration

The effect of reagent concentration was examined by measuring the absorbance at 511 nm of solutions containing 4 μg/ml of osmium(VIII) or 6 μg/ml of osmium (VI) and varying amounts of PPC. A 30-fold molar excess of PPC over osmium (VIII) or a 20-fold molar excess of PPC over osmium(VI) was needed for completeness of the reaction. Addition of more reagent had no effect on the absorbance reading. An optimum amount of 3 ml of 0.2 % reagent solution was used.

21.3.3 Effects of Sequence of Addition of Reactants and Temperature

There was no change in the absorbance reading when the order of addition of reactants was changed. The absorbance was not affected by temperature in the range of 8–52 °C, afterwards the absorbance gradually decreased with increase in temperature.

21.3.4 *Calibration, Range, Sensitivity and Precision*

Beer's law was obeyed over the concentration range 0.10–8.0 ppm of osmium(VIII) or 0.20–10.0 ppm of osmium(VI). The optimum concentration ranges evaluated by Ringbom's method [15, 16] were 0.8–7.5 ppm for osmium(VIII) and 1.0–9.8 ppm for osmium(VI). The molar absorptivity of the orange-red colored species is 1.9×10^4 $lmol^{-1}$ cm^{-1} for osmium(VIII) and 1.41×10^4 $lmol^{-1}$ cm^{-1} for osmium (VI) at 511 nm. For log $I_o/I = 0.001$, the sensitivity of the reaction as calculated from Beer's law data is 9.4 ng/cm^2 for osmium(VIII) and 13.5 ng/cm^2 for osmium(VI). The standard deviation calculated from 10 determinations for solutions containing 2 ppm of osmium(VIII) and 3 ppm of osmium(VI) are 0.0020 and 0.0028 respectively.

21.3.5 *Effect of Diverse Ions*

In order to assess the possible analytical potential of the method, the effects of some ions which often accompany osmium were studied. For these studies different amounts of ionic species were added to solutions of 50 μg of osmium(VIII) in 6 M hydrochloric acid in 25-mL volumetric flasks and the color was developed as outlined by the standard procedure. The following amounts (μg/ml) of foreign ions were found to give less than 2 % error in the determination of 1 μg/ml of osmium (VIII): Cu(II), 450; Ni(II), 420; Co(II), 450; Fe(III), 5600; Mg(II), 6500; Ca(II), 8400; Zn(II), 4200; Ba(II), 4800; Sr(II), 4200; U(VI), 2400; Ru(III), 0.5; Pd(II), 0.5; Rh(III), 25.0; Ir(III), 35.0; Pt(IV), 20; fluoride, 2600; phosphate, 10000; bromide, 3200; iodide, 2.0; nitrate, 6000; sulphate, 10000; acetate, 10000; citrate, 1800; oxalate, 2600; EDTA, 4500. Cations such as Ag(I), V(V), Ce(IV), Au(III), dichromate and permanganate interfered in the determination at all concentrations.

21.3.6 *Applications*

Determination of Osmium in Ores

Analysed samples of osmiridium or syserkite were not available. Therefore, synthetic mixtures containing platinum group metals corresponding to osmiridium or syserkite were prepared and the osmium content was determined using the standard procedure. The results are presented in Table 21.1.

21.4 Conclusions

The sensitivity of the proposed method is greater than that of *o*-(β-benzoyl-thiourido) benzoic acid [17], 2-mercaptobenzothiazole [18], tiron [19], 1-carbamidine-3-methyl-5-pyrazolene [20], β-benzoyl-α-(o-tolyl)thiourea [21], thiobenzohydrazine [22],

Table 21.1 Determination of osmium in synthetic mixtures corresponding to osmiridium or syserkite

Osmium present (ppm)	Ir (ppm)	Ru (ppm)	Pt (ppm)	Osmium found[a] (ppm)
0.8	0.200	0.120	0.040	0.82
1.5	0.375	0.225	0.075	1.48
2.4	0.600	0.360	0.120	2.39
3.0	0.750	0.450	0.150	2.99
5.0	1.250	0.750	0.250	5.00

[a]Mean of five determinations

isoniazid [23] which have been proposed as sensitive spectrophotometric reagents for osmium. In these methods, however, the formation of complexes is quite slow in aqueous medium at room temperature and conditions for obtaining reproducible color development are usually critical.

The proposed reagent is sensitive and selective for the spectrophotometric determination of microgram amounts of osmium. The application of the proposed method in the determination of osmium content in synthetic ores has been explored.

Acknowledgements The author thanks the INSA, New Delhi for financial assistance and Dr L. Julou and J. Molle of Rhone-Poulenc Centre, Nicolas Grillet, Paris, France, for the generous supply of pure PPC.

References

1. Beamish FE, Van Loon JC (1972) Recent advances in the analytical chemistry of noble metals. Pergamon Press, Oxford, pp 306–345
2. Beamish FE (1966) The analytical chemistry of noble metals. Pergamon Press, Oxford, pp 366–374
3. Beamish FE (1965) A critical evaluation of colorimetric methods for determination of noble metals-III. Rhodium, Iridium, Ruthenium, Osmium and Gold. Talanta 12:789–816
4. Boltz DF, Mellon MG (1968) Light absorption spectrometry. Anal Chem 40:255R–273R
5. Howell JA, Hargis LG (1982) Ultraviolet and light absorption spectrometry. Anal Chem 54:171R–188R
6. Goldstein G, Manning DL, Menis O, Dean JA (1961) Spectrophotometric method for the determination of osmium: extraction and determination of osmium in situ with 1,5-diphenylcarbohydrazide. Talanta 7:300–306
7. Kalyanaraman S, Khopkar SM (1975) The extraction of osmium with tributyl phosphate and photometric determination with thiourea. Anal Chim Acta 78:231–233
8. Agarwala BV, Ghose AK (1973) Spectrophotometric determination of osmium with 2R-acid in the presence of other platinum metals. Talanta 20:129–130
9. Gowda TA, Gowda SH, Gowda MNM (1983) Propericiazine as a reagent for the spectrophotometric determination of ruthenium(III). Anal Chim Acta 154:347–350
10. Gowda TA, Gowda SH, Gowda MNM (1983) Spectrophotometric determination of palladium (II) with propericiazine. Anal Chem 55:1816–1817
11. Ayres GH, Wells WN (1950) Spectrophotometric determination of osmium with thiourea. Anal Chem 22:317–320

12. Dwivedi PC, Gurudath Rao SN, Bhat SN, Rao CNR (1975) Spectrophotometric studies of electron donor properties and radical cations of phenothiazine derivative. Spectrochim Acta 31A:129–135
13. Forrest IS, Forrest FM, Berger M (1958) Free radicals as metabolites of drugs derived from phenothiazine. Biochim Biophys Acta 29:441–442
14. Stan M, Dorneanu V, Ghimicescu GH (1977) Phosphomolybdic acid as a reagent for the spectrophotometric determination of certain phenothiazine derivatives of pharmaceutical products. Talanta 24:140–142
15. Ayres GH (1949) Evaluation of accuracy in photometric analysis. Anal Chem 21:652–657
16. Ringbom A (1939) Accuracy of colorimetric determinations. Z Anal Chem 115:332–343
17. Majumdar AK, Bhowal SK (1966) Spectrophotometric determination and solvent extraction of osmium with o-(β-benzoylthiourido)benzoic acid as reagent. Anal Chim Acta 35:479–483
18. Bera BC, Chakrabartty MM (1966) Spectrophotometric determination of osmium with 2-mercaptobenzothiazole. Microchem J 11:420–429
19. Majumdar AK, Savariar CP (1959) Tiron as a spectrophotometric reagent for the estimation of osmium. Anal Chim Acta 21:146–150
20. Poddar SN, Adya JN (1964) Spectrophotometric determination of osmium with pyrazolone derivatives. Sci Cult Part I 30:50–51
21. Bhowal SK (1975) Spectrophotometric determination of osmium with 1-benzyl-3-(o-tolyl) thiourea. Curr Sci 44:157–158
22. Shome SC, Gangopadhyay PK, Gangopadhyay S (1976) Extraction and spectrophotometric determination of osmium(VIII). Z Anal Chem 282:141–142
23. Sindhwani SK, Singh RP (1973) Spectrophotometric determination of osmium by using acenapthenequinonemonoxime. Microchem J 18:627–635

Chapter 22
An Assessment of Physico-Chemical Parameters of Ganga Water Using Multivariate Analysis

Sukarma Thareja

Abstract An extensive investigation of the physico-chemical parameters of water samples of Ganga river, surface water, treated water and ground water at Kanpur during pre-monsoon, monsoon, and post-monsoon seasons of India is presented. Correlation coefficients were calculated between different pairs of parameters to identify the highly correlated and interrelated water quality parameters and *t*-test was carried out for checking significance. Principal Components Analysis (PCA) was done to ascertain the parameters responsible for variability in water quality variance. Multiple linear regression analysis is also discussed.

22.1 Introduction

It is known that good water quality produces healthier humans. River Ganga is the lifeline of Kanpur and its water is used for domestic and agricultural purposes; therefore, effective maintenance of water quality is required through appropriate measurements. River Ganga is of particular importance in the study of surface water pollution because effluents from ternaries, municipal sewage, agricultural and urban runoff are discharged into it during different seasons leading to considerable change in water quality. Physico-chemical and micro-biological characteristics describe the quality of water; therefore, an analysis of physico-chemical parameters of Ganga water has been reported by many workers [1–3]. Regular monitoring of all the parameters is a very difficult and laborious task even if adequate manpower and laboratory facilities are available. Therefore, statistical correlation technique has been used for comparison of physico-chemical parameters. The first part of present work deals with the study of 15 physico-chemical parameters like pH, temperature, turbidity (Tu), total hardness (TH), Fe^{3+}, Cl^{-}, total dissolved solids (TDS),

S. Thareja (✉)
Department of Chemistry, Christ Church College, CSJM Kanpur University, Kanpur, Uttar Pradesh, India
e-mail: sukarmathareja@yahoo.com

M. Gupta Bhowon et al. (eds.), *Chemistry: The Key to our Sustainable Future*, DOI 10.1007/978-94-007-7389-9_22, © Springer Science+Business Media Dordrecht 2014

Ca^{2+}, SO_4^{2-}, NO_3^-, F^-, total alkalinity (TA), Mg^{2+}, oxygen consumption (OC) and suspended solids (SS) of Ganga river water in Kanpur. The objective is to minimize the complexity and dimensionality of large set of data. Systematic calculation of correlation coefficient between physico-chemical parameters has been carried out and significant correlation has been further verified by using *t*-test [4]. Physico-chemical properties of Ganga water at various places in UP like Buxar (Unnao), Narora and Kannauj in and around Haridwar have been reported [1, 5]. The seasonal analysis of Kanpur water showed that the extent of pollution varied in different seasons [6a, 6b]. The steep growth in population due to rapid urbanization and industrial development of Kanpur city has increased the demand of water manifold. In the present work we consider the drinking water quality of the filtration plant and other sources at Kanpur.

The particular problem in the case of water quality monitoring is the complexity associated with analyses of a large number of measured variables [7a, b]. The data sets contain rich information about the behavior of the water resources. The classification, modeling and interpretation of monitoring data are the most important steps in the assessment of water quality. Surface water, groundwater quality assessment and environmental research employing multi-component techniques are well described in the literature [8a, b]. Multivariate statistical approaches allow deriving hidden information from the data set about the possible influence of the environment on water quality [9, 10].

Principal component analysis (PCA) method provides a unique solution so that the original data can be reconstructed from the results. Principal Components (PCs) actually take the cloud of data points and rotate it such a way that maximum variability is visible. In other words, it identifies the most important gradients. In recent years many studies have been done using PCA in the interpretation of water quality parameters [11]. In the present study we have also carried out PCA of physico-chemical parameters of water for assessing the water quality of Ganga river.

Water quality index (WQI) [12] is an effective tool for assessment and management of water quality. We have calculated WQI from the point of view of suitability of surface water, treated water and ground water.

22.2 Methodology

22.2.1 Experimental

Water samples were collected from Kanpur sampling stations during pre-monsoon (April-May), monsoon (July-August) and post- monsoon (October-November) phase in the year 2008. During sampling, pH, temperature and turbidity were determined using digital pH meter, thermometer and turbidity meter respectively. $F^-$and nitrate were estimated using colorimetric method. The laboratory analysis of samples was done using standard methods (APHA, 1998) [13]. Titrimetric method was used for

Table 22.1 Average values of physico-chemical parameters of Ganga river water for the year 2008 at Kanpur. HDL: Highest Desirable Limit; MPL: Maximum Permissible Limit

			Drinking water WHO standard		
No.	Parameter	Units	HDL	MPL	Experimental values (range)
1	Temperature	°C			22–30
2	Turbidity	NTU	5	10	18–470
3	pH value	–	6.5–8.5	No relaxation	8.4–8.9
4	Total hardness (as $CaCO_3$)	mg/l	300	600	122–212
5	Iron	mg/l	0.3	1.0	0.2–0.8
6	Chlorides	mg/l	250	1,000	7–26
7	Dissolved solids	mg/l	500	2,000	256–500
8	Calcium	mg/l	75	200	28–48
9	Sulphate	mg/l	200	400	50–91
10	Nitrate	mg/l	50	No relaxation	0–1.77
11	Fluoride	mg/l	1.0	1.5	0–0.8
12	Total alkalinity	mg/l	200	600	13–246
13	Magnesium	mg/l	30	150	9.23–26.24
14	Oxygen consumption[a]	mg/l	3.0	No relaxation	2.4–7.8
15	Suspended solids	mg/l	20	150	70–280

[a]Observed from $KMnO_4$ at 37 °C in 3 h

the determination of total alkalinity and gravimetric method for total dissolved solids and total suspended solids. Mohr's argentometric titration method was used for chloride [14], sulphate was estimated using turbidometric method [14] whereas Ca^{2+}, Mg^{2+} and TH were determined by EDTA titrimetric method [14]. Atomic absorption spectrophotometer was used for the determination of Fe contents. All the chemicals used were of analytical reagent (AR) grade. The average values of physico-chemical parameters of Ganga river water for the year 2008 at Kanpur are listed in Table 22.1. The observed values of various physico-chemical parameters of water samples were also compared with standard values recommended by the World Health Organization (WHO) [15a].

22.2.2 *Sampling*

In the present work we report the quality of water taken from 200 MLD filtration plant sites and ground water resources. The sites are GW, CGW, TW and the ground water sampling site is named as PSP i.e. postal station pumps (hand pumps). These PSP and surface water samples are Grab samples collected from six different zones of Kanpur during monsoon (July and August), autumn (September and October), winter (November to January), spring (February and March) and summer (April to June) for the period July 2008-June 2009.

Experimental values of the various (15) physico-chemical (WT, pH, turbidity, TA, TDS, TH, Ca^{2+}, Mg^{2+}, Cl^-, NO_2^-, NO_3^-, SO_4^{2-}, Cr^{6+}, Fe^{3+}, F^-) parameters in

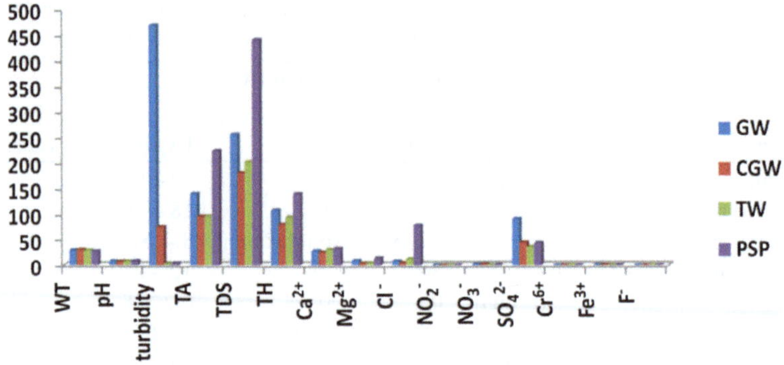

Fig. 22.1 Physico-chemical parameters in monsoon season

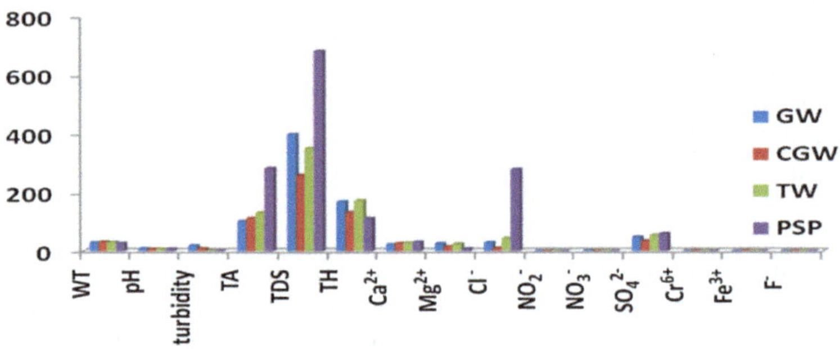

Fig. 22.2 Physico-chemical parameters in summer season

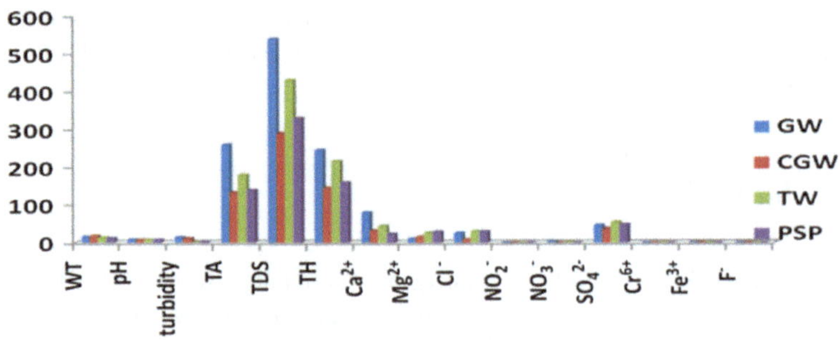

Fig. 22.3 Physico-chemical parameters in winter season

different seasons for Ganga water (GW), lower canal Ganga (CGW), treated water from filtration plant (TW), ground water sampling site postal station pumps (PSP) for April 2008-June 2009 are reported in Figs. 22.1, 22.2, 22.3, 22.4 and 22.5 respectively. For seasonal variation study at GW, CGW, TW and PSP, we concentrated on the above mentioned parameters.

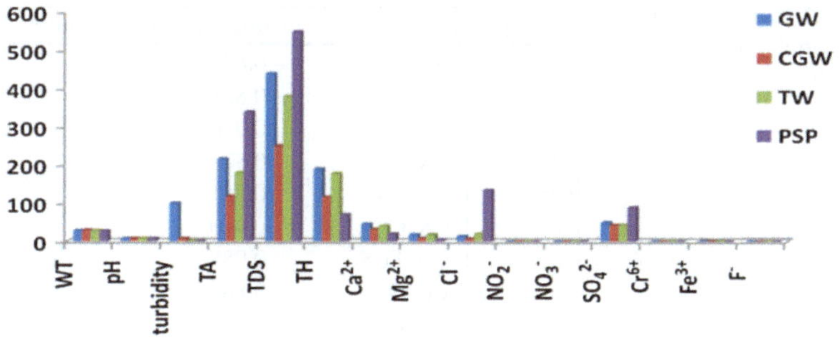

Fig. 22.4 Physico-chemical parameters in autumn season

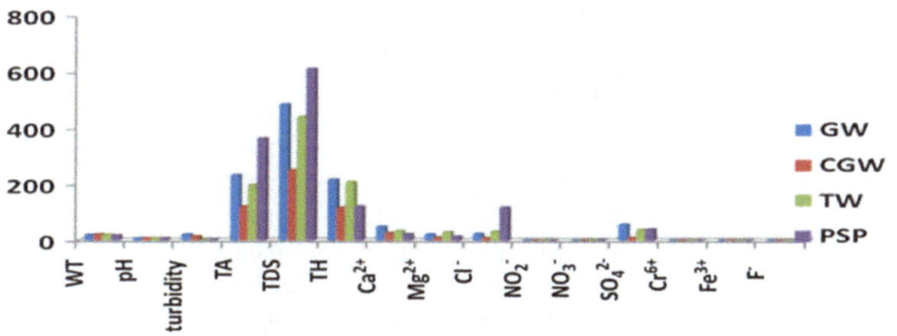

Fig. 22.5 Physico-chemical parameters in spring season

22.2.3 Correlation Study

In order to calculate the correlation coefficients, a correlation matrix was constructed for the 15 physico-chemical parameters (pH, temperature, turbidity (Tu), total hardness (TH), Fe^{3+}, Cl^-, total dissolved solids (TDS), Ca^{2+}, SO_4^{2-}, NO_3^-, F^-, total alkalinity (TA), Mg^{2+}, oxygen consumption (OC) and suspended solids (SS)) for Ganga water for the year 2008, by calculating the coefficients of different pairs of parameters and correlation for significance was further tested by applying *t*-test [15b]. The results are listed in Table 22.2.

22.2.4 Data Processing by Principal Components Analysis (PCA)

For studying PCA for the year 2008 we concentrated on 14 parameters (Ca^{2+}, Mg^{2+}, SO_4^{2-}, WT, F^-, pH, Fe, OC, Cl^-, TH, TA, Tu, SS, TDS) for 12 months. The 168 (14×12) parameters obtained from the laboratory analysis were used as variable inputs for Principal Components Analysis (PCA). PCA was performed using the SPSS package. Prior to the analysis, the data were standardized to produce

Table 22.2 Different paired sample correlations (Year 2008)

Paired parameters		Correlation coefficient (r)	[*t*]
Pair 1	Temp & NO_3^-	−0.829	21.514
Pair 2	pH & OC	0.821	4.762
Pair 3	TA & TH	0.970	4.694
Pair 4	TA & Ca^{2+}	0.821	9.145
Pair 5	TA & Mg^{2+}	0.851	10.179
Pair 6	TA & Cl^-	0.910	10.511
Pair 7	TA & TDS	0.988	8.233
Pair 8	TA & OC	0.880	9.942
Pair 9	TA & F^-	0.836	9.850
Pair 10	TH & Mg^{2+}	0.916	9.973
Pair 11	TH & Cl^-	0.870	10.091
Pair 12	TH & TDS	0.993	8.653
Pair 13	TH & F^-	0.912	9.516
Pair 14	Ca^{2+} & TDS	0.816	8.711
Pair 15	Mg^{2+} & Cl^-	0.872	2.175
Pair 16	Mg^{2+} & TDS	0.889	9.179
Pair 17	Mg^{2+} & SS	−0.935	3.261
Pair 18	Mg^{2+} & F^-	0.977	6.239
Pair 19	Cl^- & TDS	0.877	9.256
Pair 20	Cl^- & OC	0.985	4.639
Pair 21	Cl^- & SS	−0.923	3.348
Pair 22	TDS & OC	0.827	9.070
Pair 23	TDS & SS	−0.971	3.049
Pair 24	TDS & F^-	0.890	9.051
Pair 25	OC & SS	−0.894	3.817
Pair 26	SS & F^-	−0.928	4.016

Note: Significant if value of *t* using the 0.05 significance level is >2.18

a normal distribution of all variables. In PCA, the data cluster is rotated by subtracting the mean of the data and dividing by the standard deviation. So the centroid of whole data set is zero and the relative location of all the points remains the same. This type of ordination reduces the dimensionality of the data set and minimizes the loss of information caused by reduction of Eigen values and the corresponding Eigen vectors of standardized covariance matrix were calculated. Then a number of PCs were selected from the initial PC according to their Eigen values and scree diagram. The Eigen value associated with each principal component tells us how much variation in the data set it explains. They are usually expressed as a percentage of the total variation in the data set and percent of variance contributed in each PC.

22.2.5 Calculation of the Water Quality Index

The Water Quality Index (WQI) was calculated using weighted arithmetic index method. The method involved three steps.

In the first step, the quality rating scale qi for each parameter was assigned by dividing the Ci, the concentration of each chemical parameter in water sample in mg/l to its respective standard, Si, where Si, is the Indian drinking water standard for each chemical parameter according to guidelines of the BIS 10500, 1991 [16] and the results are multiplied by 100.

$$\mathrm{qi} = (\mathrm{Ci}/\mathrm{Si}) \times 10 \tag{22.1}$$

In the second step, wi the unit weight factor was calculated by a value inversely proportional to the recommended standard Si of the corresponding parameter where qi is the quality rating.

$$\mathrm{wi} = 1/\mathrm{Si} \tag{22.2}$$

The overall Water Quality Index (WQI) was calculated by aggregating the quality rating qi with unit weight wi linearly.

$$\mathrm{WQI} = \mathrm{wiqi} = \mathrm{SIi} \tag{22.3}$$

where SIi is the sub index of ith parameter.

$$\text{Overall WQI} = \sum \mathrm{WQI}/\sum \mathrm{wi} \tag{22.4}$$

22.3 Results and Discussion

22.3.1 Physico–Chemical Characteristics of Ganga River Water at Kanpur for 2008 by Using Correlation Study

The observed pH values ranging from 8.4 to 8.9 show that the water samples are slightly alkaline in pre-monsoon season. These values are within the maximum permissible limit prescribed by WHO [15a]. Other parameters like turbidity (18–470 NTU), TH (122–212 mg/l), Fe^{3+} (0.2–0.8 mg/l), chloride (7–26 mg/l), TDS (256–500 mg/l), Ca^{2+} (28–48 mg/l), $SO_4^{\ 2-}$(50–91 mg/l), $NO_3^{\ -}$ (0–1.77 mg/l), F^- (0–0.4 mg/l), TA (13–246 mg/l), Mg^{2+} (9.23–26.24 mg/l), OC (2.4–7.8 mg/l), SS (70–280 mg/l) are found within the highest desirable or maximum permissible limit set by WHO. However, turbidity and Fe contents are observed to be on the higher side compared to the maximum permissible limit set by WHO in all the seasons. For the year 2008, a significant positive correlation was found between total alkalinity and total hardness ($r = 0.970$, $t = 4.694$), Ca^{2+} ($r = 0.821$, $t = 9.145$), Mg^{2+} ($r = 0.851$, $t = 10.179$), Cl^- ($r = 0.91$, $t = 10.511$), TDS ($r = 0.988$, $t = 8.233$), OC ($r = 0.880$, $t = 9.942$) and F^- ($r = 0.836$, $t = 9.850$). This shows that with increase or decrease in the values of TA, TH, Ca^{2+}, Mg^{2+}, Cl^-, total dissolved solids, OC and F^- also exhibit increase or decrease in their values.

Total hardness bears a positive correlation with Mg^{2+} (r = 0.916, t = 9.973), Cl^- (r = 0.870, t = 10.091), TDS (r = 0.993, t = 8.653) and F^- (r = 0.912, t = 9.516). The correlation between Mg^{2+} and Cl^- is high (r = 0.872, t = 2.175) as well as the correlation between Mg^{2+} and F^- (r = 0.977, t = 6.239), suggesting that the total hardness of water samples is mainly due to the presence of $MgCl_2$ and MgF_2 [17]. Chloride ion bears significant positive correlation with TDS (r = 0.877, t = 9.256), Mg^{2+} (r = 0.872, t = 2.175) and OC (r = 0.985, t = 4.639). It reveals that Mg^{2+} mainly remains present as $MgCl_2$. A significant positive correlation was found between Ca^{2+} and TDS (r = 0.816, t = 8.711). The magnesium content of Ganga water increases or decreases with increase or decrease in the value of TDS (r = 0.889, t = 9.179). OC and SS are significantly negatively correlated showing that when SS increases, OC decreases. High levels of suspended solids (SS) in the water systems increase the biological and chemical oxygen demand, which deplete the dissolved oxygen level in the aquatic system. High negative correlations were also found between SS and TDS, Cl^- and Mg^{2+}.

22.3.2 *Seasonal Variation of Water Physico-Chemical Parameters at Different Sites for 2008–2009*

Water Temperature

In the present study, the temperature varied from 14 to 32 °C. In 2002 and 2003 the mean monsoon temperature was 25.8 °C and 26.1 °C respectively. However, for the same seasons in 2008–2009, mean temperature was found to be 30 °C. Mean temperatures of 20 and 30 °C for spring and summer seasons were observed. It is found that over the passage of time from 2002 to 2008–2009 the monsoon season temperature has increased significantly.

Hydrogen Ion Concentration (pH)

The pH of water is important because many biological activities can occur only within a narrow range. Thus, any variation beyond an acceptable range could be fatal to a particular organism. In the present study, the mean pH of GW samples for monsoon season was found to be 8.4 (Fig. 22.1). However, for spring and summer seasons it is 8.5 and 8.9 respectively. The present study also shows that the pH is alkaline in most of the samples and it ranges from 7.7 to 8.9.

Total Alkalinity

A mean TA of 140 mg/l in surface GW was obtained for the monsoon season. However, for the five seasons TA varied from 104 to 260 mg/l. Total alkalinity for all the seasons for treated water (TW) and GW is within the permissible limit of WHO which is 200 mg/l except in the winter season for GW samples when the TA is greater than the HDL prescribed by WHO. TA for GW is lowest during summer and highest during winter as shown in Figs. 22.2 and 22.3 respectively.

Total Hardness

Hardness is an important parameter in decreasing the toxic effect of poisonous element. The measured value of TH for the monsoon season increased to 108 mg/l compared to 81.70 and 97.45 mg/l respectively for GW samples in 2002 and 2003 [2]. The TH of surface water in GW and CGW and ground water samples at PSP and treated water was found to be in the range of 80–246 mg/l, within the prescribed limit of 300 mg/l set by WHO.

Turbidity

The present study results show that the mean turbidity of GW samples in the monsoon season has increased tremendously to 470 NTU. Values of turbidity for TW samples and ground water samples at PSP for all the seasons were found to be lower than the HDL prescribed by WHO. However the surface water samples in GW and CGW showed higher values than the HDL. The immersion of idols could be the reason of the high value of turbidity shown by surface GW samples during the festival season in autumn (September and October) as depicted in Fig. 22.4.

Total Dissolved Solids

TDS indicate the total amount of inorganic chemicals in solution. TDS of GW, CGW and PSP showed seasonal fluctuation for the study period. TDS values of ground water samples at PSP in summer, autumn and spring (Figs. 22.2, 22.4 and 22.5) are higher than the HDL of 500 mg/l prescribed by WHO.

Chloride

The chloride concentration in water indicates the presence of organic waste particularly of animal origin [18]. The maximum Cl^- contents are found to be 280 mg/l in the summer season in PSP water samples. The mean chloride contents of GW samples in the monsoon season are found to be 8 mg/l. For surface water samples in GW, Cl^- concentration increases from monsoon, autumn, spring, winter to the summer season in the range of 8–30 mg/l. The high chloride content in PSP water makes it taste salty and also promotes pipe corrosion.

Nitrate

In the present study in PSP samples the NO_3^- levels are below 1 mg/l in the monsoon, autumn, winter and summer season, but in the spring season its mean value is 1.772 mg/l. In GW samples in the autumn and spring seasons the nitrate level is less than 1 mg/l, but in the monsoon, winter and summer season it is more than 1 mg/l with the highest value observed in the winter season of 3.544 mg/l. In CGW samples the level is less than 1 mg/l in all the seasons except in monsoon.

Fluoride

The fluoride contents are not detectable in PSP and TW samples in all the seasons. In GW and CGW samples, the fluoride contents ranged from 0 to 0.4 mg/l which are less than the HDL of 1 mg/l prescribed by WHO for good health.

Sulphate

The mean value of SO_4^{2-} contents for surface water in GW, CGW, ground water at PSP and for TW samples is far below the maximum allowable concentration for sulphate ions in drinking water prescribed by WHO which is 250 mg/l.

Iron

Excessive iron content makes the water turbid and discoloured, and imparts an astringent taste to water. The present study shows that in monsoon, spring and autumn seasons, the iron content of CGW is greater than GW and the values are greater than the HDL prescribed by WHO which is 0.3 mg/l. The Fe contents of TW samples in the autumn season and the ground water samples at PSP in summer are also higher than the prescribed limit of WHO. It indicates that the filtration plant is not effective in reducing the iron contents of surface water during the autumn season.

Ca^{2+} and Mg^{2+} Contents

Ca^{2+} and Mg^{2+} are important contributors to water hardness. For all the seasons, the surface water at GW and CGW, ground water at PSP and for TW, the content of Ca^{2+} is greater than Mg^{2+} except for the summer season where the content of Mg^{2+} in the surface water samples at GW is greater than Ca^{2+}. The values of Ca^{2+} and Mg^{2+} obtained from surface water samples in GW and CGW, ground water samples at PSP and for TW for all the seasons except winter are within the HDL prescribed by WHO.

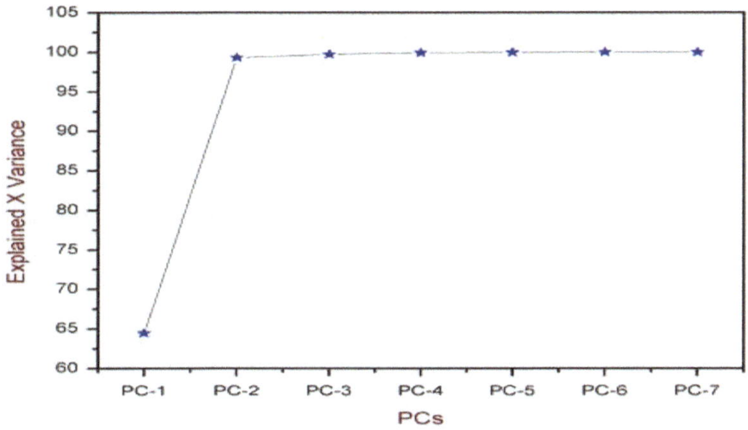

Fig. 22.6 Scree plot showing X-Variance vs PCs

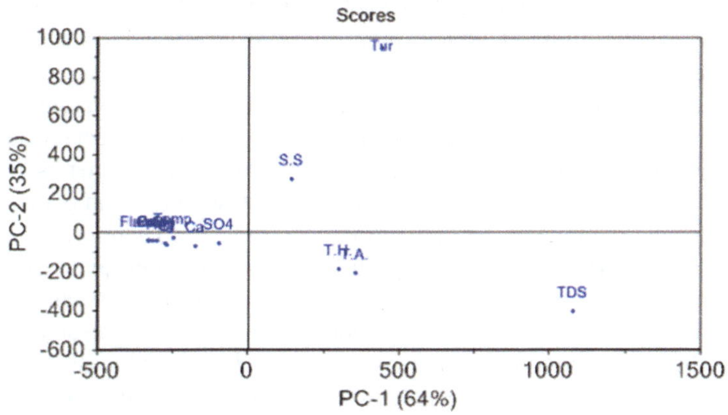

Fig. 22.7 Score Plot PC-1 vs PC-2

22.3.3 PCA of Physico-Chemical Parameters of Ganga Water

The PCA results of Ganga water are shown in Fig. 22.6, depicting a scree plot of X-Variance (also represents eigenvalue) for each component. The scree plot shows that the first two PCs are the most significant components representing more than 99.32 % of the variance in water quality of Ganga river, 64.47 % by PC-1 and 34.85 % by PC-2, showing that the first two eigen values are significant as compared to others.

As the principal components, PC-1 and PC-2 contribute about 99 % of the total variance in the data, they are sufficient to interpret our analysis. Figure 22.7 shows the 64.5 % variance along PC-1 axis where TDS lies at the rightmost position.

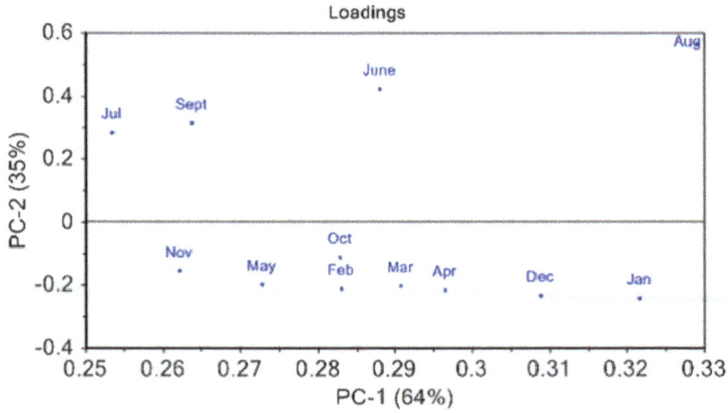

Fig. 22.8 Loading plot PC-1 vs PC-2

Principal Component one (PC-1) has a high loading of Oct, Nov, March, May, April, Dec, Feb and Jan. Figure 22.8 explains 64.5 % of the total variance. This component can be ascribed as post-monsoon season effect on water quality of Ganga river.

Principal Component two (PC-2) explains 34.8 % of the total variance, has high loading of June, July, Aug and Sep and it reflects the monsoon effect on Ganga river within Kanpur city. PCA study on the physico-chemical parameters found that TDS, TH and TA (Fig. 22.7) contribute to the construction of component one and are significant parameters in post-monsoon season. Therefore these three parameters which are responsible for main variability in water quality were monitored during post-monsoon season. On the other hand, turbidity and suspended solids are significant parameters in monsoon season as they are responsible for the main variability in water quality in monsoon season. Figures 22.7 and 22.8 show the position of Ca^{2+}, Mg^{2+}, SO_4^{2-}, temperature, F^-, pH, Fe, OC and Cl^- representing their least contribution with season change during months, and their position near to zero along PC-2 axis shows almost their no contribution at all. The frequency of sampling for these insignificant parameters may thus be reduced. Turbidity and suspended solids are monitored for monsoon season as they are responsible for variability in water quality. Variation in hardness of river water is probably due to the regular addition of sewage detergents arising from huge human activities called Anthropogenic Factor. Anthropogenic Factor is identified as responsible for explaining the physical nature of component one accounting for 64.5 % of the total variance. Turbidity and SS contribute to the construction of component two accounting for 34.9 % of total variance. Urbanized areas contribute large amounts of turbidity to nearby waters, through storm water pollution from paved surfaces such as roads, bridges and parking lots. Certain industries such as leather tanneries near the bank of Ganga have generated very high levels of turbidity from colloidal particles from their wastes. Therefore, the industrial factors are identified as responsible for explaining the physical nature of component two.

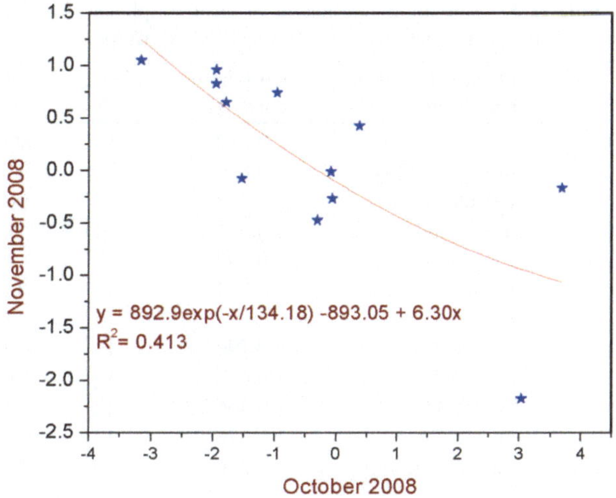

Fig. 22.9 Relationship between October and November 2008 for data of all the 14 physico-chemical parameters

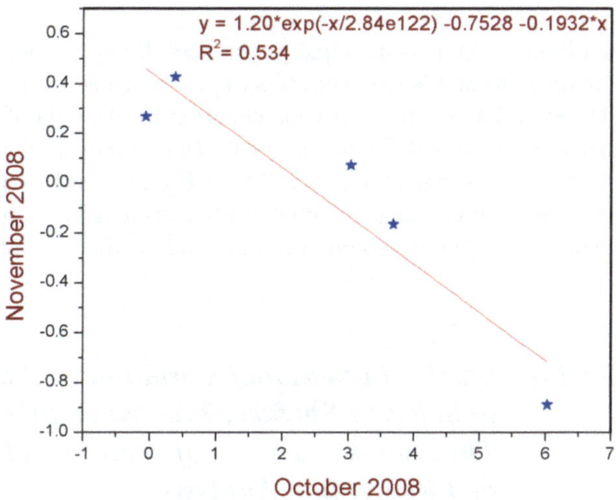

Fig. 22.10 Relationship between October and November 2008 for the 5 principal parameters

Validation of PCA Results

Before applying the above finding, its scientific reliability must be validated using other independent methods. One way to achieve this goal is to compare the water quality data with non-principal physico–chemical parameters. In this study we developed the comparison between two cases. In the first case, data from the physico-chemical parameters were used to formulate relationship by regression analysis for the months of October and November 2008. Comparison of the relationship for October-November 2008 using the data of all the 14 physico-chemical parameters (Ca^{2+}, Mg^{2+}, SO_4^{2-}, WT, F^-, pH, Fe, OC, Cl^-, TH, TA, Turbidity, SS and TDS) (Fig. 22.9) with that obtained using data of the five principal physico-chemical parameters (TH, TA, Turbidity, SS and TDS) (Fig. 22.10) showed that the

Table 22.3 Correlation coefficients of different physico-chemical parameters of GW at Kanpur along with their equations showing linear relationships

S.N.	Physico-chemical parameters	Correlation coeff (r)	Developed linear relationship	Eq No.
1.	Ca^{2+} and TDS	0.810	$Ca^{2+} = 0.06$ TDS + 17.03	5
2.	Mg^{2+} and TDS	0.890	$Mg^{2+} = 0.064$ TDS − 7.5	6
3.	Cl^- and TDS	0.880	$Cl^- = 0.068$ TDS − 10.6	7
4.	SO_4^{2-} and TDS	0.081	$SO_4^{2-} = 0.0164$ TDS + 63.16	8
5.	F^- and pH	0.390	$F^- = -2.923 + 0.38$ pH	9
6.	F^- and Turbidity	−0.730	$F^- = 0.417 - 0.0005$ Turbidity	10
7.	F^- and TA	0.700	$F^- = -0.3046 + 0.003$ TA	11
8.	F^- and TH	0.676	$F^- = -0.269 + 0.0031$ TH	12
9.	Fe and TH	−0.366	Fe = 0.91 − 0.0016 TH	13
10.	Fe and TA	−0.376	Fe = 0.9264 − 0.0015 TA	14
11.	Fe and Turbidity	0.380	Fe = 0.553 + 0.0028 Turbidity	15
12.	Fe and pH	−0.390	Fe = 3.78 − 0.37 pH	16

addition of the non-principal parameters did not improve the curve fitting between October (X) and November 2008 (Y) as indicated by the correlation coefficient (R^2). The R^2 value for the regression equation for the data of all the 14 physico-chemical parameters was 0.413 whereas the R^2 value for the regression equation for the data of the 5 principal parameters was 0.534. The latter is better than the former. Therefore, the 9 physico-chemical parameters are considered non-principal, since the addition of data of these 9 physico-chemical parameters did not improve the curve-fitting.

22.3.4 Studies in Seasonal Variation in Water Quality of Selected Surface, Treated and Ground Water Through Assessment of Water Quality Index and Regression Analysis

The degree of a linear association between any two of the water quality parameters as measured by simple correlation coefficient (r) is presented in Table 22.3. Simple regression analysis was done to determine the nature of correlation and significance of the quality parameters on TDS, F^- and Fe content of Ganga water. An overall correlation analysis shows a positive correlation between TDS with Ca^{2+}, Mg^{2+}, Cl^- and SO_4^{2-}. The developed linear relationships for TDS with Ca^{2+}, Mg^{2+}, Cl^- and SO_4^{2-} are given in the form of equations along with correlation coefficients between the above mentioned physico-chemical parameters in Table 22.3.

F^- varies in the range of 0–0.8 mg/l. An overall correlation analysis shows positive correlation of F^- with pH, TA and TH. Correlation coefficient (r) of turbidity with F^- is −0.73. The developed linear relationships for F^- with pH,

Turbidity, TA and TH are given in Table 22.3. Fe concentration varies between 0.2 and 0.8 mg/l (Table 22.1). Fe is negatively correlated with TH, TA and pH while it is positively correlated with turbidity. The developed linear relationships for Fe with pH, Turbidity, TA and TH are given in Table 22.3. The correlation of Fe content of Ganga water with individual chemical parameters in the form of Eqs. 22.1, 22.2, 22.3 and 22.4 is not satisfactory (correlation coefficient is less than 0.4). Such a poor correlation indicates that no single chemical parameter would sufficiently explain the existence of iron content values. So the Fe contents data were subjected to multiple regression analysis in the form of Eq. 22.5.

The developed multiple linear relationship of Fe with F^-, $SO_4{}^{2-}$, TDS, Cl^-, Mg^{2+}, Ca^{2+}, TH, TA and pH is given below:

$$\begin{aligned} Fe = {} & 4.989 \times F^- - 0.002 \times SO_4{}^{2-} - 0.1872 \times TDS - 0.457 \\ & \times Cl^- - 377.31 \times Mg^{2+} - 229.268 \times Ca^{2+} + 91.95 \times TH + 0.223 \\ & \times TA - 0.835 \times pH - 0.207 \end{aligned} \tag{22.5}$$

Considering a known value of iron and F^-, the percentage contribution of each parameter i.e. TH, TA, turbidity and pH can be obtained by substituting an average value for study water source/studied area for a particular season from Eqs 13, 14, 15, 16 and Eqs. 9, 10, 11 and 12 (Table 22.3) respectively.

The overall WQI of all the samples taken during summer, winter, spring, autumn and monsoon for GW, CGW and TW were calculated according to procedure explained above (Sect. 22.2.5). Ground water samples in the winter season are found to have WQI = 5.7 and are of excellent quality. TW-WQI lie in the range of 50–75 indicating good quality water for all the seasons except for autumn when the WQI is 142.6 indicating poor quality water. For summer Ground water WQI is 143.5, and for the four other seasons Ground water WQI range is 5.7–54.0 indicating excellent and good quality water in all the seasons except for summer. GW-WQI values show poor quality water for all the seasons except for spring, when the water quality of GW is good with WQI equal to 65.95. The poor quality water could be attributed to improper disposal of wastes, large quantity of agricultural and urban run-off, sewage, over-application of inorganic fertilizers, improper operation and maintenance of septic systems (World Health Organisation, 1993) [15a]. For GW, WQI ranges from 117 to 192 for all the seasons except for monsoon where the WQI value is 579.3 indicating that water is unsuitable for drinking and cattle purposes.

Table 22.4 shows the water sampling sites with the seasons that fall under different quality. In monsoon and winter season the ranking of WQI values is in the order GW > CGW > TW > PSP water, while in spring and summer seasons the ranking of WQI values is in the order GW > CGW > TW, PSP. The high value of WQI is due mainly to higher values of turbidity and iron contents in the water samples.

Table 22.4 Water quality classification based on WQI value

WQI value	Water quality	Water station and seasons
< 50	Excellent	PSP in winter
50–100	Good	PSP in monsoon, autumn, spring and summer
		TW in monsoon, winter, spring and summer
		CGW in spring
100–200	Poor	PSP in summer, TW in autumn
		CGW in monsoon, winter, autumn and summer
		GW in winter, autumn, spring and summer
> 300	Unsuitable for drinking	GW in monsoon

22.4 Conclusions

A large number of factors and geological conditions influence the correlations between different pairs of physico-chemical parameters of water samples directly or indirectly. For the year 2008 Ganga water recorded an appreciable significant positive correlation between TA and TH, Ca^{2+}, Mg^{2+}, Cl^-, TDS, OC and F^-, and Cl^- with Mg^{2+}, TDS, OC and TH. Total hardness has significant positive correlation with Mg^{2+}, Cl^- and F^- and TDS with Mg^{2+} and F^-. SS show significant negative correlation with Mg^{2+}, TDS, Cl^-, OC and F^-.

Significant seasonal variations in the physico–chemical parameters of surface water of Kanpur city were observed during the study period April–December, 2008 and January–March, 2009. PSP ground water source was found to be contaminated with Cl^- in summer, TA in monsoon and summer, and TDS in autumn, spring and summer seasons as their values were found to be higher than the HDL prescribed by WHO for drinking purposes.

The first two principal components explain more than 99.0 % of the total variance of the original data set of the physico-chemical parameters of Ganga river water, and are sufficient to explain the monitoring area as well as the physico-chemical water quality parameters. PCA results also show that TA, TH and TDS are significant parameters in post-monsoon seasons while Turbidity and SS are significant parameters in the monsoon season. Seasonal variations of WQI for the samples at different stations range from 5.62 to 587.49. At the outset, the study clearly indicates that the water of GW and CGW cannot be used for public consumption in all the seasons except CGW in spring. The analysis reveals that the groundwater of the area in summer needs some degree of treatment before consumption, and it also needs to be protected from the perils of contamination.

Acknowledgement ST is grateful to Dr. Pervez E. Dean, Principal Christ Church College, Kanpur for his encouragement throughout the work.

References

1. Sinha UK (1986) Ganga pollution and health hazard. Inter–India Publication, New Delhi
2. Mehrotra MN (1990) The role of sediments in environmental pollution. A case study of the Ganga at Varanasi. J Indian Assoc Sedimentol 9:1–14
3. Sinha AK, Singh VP, Srivastava K (2000) Physico-chemical studies on river Ganga and its tributaries in Uttar Pradesh – the present status. In: Trivedi RK (ed) Pollution and biomonitoring of Indian rivers. ABD Publ, Jaipur, pp 1–29
4. Bhandari NS, Nayal K (2008) Correlation study on physico-chemical parameters and quality assessment of Kosi river water. Uttarakhand. J Chem 2:342–346
5. Kaur S, Joshi BD (2003) Seasonal variation of some physico-chemical parameters of river Ganga in and around Haridwar. Him J Zool 17:45–55
6. (a) Zafar A, Sultana N (2008) Seasonal analysis in the water quality of the river Ganga. J Curr Sci 12:210–217; (b) Trivedi P, Bajpai A, Thareja S (2009) Ganga river at Kanpur by using correlation study. Nat Sci 1:91–94
7. (a) Saffran K (2001) Canadian water quality guidelines for the protection aquatic life, CCME water quality Index. Users manual Excerpt from Publication No.1299:ISBN1-896997-34; (b) Trivedi P, Bajpai A, Thareja S (2010) Comparative study of seasonal variation in physico-chemical characteristics in drinking water quality of Kanpur, India with reference to 200 MLD filtration plant and ground water. Nat Sci 8:11–17
8. (a) Praus P (2006) Water quality assessment using SVD-based principal component analysis of hydrological data. Water SA 31:417–422; (b) Thareja S, Choudhury S, Trivedi P (2011) Assessment of water quality of Ganga river in Kanpur by using principal components analysis. Adv Appl Sci Res 2:84–91
9. Agarwal A, Saxena M (2011) Assessment of pollution by physico-chemical water parameters using regression analysis: a case study of Ganga River at Moradabad. Adv Appl Sci Res 2:185–189
10. Sneyers R, Vandiepenbeeck M, Vanlierde R (1989) Principal component analysis of Belgian rainfall. Theor Appl Climatol 39:199–204
11. Lohani BN, Todino G (1984) Water quality index of Chao Phraya river. J Environ Eng 11:1163–1176
12. Shihab AS (1993) Application of multivariate methods in the interpertation of water quality monitoring data of Saddam Dam Reservior. Confidential, The Social Research and Development Center (SRDC), Mousal University
13. (1998) Standard methods for the examination of water and waste water, 20th edn. APHA/AWWA/WEF, Washington, DC
14. Vogel AI (1978) A text book of quantitative inorganic analysis including elementary instrumental analysis, 4th edn. The English Language Book Society and Longmans Green Co Ltd., London (a) p 837 (b) p 328–329 (c) pp 504–506
15. (a) World Health Organization (1993) Guidelines for drinking water quality. I. Recommendations, 2nd edn. Geneva www.lenntech.com/drinking-water-standards.htm; (b) SPSS Advanced Models 12.0, http://www.spss.com
16. BIS:10500 (1991) Guidelines for drinking water quality standards. Bureau of Indian Standards, New Delhi
17. Bhoi DK, Raj DS, Mehta YM, Chauhan MB, Machhar MT (2005) Physico-chemical analysis of bore wells drinking water of Nadiad territory. Asian J Chem 17:404–408
18. Singh MR, Beeteswari KH, Gupta A (2009) Water quality status of the Iril River, Manipur. J Curr Sci 14:173–180

Chapter 23
Toxicity Studies of *Trachyspermum ammi* (L.) Sprague ex Turrill and Its Smooth Muscles Effects

Noor Jahan, Mansoor Ahmad, and Mehjabeen

Abstract Spices are utilized for the treatment of different disorders since the ancient age. *Trachyspermum ammi* (*T. ammi,* family Apiaceae, commonly known as ajowan) is an annual herb and is widely distributed in Pakistan. It is one of the spices that is used traditionally to relieve certain complaints. It is also used world-wide as a flavoring agent in curries, pickles, sauces and cakes. The seeds of the ajowan are widely used for medicinal purpose. Major active constituents of *T. ammi* are thymol, *p*-cymene and γ-terpinene. The present study is conducted to find out the effect of *T. ammi* on gastrointestinal tract. The in-vitro study was conducted on rabbit jejunum. The volatile oil of *T. ammi* seed extract produced interesting effects on intestinal smooth muscles. Anti-diarrhoeal effect was observed at different concentrations of seed extract of *T. ammi*. Absence of insect infestation and contamination has become an important consideration in storage of grain and to maintain a high quality product. Toxicity of *T. ammi* was assayed by performing insecticidal and wormicidal activity. Insecticidal activity was determined on *Tribolium castaneum* and *Sitophilus oryzae* at different concentrations, that is, 1, 5, 10, 25, 50, 75 and 100 mg. 100 % mortality was obtained at concentrations 25, 50, 75 and 100 mg and no effect was observed at low concentration. Wormicidal potential was determined on earth worm (*Lumbricus terrestris*). The extract of *T. ammi* has no toxicity at low doses, so it can be utilized safely in the treatment of diarrhoea.

N. Jahan (✉)
Dow College of Pharmacy, Dow University of Health Sciences, 74200 Karachi, Pakistan
e-mail: jahann7816@yahoo.com

M. Ahmad
Research Institute of Pharmaceutical Sciences, Department of Pharmacognosy,
University of Karachi, 75270 Karachi, Pakistan
e-mail: herbalist53@yahoo.com

Mehjabeen
Department of Pharmacology, Federal Urdu University for Science,
Arts & Technology, Karachi, Pakistan
e-mail: mehjbn1@gmail.com

M. Gupta Bhowon et al. (eds.), *Chemistry: The Key to our Sustainable Future*,
DOI 10.1007/978-94-007-7389-9_23,

23.1 Introduction

T. ammi is widely distributed in India, Europe, Persia, Egypt and Afghanistan [1] and is also cultivated in different areas of Pakistan [2]. *T. ammi* contains different constituents and the active constituents include thymol (39.1 %), *p*-cymene (30.8 %) and γ-terpinene (23.2 %) [3]. Ajowan fruits contain proteins, resins, thiamine and riboflavin [4]. With its characteristic aromatic smell and pungent taste, Ajowan is widely used as a spice and is employed either alone or in mixture with other spices and condiments. It is also used in pickles, certain types of biscuits, confectionery, beverages and pan mixtures [5].

Fruits of *T. ammi* are used as carminative, spasmolytic and as flavouring agent [3]. *T. ammi* also acts as an antioxidant, antipyretic, cardiac depressant and antibacterial as well as antifungal agent [6]. Fruits are also effective against diarrhoea, flatulence, cholera and colic [5]. Other reported traditional therapeutic uses of *T. ammi* fruits include galactogogue and stomachic [7]. Seeds are fried in oil and used as a thin soup as a galactogogue [8]. It is also used against diarrhoea and as a parasiticidal agent [9]. Seeds soaked in lemon juice with *Prunus amygdalus* (Badam) are effective against amenhorroea [10], it is a cooling (antipyretic, febrifugal) drink [11] and used in typhoid fever [12]. The present report describes the in-vitro toxicity and identification of receptors involved in gastrointestinal effects.

23.2 Materials and Methods

The fruits of *T. ammi* (herbarium number, 001006–11) were purchased from local market in Nov-Dec 2006. The fruits/seeds were reduced in size and macerated with ethanol for 15 days in a closed amber glass. The glass chamber was shaken at different intervals. After 15 days the extract was filtered and dried in a rotary evaporator. The extract was stored at room temperature for further use [13].

23.2.1 *Insecticidal Activity*

Insecticidal activity was performed using filter method. The extract in the doses of 1–100 mg/ml was administered on the filter paper [14]. Filter papers of 90 mm diameter were used on the petri dishes. 10 insects each of *Tribolium castaneum* and *Sitophilus oryzae* were placed in each plate and 24 h readings were observed. Permethrin (235.9 µg/cm^3) was used as standard drug [14].

The crude extract dissolved in distilled water was spread uniformly on the filter paper; then after air drying of filter paper, insects were placed. Ten insects were placed on each filter paper placed in seven petri dishes containing dilutions 1–100 mg/ml respectively. The movement of the insects attracted towards the extract or repelled by sticking to the walls of the petri dish was observed every 5 h during 24 h. Convulsions, faintness or paralysis, if any, in the insects were also observed. The numbers of insects dying immediately or within 5 or 12 or 24 h were noted. The drug showing high mortality was more toxic as compared to the drug that caused no death or lower number of insects' death. The percentage of mortality determines the higher or lower toxicity of the drug. The results were compared with the positive as well as negative standard. Permethrin was used as standard causing high toxicity.

23.2.2 Wormicidal Activity

Wormicidal activity was performed on earth worms (*Lumbricus terrestris*) at different concentrations (1–100 mg/ml). Vermox and Zental were used as standard drugs [15]. The different concentrations of the *T. ammi* ethanolic extract were prepared in distilled water. In each petri dish three worms were placed and when the motility became normal, the drug was administered in different concentrations and their movements were observed for 5, 12 and 24 h. The pattern of movement, as increased or decreased, irritating or normal, was observed. The worms were also observed for the paralysis of any type of body secretions as either blood, yellowish or white secretions that indicated the toxicity or the hemorrhages. The worms were also observed for whether the body was stiff like stone or it was soft indicating that the drug has either muscle relaxant or constriction effect and the death either occurs due to severe constriction or paralysis. The standard drugs utilized were Vermox and Zental which are locally available in the market and known for having anthelmintic activity.

23.2.3 Smooth Muscles Activity

Smooth muscles activity was performed on rabbit intestine (Jejunum). The rabbits weighing 1 kg were sacrificed by neck blow and then the intestine was removed and placed in Tyrode solution. Carbogen was provided by gas tubing and the temperature was maintained at 37 °C. Intestinal movements were recorded isotonically using Harvard transducers and Harvard Student Oscillograph. Pieces of 3 cm of intestine were cut and tied onto the water bath containing 5 ml of Tyrode solution. After equilibration of intestinal movement, the effects of crude drug were observed. The results were compared with standards such as atropine, acetylcholine and adrenaline [16–18]. Crude extract was administered in different doses from 1 to 25 mg dissolved in distilled water and the pattern of movement was observed using an oscillograph. Each observation was made three times after washing the intestinal segment with Tyrode solution. The drugs inducing relaxing effects give smaller peaks indicating that they have muscle relaxant or antidiarrhoeal effect and the drugs inducing contraction cause increase in peak length indicating the effect of contraction of smooth muscles or a diarrhoea inducing effect. The results were analysed by using the student *t*-test, and significance was determined at 0.05 *p*-value.

23.3 Results and Discussion

T. ammi is an important remedial agent for flatulence, atonic dyspepsia, and diarrhoea [5]. The seed of ajowan is bitter, pungent, and it acts as anthelmintic, carminative, laxative, and stomachic. It also cures abdominal tumors, abdominal pains, and piles [19]. Ajowan seeds contain an essential oil containing about 50 % thymol which is a strong germicide, anti-spasmodic and fungicide.

The study was carried out on aerial parts of fruits/seeds of *T. ammi*. The ethanolic crude extract was used for the toxicological and gastrointestinal investigations. The results show the death or survival of insects after exposure with crude extract of ajowan seeds. The results were observed for a period of 24 h. The repellant or attractive movements of the insects for the extracts were observed. The decreased or increased motility as well as the decrease in response or paralysis were also noted. Chaubey worked on isolated essential oil from seven common spices, *Anethum graveolens, Cuminum cyminum, Illicium verum, Myristica fragrans, Nigella sativa, Piper nigrum* and *T. ammi* and determined their insecticidal and other activities [20]. Pandey et al. worked on essential oil of seeds of *T. ammi* (Linn.) Sprague and its pure constituent thymol showed promising results when evaluated for larvicidal, oviposition deterrent, vapor toxicity, and repellent activity against the malarial vector, *Anopheles stephensi* [21]. Thymol is the main constituent of ajowan. The present study confirmed the insecticidal and wormicidal effect of ajowan. The observations showed that the crude ethanolic extract of ajowan is toxic against the insects as it increased agitation in the insects and gave a repellent effect. At lower concentration of 1 mg also, the death of insects was observed (Tables 23.1 and 23.2). The results observed with *Tribolium castaneum* show 20 % death at 1 mg dose, 50 % death at 5 and 10 mg dose and 100 % death at 25–100 mg doses. *Sitophilus oryzae* were found to be more resistant as at 1 mg dose 10 % death was observed while at 5, 10, 25, 50 and 100 mg the death ratio was 30 %, 50 %, 70 %, 80 % and 100 % respectively. The standard drug Permethrin showed 100 % death indicating highest toxicity. High doses of *T. ammi* are toxic but it is less than standard drug (Permethrin).

Anthelmintic activity of seed extract of *T. ammi* was investigated on earth worms (*Lumbricus terristris*) (Table 23.3). *T. ammi* showed high wormicidal activity in comparison with standards Vermox and Zental. At 0.5 mg, death was observed within 24 h. Similar results were observed at 1 mg dose of seed extract. At 5 mg dose, the death was observed after 2 h. At 10 and 25 mg dose the death was observed after 15 min whereas at 50, 75 and 100 mg dose, 100 % mortality was observed within 1 min. Vermox and Zental were used as standard drugs for the

Table 23.1 Insecticidal activity of *T. ammi* against *Tribolium castaneum*

Dose (mg/ml)	No. of survivors	Time of onset of drug action (immobility time/h)	% mortality
1	8	24 (10P, 8R, 2D)	20
5	5	12 (10P, 5R, 5D)	50
10	5	12 (5D)	50
25	0	12 (10P, 10D)	100
50	0	12 (10P, 10D)	100
75	0	12 (10P, 10D)	100
100	0	12 (10P, 10D)	100
+ve control[a]	0	0.5 (10D)	100
−ve control[b]	100	(0D)	0

The results are expressed in percentage; No of insects (n) = 10; *P* paralysed; *R* recover; *D* death
[a]Permethrin (copex 235.9 μg/cm^2)
[b]Solvent

Table 23.2 Insecticidal activity of *T. ammi* against *Sitophilus oryzae*

Dose (mg/ml)	No. of survivors	Time of onset of drug action (immobility time/h)	% mortality
1	9	24 (10P, 9R, 1D)	10
5	7	12 (10P, 7R, 3D)	30
10	5	12 (5D)	50
25	3	12 (10P, 7D)	70
50	2	12 (10P, 8D)	80
75	0	12 (10P, 10D)	100
100	0	12 (10P, 10D)	100
+ve control[a]	0	0.5 (10D)	100
−ve control[b]	100	(0D)	0

The results are expressed in percentage; No of insects (n) = 10; *P* paralysed; *R* recover; *D* death
[a]Permethrin (copex 235.9 μg/cm^2)
[b]Solvent

Table 23.3 Assessment of anthelmintic activity of *T. ammi*

Treatment dose	Time									
of *T. ammi* (mg)	≤ 1 min	5 min	10 min	15 min	30 min	1 h	2 h	4 h	12 h	24 h
Control	N	N	N	N	N	N	N	N	N	N
0.5	0+	0+	0+	0+	0+	0+	0+	2	2	↓2D
1	0+	0+	0+	0+	0+	2	1D	3	4	↓D
5	0+	0+	0+	0+	2	↓1D	D	–	–	–
10	0+	0+	0+,B	↓D	–	–	–	–	–	–
25	0+	0+	0+,S,↓	D	–	–	–	–	–	–
50	↓D	–	–	–	–	–	–	–	–	–
75	↓D	–	–	–	–	–	–	–	–	–
100	↓D	–	–	–	–	–	–	–	–	–

Number of worms used for each dose = 3; *N* Normal; 0, 1, 2, 3, 4 = intensity of movement; + = movement present; *S* and *B* secretions and blood; *↓D* decreased activity and death

Table 23.4 Dose related response of crude extract of *T. ammi* on isolated rabbit intestine

Dose (mg/ml)	Control (cm)	Response (cm)	Response (%)	*t*-value	*p*-value
1	0.60 ± 0.06	0.10 ± 0.03	83.33	7.745[a]	0.0007
5	0.60 ± 0.12	−0.10 ± 0.04	116.66	5.797[a]	0.0020
10	0.70 ± 0.10	−0.10 ± 0.06	114.28	12.499[a]	0.0001
15	0.70 ± 0.10	−0.10 ± 0.03	114.28	7.905[a]	0.0006
20	0.50 ± 0.09	−0.10 ± 0.03	120.00	6.824[a]	0.0012
25	0.60 ± 0.09	−0.10 ± 0.03	116.66	7.902[a]	0.0006

The results are expressed at $p \leq 0.05$ and $p \geq 0.05$
[a]Highly significant

comparison of results of *T. ammi*. The results of wormicidal activity were similar to insecticidal activity indicating the irritating action of the drug as well as toxicity of high doses of the drug as at low dose of 1 mg, the death was observed after a long time period of exposure of 24 h. Worms also showed secretions at 10 mg dose of extract. The blood secretion was observed at 25 mg dose indicating hemorrhage. After the death of worms, the body was observed for paralysis or convulsions that indicated the flaccidity of the body muscles. At higher doses, very irritating movement and immediate death of worms were observed. The pin prick response was also reduced in a dose-dependent manner. With standard drugs, no death was observed but the pin prick response was reduced and worms were paralyzed. No death but reduced activity was observed with both standard drugs Zental and Vermox.

The effect on isolated intestinal segment of rabbit jejunum was observed on oscillograph and the results showed the relaxant effect but this relaxant effect was not dose dependent. Aftab et al. [22] evaluated blood pressure lowering action of active principle thymol by calcium channel blocker. Sublethal exposure to thymol and [5]-gingerol reduced the levels of 5-hydroxytryptamine and dopamine in the nervous tissue of *L. acuminata*. Boskabady et al. [23] described several therapeutic effects including anti-asthma and dyspnea for the seeds of *Carum copticum*. In previous studies on the relaxant and anticholinergic (functional antagonism) effect, histamine (H1) inhibitory effect of *Carum copticum* has been demonstrated on guinea pig tracheal chains [23]. In present studies the inhibitory effect of *T. ammi* on isolated rabbit jejunum was observed due to its anticholinergic, antihistaminic, antidopaminic and adrenergic effects.

The effect was observed at different concentrations. At 0.5 mg the effect was 0.1 mm and at 1–25 mg same effect as −0.1 mm was observed that was below the base line. This peak size indicates the complete relaxing effect of extract. The relaxant effect of *T. ammi* is responsible for its spasmolytic effect (Table 23.4, Fig. 23.1).

The spasmolytic effect of ajowan was analysed by comparison with the standard drugs and the involvement of muscarinic receptors was found (Figs. 23.2, 23.3, 23.4 and 23.5). The receptors were determined by comparing the effect of crude extract along with the standard drugs, acetylcholine, aderanaline, atropine and avil in three

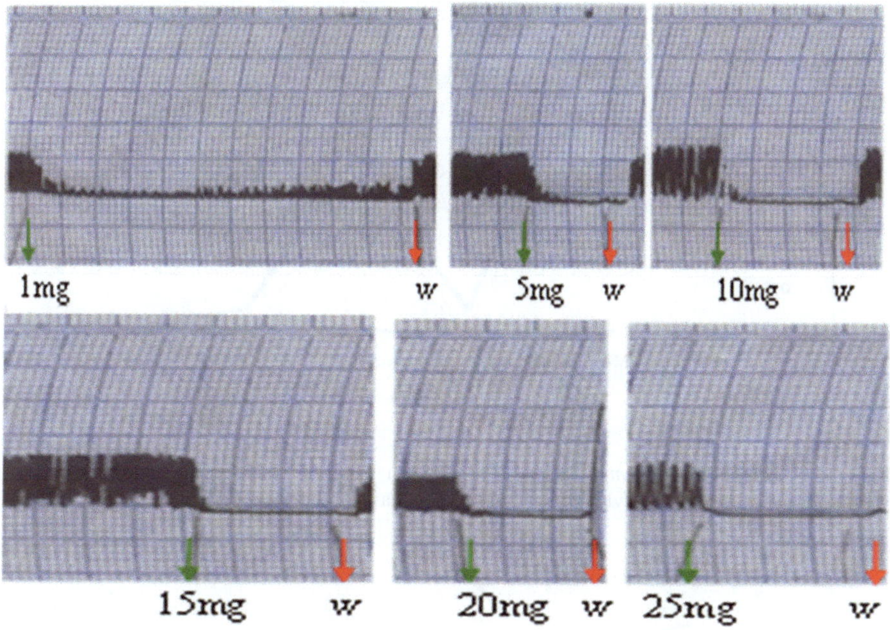

Fig. 23.1 Tracing showing effect of *T. ammi* extract on smooth muscles of isolated rabbit intestine (Dose concentration of extract = 1, 5, 10, 15, 20 and 25 mg; *W* = washing of intestinal segment)

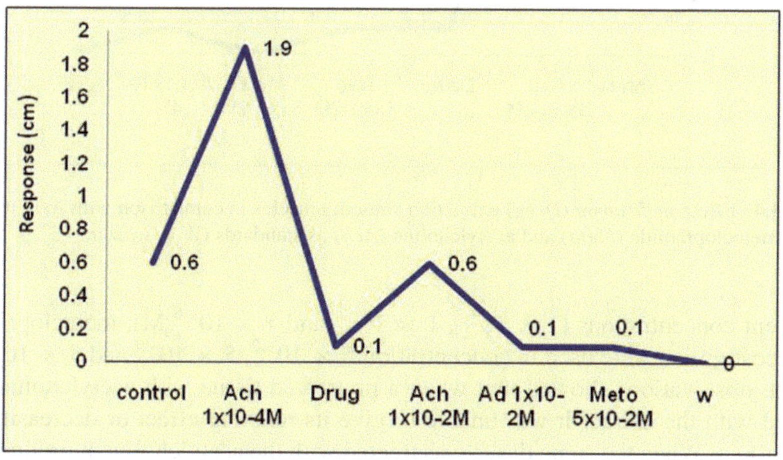

Fig. 23.2 Effect of *T. ammi* (*Drug*) extract on smooth muscles in comparison with acetylcholine (*Ach*), adrenaline (*Ad*) and metoclopramide (*Meto*) as standards (*W* = *washing*)

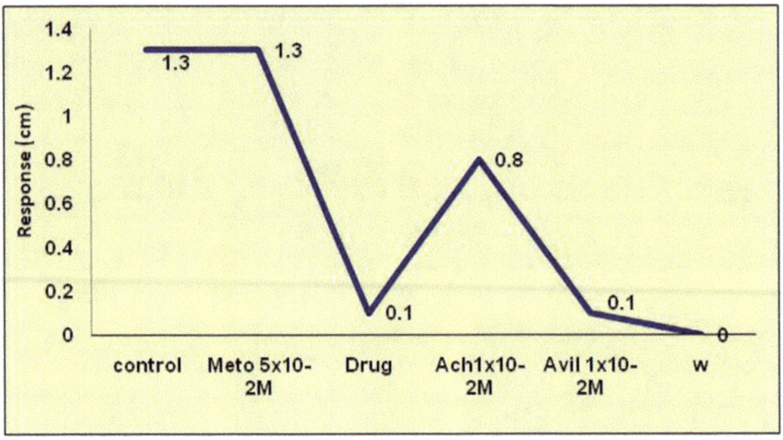

Fig. 23.3 Effect of *T. ammi* (*Drug*) extract on smooth muscles in comparison with metoclopramide (*Meto*), acetylcholine (*Ach*) and *Avil* as standards (*W* = *washing*)

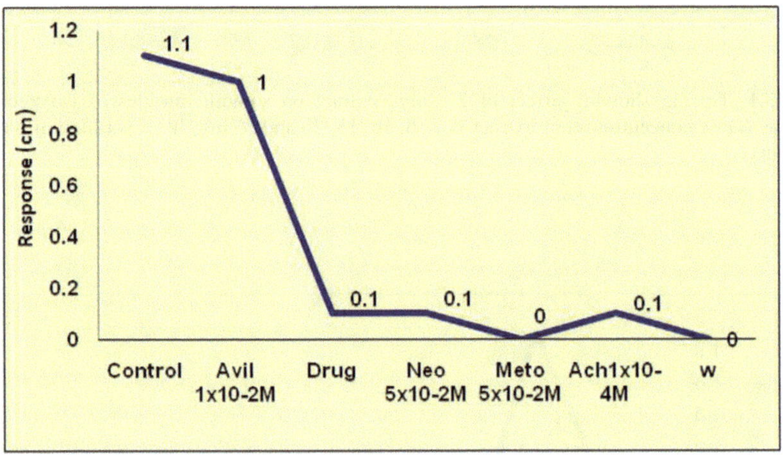

Fig. 23.4 Effect of *T. ammi* (*Drug*) extract on smooth muscles in comparison with *Avil*, neostigmine, metoclopramide (*Meto*) and acetylcholine (*Ach*) as standards (*W* = *washing*)

different concentrations (1×10^{-2}, 1×10^{-4} and 1×10^{-6} M); metoclopramide and neostigmine were used in concentration 5×10^{-2}, 5×10^{-4} and 5×10^{-6} M.

The observations showed that when a pretreated tissue with acetylcholine was treated with the extract it was unable to give its relaxing effect or decreasing the peak effect. When the same tissue was treated with the acetylcholine it produced no response as the receptors were occupied already. However, when the same tissue was again treated with the metoclopramide standard drug, it did not produce its response, that is, increase in gastrointestinal motility, clearly indicating that some

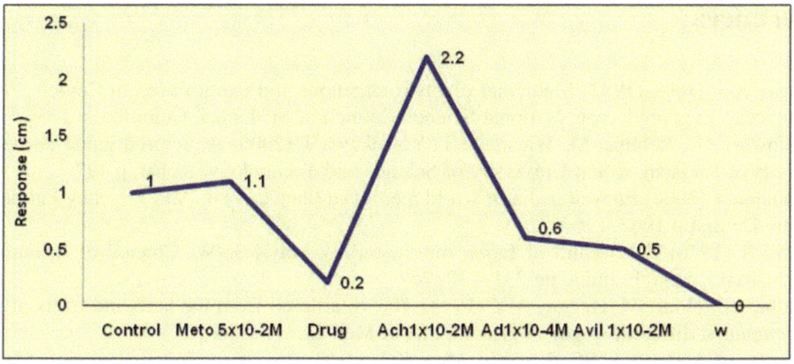

Fig. 23.5 Effect of extract of *T. ammi* (*Drug*) on smooth muscles in comparison with metoclopramide (*Meto*), acetylcholine (*Ach*) and adrenaline (*Ad*) and Avil as standards (*W = washing*)

dopamine receptors were also occupied. When a pretreated tissue with metoclopramide was treated with extract, the latter was unable to give its full effect, which is gastrointestinal relaxant effect that indicates the receptor occupancy while when the same tissue was treated with Avil, one of the standard drugs, it produced its full effect, that is, gastrointestinal relaxant effect, indicating that histamine receptors were not involved.

When a pretreated tissue with Avil was treated with extract, it produced its relaxing effect gradually while after treatment with metoclopramide as well as neostigmine, it induced no response confirming the involvement of muscarinic receptors.

When a pretreated tissue with drug was treated with adrenaline, the latter did not produce its full smooth muscle relaxant effect indicating the adrenergic receptors involvement in the relaxing movement of intestine by the extract.

23.4 Conclusions

The present work was aimed to find out the toxicity as well as the identification of receptors involved in gastrointestinal effects of *T. ammi*. In this study it is concluded that *T. ammi* induces its gastrointestinal relaxant effect by involvement of muscarinic as well as adrenergic and somewhat dopamine receptors. Histamine receptors involvement was not observed when compared with standards. *T. ammi* seed extract can be utilized as antidiarrhoeal agent and to relieve abdominal cramps as it gives gastrointestinal relaxant effect. It also has good insecticidal and wormicidal activity. Its use is safe as it induces toxicity at high concentration.

References

1. Jayaweera DMA (1982) Medicinal plants (indigenous and exotic) used in Ceylon. Part V. - Rutaceae-Zygophylaceae. National Science Council of Sri Lanka, Colombo, p 133
2. Shinwari ZK, Rehman M, Watanabe T, Yoshikawa T (2006) A pictorial guide to medicinal plants of Pakistan. Kohat University of Science and Technology, Kohat, p 492
3. Pullaiah T (2006) Encyclopedia of world medicinal plants, vol 4. A.P. Regency Publications, New Delhi, pp 1962–1964
4. C.S.I.R (1976) The wealth of India, raw materials, vol X:Sp-W, Council of Scientific and Industrial Research, India, pp 231–232, 267
5. Balbaa SI, Hilal SH, Haggag MY (1973) The volatile oil from the herb and fruits of *Carum copticum* at different stages of growth. Planta Med 23:312–320
6. Chatterjee A, Pakrashi SC (2003) In: Marg KS (ed) The treatise on Indian medicinal plants, vol 4. NISCAIR Press, National Institute of Science Communication and Information Resources/ CSIR, New Delhi, pp 45–47
7. Chialva F, Monguzzi F, Manitto P, Akgiil A (1993) Essential oil constituents of *Trachyspermum copticum* (L.) Link fruits. J Essent Oil Res 5:105–106
8. Howard MP, Reynolds RD, Moser PB, Andon MB, McConnell W, Acharya S (1985) Special foods of high nutritive value consumed by Nepalese lactating women. Fed Proc 44:1505
9. Jansen PCM (1981) Spices, condiments and medicinal plants in Ethiopia, their taxonomy and agricultural significance. Centre for Agricultural Publication and Documentation, Wageningen
10. Shome U, Rawat AKS, Mehrotra S (1996) Time-tested household herbal remedies. In: Jain SK (ed) Ethnobiology in human welfare. Deep Publications, New Delhi, pp 96–100
11. Umadevi I, Daniel M (1990) Phenolics of some fruit spices of the Apiaceae. Natl Acad Sci Lett 13:439–441
12. Vedavathy S, Rao DN (1995) Herbal folk medicine of Tirumala and Tirupati region of Chittoor district, Andhra Pradesh. Fitoterapia 66:167–171
13. Ahmad M, Muhammad N, Ahmad M, Lodhi MA, Mehjabeen, Jahan N, Khan Z, Ranjit R, Shaheen F, Choudhary MI (2008) Ureaese inhibitor from *Datisca cannabina* Linn. J Enzym Inhib Med Chem 23:386–390
14. Ahn YJ, Kim GH, Cho KY (1995) Bioassay system for insecticidal compounds. In: Proceedings of the third symposium on the biochemical methodology for the research and development of the bioactive substances, Seoul, pp 495–506
15. Syed S, Ahmad M, Mehjabeen, Jahan N, Fatima N (2010) Assaying insecticidal and wormicidal activities in crude extracts of leaves and seeds of *Portulaca oleracea* L. Int J Biol Biotech 7:439–443
16. Jahan N, Mehjabeen, Ahmad M, Chishti KA, Hamid S, Rehman AB (2004) Spasmolytic activity of *Thuja occidentalis* (Cupressaceae). Int Chem Pharm Med J 1:135–140
17. Mehjabeen, Jahan N, Ahmad M, Chishti KA, Shaista H, Rehman AB (2004) Comparative pharmacological study on red and black fruits and leaves of *Solanum nigrum* (Black nightshade). Int Chem Pharma Med J 1:39–47
18. Ahmad M, Mehjabeen, Jahan N (2012) Exploration of signal transduction mechanism of therapeutic action of *Solanum nigrum*. Afr J Pharm Pharmacol 6:322–329
19. Krishnamoorthy V, Madalageri MB (1999) Bishop Weed (*Trachyspermum ammi*): an essential crop for north Karnatka. J Med Arom Plant Sci 21:996–998
20. Chaubey MK (2008) Fumigant toxicity of essential oils from some common spices against pulse beetle, Callosobruchus chinensis (Coleoptera: Bruchidae). J Oleo Sci 57:171–179
21. Pandey SK, Upadhyay S, Tripathi AK (2009) Insecticidal and repellent activities of thymol from the essential oil of *Trachyspermum ammi* (Linn) Sprague seeds against *Anopheles stephensi*. Parasitol Res 105:507–512
22. Aftab K, Atta-Ur-Rahman K, Usmanghani K (1995) Blood pressure lowering action of active principle from *Trachyspermum ammi* (L.) Sprague. Phytomedicine 2:35–40
23. Boskabady MH, Jandaghi P, Kiani S, Hasanzadeh L (2005) Antitussive effect of *Carum copticum* in guinea pigs. J Ethnopharmacol 97:79–82

Chapter 24
Metal Levels in Traditional Chinese and Ayurvedic Medicines

Henri Li Kam Wah, Kanisha Ramchurn, and Safeenaz B. Alladin

Abstract The calcium, magnesium, nickel and zinc contents in seven Traditional Chinese Medicines (TCM) and four Ayurvedic Medicines (AM) purchased locally were determined using flame atomic absorption spectroscopy following five different digestion methods: four open wet hot plate with various mixtures of acids and/or H_2O_2, and microwave oven (**MWO**) with conc HNO_3 and H_2O_2. Sodium and potassium levels were analysed using flame photometry after **MWO** digestion. Potassium had the highest mean level among the six metals analysed. **MWO** was found to be more effective for Mg, method **M1** [conc HNO_3 + H_2O_2 (30 %)] for Ca and Zn, and method **M4** (aqua regia) for nickel. From replicate measurements, the inter-day precision of the flame photometric measurements for sodium and potassium was found to be 3.0–9.0 % and 1.3–9.0 % respectively. Comparison of results obtained from standard calibration curves and standard addition indicated that the percentage difference of sodium content in the TCM *Niuhuangchiehdu Pien* and the potassium content in AM, *Diabecon* was 3.9 % and 0.22 % respectively, indicating that the matrix did not have much influence on the metal content.

24.1 Introduction

The World Health Organisation (WHO) defines traditional medicine as: "the sum total of the knowledge, skills, and practices based on the theories, beliefs, and experiences indigenous to different cultures, whether explicable or not, used in the maintenance of health as well as in the prevention, diagnosis, improvement or treatment of physical and mental illness" [1].

There exist several traditional systems of medicine around the world, each with different associated philosophies and cultural origins, of which the two most

H. Li Kam Wah (✉) • K. Ramchurn • S.B. Alladin
Department of Chemistry, Faculty of Science, University of Mauritius, Réduit, Mauritius
e-mail: lkwah@uom.ac.mu

M. Gupta Bhowon et al. (eds.), *Chemistry: The Key to our Sustainable Future*,
DOI 10.1007/978-94-007-7389-9_24,

prevalent are the Traditional Chinese Medicines (TCM) and Ayurvedic Medicines (AM). During the last two decades, there has been increased global interest in traditional medicines, especially those based on herbal products and extracts. These medicines contain a wide range of inorganic elements, both toxic and essential. Toxic metals of concern are mainly As, Cd, Hg and Pb. Several studies have reported potentially harmful levels of these heavy metals in both TCM and AM [2–24], thus emphasizing the need for more stringent quality standards and regulations for these products [8, 9, 21, 25, 26]. However, some authors highlighted that detailed chemical speciation be conducted to assess the effective toxicity of the medicines as the toxicity of elements varies with their chemical forms [3, 7, 10, 27].

On the other hand, TCM and AM are also known to have many essential and nutritional elements such as Ca, Cu, K, Mg, Mn, Na, Ni and Zn, whose excess or deficiency may disturb the normal biochemical functions of the human body. Several studies have reported elemental contents in traditional medicines or medicinal plant extracts [3, 11, 12, 17, 28–35].

Considering the importance of trace elements in various human metabolic processes and also their curative properties, the analysis of six essential and trace elements, namely calcium, magnesium, nickel, potassium, sodium and zinc, in 7 TCM and 4 AM commercially available in the Mauritian market was carried out.

Calcium is an essential plant nutrient and is omnipresent in all living things (animals and plants), water, soils and sediments. It is essential for healthy bones, teeth and blood. Deficiency of calcium causes rickets, osteomalacia and scurvy [36]. Magnesium is the fourth most abundant mineral in the human body and is needed for more than 300 biochemical reactions. It helps in maintaining a healthy immune system, strong bones and regulating the levels of several vitamins and minerals. Magnesium deficiency can cause amongst others cramps, muscle tension, poor memory, irritability, depression, and can result in low levels of calcium and potassium in the blood [37]. Nickel, despite being toxic, is found in minute amounts in the human body and is believed to play a role in the stabilisation of the RNA molecule, in the normal bone functioning, in the breakdown of glucose and human breast milk production. Deficiency of nickel is normally rare but can lead to liver, skin and kidney diseases. However, at high levels of accumulation, nickel is capable of causing serious health problems such as paralysis of the spinal column, tumours, thyroid and heart diseases [38]. Zinc is an essential element for the body and is involved in numerous aspects of cellular metabolism. Zinc plays important roles in normal growth and development during pregnancy, childhood and adolescence, and nearly 100 specific enzymes depend on zinc for their catalytic activity. Severe zinc deficiency can result in the slowing of growth and development, delayed sexual maturation, chronic and severe diarrhea, immune system deficiencies, impaired wound healing, reduced appetite, impaired taste sensation and night blindness [39]. Both potassium and sodium are essential body minerals. The body depends on the electrolytic properties of potassium for conduction of nerve impulses, muscle contractions and regulation of balance of acids and bases in the blood. Potassium is also involved in various biochemical processes in the human body. Some common problems associated with low potassium levels include hypertension, congestive heart failure, cardiac arrhythmia, fatigue and osteoporosis [40]. On the other hand, the body uses sodium to regulate blood pressure and blood volume. Sodium is also essential for the transmission of nerve impulses, heart activity, and certain metabolic functions. Low sodium levels (hyponatremia) are uncommon but can occur whenever there is an increase in the amount of body water relative to sodium. On the contrary, hypernatremia (high sodium levels) is common, caused principally with high sodium content in food. Other causes include kidney diseases, low water intake, severe diarrhea and vomiting [41].

24.2 Methodology

24.2.1 Materials and Reagents

7 TCM (*Jin Qian Chao, Leung Pui Kee, Po Chai Pills, Quiang Li Tien-Ma, Saplingtan, Watermelon Frost* and *Niuhuangchiehdu Pien*) imported from China and 4 AM (*Diabecon, Pilex, Septilin* and *Softovac*) imported from India were purchased locally.

The standard stock 1,000 ppm solutions of potassium and sodium were bought from Sherwood Scientific Ltd, calcium and zinc from Fisher Scientific, magnesium and nickel from Fisons Scientific. Concentrated HNO_3, HCl and H_2SO_4 were obtained from SD Fine Chemicals Ltd while H_2O_2 (30 %) from Fisons Scientific.

All glassware used (burettes, pipettes, volumetric flasks, beakers) were of grade A. Prior to use, all glassware were acid-washed.

24.2.2 Sample Preparation

The traditional medicines which were mostly in the form of pills were ground using a mortar and pestle in order to obtain fine and homogeneous solids.

24.2.3 Microwave Oven Digestion

The samples were digested using the microwave digester Milestone High Performance Microwave Unit mls 1,200 M. Prior to use, the Teflon flasks were acid washed. Approximately 1 g of each powdered sample was weighed in duplicate in dry labelled Teflon flasks to which were added 7 ml of concentrated nitric acid (72 %) followed by 5 ml of H_2O_2 (30 %). The samples were digested using the program shown in Table 24.1.

After cooling for about 30 min, the resulting digested samples were filtered through fluted filter paper into 25 mL volumetric flasks and the filtrates were made up to the mark with distilled deionised water. The resulting clear solutions were then transferred to clean acid-washed labelled plastic bottles.

Table 24.1 Microwave digestion program used

Step	1	2	3	4	5
Power/W	250	0	250	400	650
Time/min	1	1	5	5	5

Table 24.2 Reagents used for wet open digestions

Method	Reagents
M1	HNO_3 (72 %): 16 ml & H_2O_2 (30 %): 4 ml
M2	HNO_3 (72 %): 20 ml
M3	HNO_3 (72 %): 16 ml & H_2SO_4 (98 %): 4 ml
M4	HCl (35 %): 15 ml & HNO_3 (72 %): 5 ml

Table 24.3 Concentration of standards and wavelengths used

Metal	Wavelength/nm	Concentration of prepared standards/ppm
Calcium	422.7	1, 2, 3, 4 and 5
Magnesium	202.6	5, 10, 15, 20 and 25
Nickel	232.0	2, 4, 6, 8 and 10
Zinc	213.9	0.2, 0.4, 0.8, 1.0 and 1.2
Potassium	766.0	10, 15, 20, 25 and 30
Sodium	589.0	10, 15, 20, 25 and 30

24.2.4 Wet Open Digestions

Wet open digestion of the samples was performed using four different methods as given in Table 24.2.

Approximately 1 g of the ground medicine was weighed in duplicate into separate 50 mL conical flasks. The appropriate reagents as detailed in Table 24.2 were pipetted into each of the flasks. The solutions were heated to boiling on a hot plate in a fume hood for 4 h before filtration into 25 mL volumetric flasks and making up to the mark with distilled deionised water. The contents were finally transferred to clean well-labelled plastic bottles.

Reagent blanks were prepared for both microwave and wet open digestions under similar conditions as those used for the medicinal samples.

24.2.5 Determination of Metal Contents

The standard calibration solutions were prepared daily from the stock solutions by proper dilution with distilled deionised water. The levels of potassium and sodium in the medicines were determined using a CIBA Corning Model 410 flame photometer whereas calcium, magnesium, nickel and zinc contents were analysed by a GBC Avanta flame atomic absorption spectrometer using air/acetylene. Table 24.3 lists the concentrations of the prepared standard solutions and wavelength used for the metals.

24.2.6 Standard Addition

Determination of Potassium Content in *Diabecon*

5 ml of the microwave oven digested medicinal sample *Diabecon* was diluted 20 times. 5 ml of this diluted sample was then transferred to each of five 25 mL volumetric flasks. To the five volumetric flasks, 0, 0.5, 1, 1.5 and 2 ml of 100 ppm potassium standard solution was respectively added. The resultant solutions were then made up to the mark with distilled deionised water and analysed by flame photometry. The concentration of the added standard for each flask was determined and a plot of emission against concentration of added standard was drawn. The concentration of potassium in the diluted sample corresponds to the x-intercept obtained by extrapolation. The concentration of potassium in the undiluted sample is then calculated by taking into consideration the dilution performed.

Determination of Sodium Content in *Niuhuangchiehdu Pien*

The experimental procedure is similar to that applied in section "Determination of Potassium Content in *Diabecon*" except that 5 ml of the undiluted microwave digested *Niuhuangchiehdu Pien* sample was pipetted in the five volumetric flasks.

24.3 Results and Discussion

24.3.1 Metal Levels in TCM and AM

Calcium

Table 24.4 gives the calcium levels obtained in the 7 TCM and 4 AM following the five methods of digestion.

All the 11 medicinal samples analysed exhibited detectable calcium levels. The highest calcium level is observed in *Niuhuangchiehdu Pien* irrespective of the applied method of digestion. The elevated concentration of calcium found in *Niuhuangchiehdu Pien* can be accounted by the presence of the gypsum selenite (26.64 %) in its formulation. Excluding the TCM *Niuhuangchiehdu Pien*, the calcium content in the four AM is higher than in the remaining six TCM. *Diabecon*, *Pilex* and *Softovac* have the highest amount of calcium amongst the AM. However, the calcium contents were considerably lower than those analysed by Giacomino et al. [12] in five products of the Parpati family where Ca was present in all the medicines with a range of 738–8,439 μg/g. Two main ingredients namely *Shilajit* and *Pterocarpus marsupium* present in *Diabecon* contain high calcium levels. *Shilajit* is an exudate that is pressed out from rock layers in sacred mountains in

Table 24.4 Calcium levels (μg/g) in the 7 TCM and 4 AM following the five methods of digestion

Medicine	M1	M2	M3	M4	MWO
Diabecon	129 ± 2	117 ± 2	93 ± 2	108 ± 6	116 ± 9
Pilex	132 ± 2	109 ± 0	44 ± 2	59 ± 6	120 ± 2
Septilin	70 ± 4	40 ± 4	53 ± 2	26 ± 6	53 ± 2
Softovac	131 ± 0	65 ± 11	47 ± 2	70 ± 4	113 ± 0
Jin Qian Chao	59 ± 2	41 ± 6	33 ± 4	40 ± 4	41 ± 6
Leung Pui Kee	76 ± 4	36 ± 4	21 ± 4	27 ± 4	53 ± 6
Po Chai Pills	67 ± 4	21 ± 4	17 ± 2	20 ± 2	56 ± 2
Quiang Li Tien-Ma	36 ± 4	32 ± 2	26 ± 2	24 ± 9	33 ± 4
Saplingtan	78 ± 2	67 ± 9	43 ± 4	46 ± 0	65 ± 2
Watermelon Frost	59 ± 2	32 ± 2	52 ± 0	50 ± 2	52 ± 0
Niuhuangchiehdu Pien	865 ± 2	912 ± 4	856 ± 19	706 ± 9	876 ± 4

Nepal and other high mountains and is rich in minerals and metals such as calcium, zinc, sodium and magnesium [42]. On the other hand, *Pterocarpus marsupium* also known as the Indian Kino tree contains 10–20 % of calcium sennoside which is a big molecule with molecular formula $C_{42}H_{36}CaO_{20}$.

Comparison between the various digestion methods showed that **M1** digestion resulted in the highest calcium levels for 10 out of the 11 medicinal samples analysed in this study, except for the TCM *Niuhuangchiehdu Pien* in which the highest calcium content was detected following digestion method **M2. MWO** digestion gave the next highest calcium levels for the medicines analysed. Least calcium level was observed in seven medicines (*Diabecon*, *Pilex*, *Softovac*, *Jin Qian Chao*, *Leung Pui Kee*, *Po Chai Pills* and *Saplingtan*) following digestion method **M3**, in three medicines (*Septilin*, *Quiang Li Tien-Mo* and *Niuhuangchiehdu Pien*) from digestion method **M4** and in one medicine (*Watermelon Frost*) from digestion method **M2**.

The statistical two-tail student's *t*-test method was applied to compare the different digestion methods to investigate whether there are any significant differences between the results obtained from all the ten possible pairs (Table 24.5). By analysing the calculated *t*-values (t_{calc}) with the t_{crit} value (2.23), it is observed that for six pairs of digestion methods, namely **M1/M2**, **M1/M3**, **M1/M4**, **M1/MWO**, **M3/MWO** and **M4/MWO**, the results are significantly different from each other at 95 % confidence level. The results obtained following the digestion method **M1** are thus significantly different from those obtained from the other four digestion methods. Similarly, calcium levels obtained following **MWO** digestion are significantly different from all the other digestion methods except **M2**. On the other hand, it can be concluded that the values obtained from digestion methods **M2**, **M3** and **M4** are comparable and hence are not significantly different.

Magnesium

The magnesium levels obtained in the 7 TCM and 4 AM following the five methods of digestion are shown in Table 24.6.

Table 24.5 Comparison of calcium content between different pairs of digestion methods using *t*-test

Digestion pair	Mean difference	Standard deviation	t_{calc}
M1/M2	20.91	28.69	2.42
M1/M3	37.91	28.69	4.38
M1/M4	47.82	42.16	3.67
M1/MWO	11.27	9.22	4.05
M2/M3	17.00	25.63	2.20
M2/M4	26.91	61.79	1.44
M2/MWO	9.64	21.88	1.46
M3/M4	9.91	48.21	0.68
M3/MWO	26.64	25.30	3.49
M4/MWO	36.55	47.95	2.53

Table 24.6 Magnesium levels (μg/g) in the 7 TCM and 4 AM following the five methods of digestion

Medicine	**M1**	**M2**	**M3**	**M4**	**MWO**
Diabecon	97 ± 4	80 ± 1	81 ± 2	83 ± 7	99 ± 2
Pilex	258 ± 2	230 ± 2	237 ± 2	242 ± 4	265 ± 2
Septilin	76 ± 1	57 ± 5	64 ± 2	67 ± 2	78 ± 2
Softovac	52 ± 1	42 ± 1	48 ± 1	48 ± 1	55 ± 4
Jin Qian Chao	128 ± 1	119 ± 2	125 ± 1	123 ± 5	137 ± 9
Leung Pui Kee	64 ± 2	48 ± 1	59 ± 4	57 ± 2	62 ± 1
Po Chai Pills	61 ± 2	48 ± 4	54 ± 2	57 ± 2	62 ± 4
Quiang Li Tien-Ma	71 ± 2	50 ± 2	61 ± 2	64 ± 2	76 ± 2
Saplingtan	42 ± 1	23 ± 2	26 ± 2	31 ± 1	42 ± 1
Watermelon Frost	36 ± 2	23 ± 2	28 ± 4	26 ± 2	38 ± 4
Niuhuangchiehdu Pien	71 ± 2	57 ± 2	62 ± 1	69 ± 4	74 ± 2

All the traditional medicines analysed were found to contain magnesium. The highest magnesium level was found in the AM *Pilex* which is due to the presence of the ingredients *Guggul* and *Shilajit*. *Guggul* is composed of a mixture of organic compounds, inorganic ions and minerals which consist mostly of calcium, magnesium and iron [43]. On the other hand, *Shilajit* is a rich source of naturally occurring minerals like magnesium [42]. The lowest magnesium level in the AM was found in *Softovac*. For the TCM, *Jin Qian Chao* was found to have the highest magnesium content. Magnesium is well-known for its calcium channel blocking effect and can reduce the formation of calcium oxalate which is the main cause for stones in body. *Herba Lysmachiae*, the only ingredient of *Jin Qian Chao*, is very effective in dissolving kidney, liver and gall stones [44] and this can explain the high level of magnesium obtained in this TCM. The range of magnesium levels obtained in this study for AM corresponds to only a certain extent to the range observed by Giacomino et al. [12] (201–4,236 μg/g). For the TCM, the range observed was within that obtained by Wang et al. [11] (ND-2,031 μg/g). The lowest magnesium levels were found in the TCM *Saplingtan* and *Watermelon Frost*.

Highest levels of magnesium were observed after **MWO** or **M1** digestion for all the 11 medicines. Digestion method **M2** gave the least amount for all the medicines

Table 24.7 Comparison of magnesium content between different pairs of digestion methods using *t*-test

Digestion pair	Mean difference	Standard deviation	t_{calc}
M1/M2	16.27	5.42	9.95
M1/M3	10.09	5.66	5.91
M1/M4	8.09	4.39	6.11
M1/MWO	2.91	3.11	3.10
M2/M3	6.18	2.96	6.93
M2/M4	8.18	3.79	7.16
M2/MWO	19.18	6.45	9.87
M3/M4	2.00	3.13	2.12
M3/MWO	13.00	6.60	6.53
M4/MWO	11.00	5.51	6.62

Table 24.8 Nickel levels (μg/g) in the 7 TCM and 4 AM following the five methods of digestion

Medicine	**M1**	**M2**	**M3**	**M4**	**MWO**
Diabecon	7.3 ± 0.3	7.0 ± 0.6	6.7 ± 0.3	7.7 ± 0.6	8.6 ± 0.3
Pilex	5.4 ± 0.3	5.4 ± 0.3	5.7 ± 0.6	6.1 ± 0.9	7.0 ± 0.6
Septilin	5.4 ± 0.3	6.1 ± 0.3	5.7 ± 0.1	6.7 ± 0.3	6.4 ± 0.6
Softovac	4.5 ± 0.6	4.8 ± 0.3	5.4 ± 0.9	6.1 ± 0.3	6.1 ± 0.3
Jin Qian Chao	6.7 ± 0.9	7.0 ± 0.6	7.6 ± 1.3	8.9 ± 0.6	8.6 ± 0.3
Leung Pui Kee	7.3 ± 0.3	8.6 ± 0.3	7.7 ± 0.6	13.1 ± 0.9	8.9 ± 1.3
Po Chai Pills	7.7 ± 0.6	8.3 ± 0.6	8.9 ± 0.6	8.9 ± 0.1	11.2 ± 0.3
Quiang Li Tien-Ma	7.7 ± 0.1	6.1 ± 0.3	6.1 ± 0.3	9.2 ± 0.3	9.2 ± 0.9
Saplingtan	3.2 ± 0.6	5.1 ± 0.6	5.7 ± 0.1	8.6 ± 0.3	7.0 ± 0.1
Watermelon Frost	3.5 ± 0.3	4.5 ± 0.1	4.8 ± 0.3	8.6 ± 0.3	8.3 ± 0.6
Niuhuangchiehdu Pien	4.5 ± 0.6	6.4 ± 0.6	7.7 ± 0.1	8.3 ± 0.6	8.0 ± 0.3

analysed. The following order of extractability of magnesium was observed: **M2** < **M3** ~ **M4** < **M1** ~ **MWO**. From Table 24.7, it was observed that the t_{calc} was greater than the t_{crit} (2.23) for all pairs of digestion methods except **M3/M4**.

Nickel

Table 24.8 gives the nickel levels obtained in the 7 TCM and 4 AM following the five methods of digestion.

Nickel was detected in all the 11 medicines. Highest level was observed in the TCM *Leung Pui Kee*. The latter is based mainly on constituents of plant origin and the only mineral present is talc (1 %). A potential source of nickel apart from talc is soils and sediments. Plants grown in soils rich in nickel absorb the metal which is then incorporated into the medicine. The plant *Aster tataricus* used in the preparation of *Leung Pui Kee* is native from the northern part of China and is grown in meadows and wetlands which are rich in minerals and metals due to the anthropogenic activities in the vicinity. The presence of nickel in the other medicines can also be explained from soil contribution.

Table 24.9 Comparison of nickel content between different pairs of digestion methods using *t*-test

Digestion pair	Mean difference	Standard deviation	t_{calc}
M1/M2	0.59	0.31	1.89
M1/M3	0.85	0.41	2.09
M1/M4	2.67	0.62	4.27
M1/MWO	2.34	0.44	5.36
M2/M3	0.26	0.19	1.40
M2/M4	2.08	0.45	4.60
M2/MWO	1.74	0.37	4.65
M3/M4	1.82	0.52	3.46
M3/MWO	1.48	0.33	4.47
M4/MWO	0.33	0.56	0.60

The nickel levels in the 11 medicines were all found in a small range (3.2–13.1 μg/g) and are comparable to those observed by Wieckowski and Mezyk [45] in TCM (1.6–11.3 μg/g), and by Akhtar et al. [46] (0.94–4.35 μg/g) in AM.

Comparison between the five different methods of digestion for nickel revealed that methods **M4** and **MWO** were the best acid mixtures for the medicinal samples analysed in this study. Highest deviations between the digestion methods **MWO** and **M4** were observed for *Leung Pui Kee* (−46.4 %), *Saplingtan* (−22.8 %) and *Po Chai Pills* (+20.0 %). The two-tail student's *t*-test was applied to compare the results obtained from all the different pairs of digestion methods and the results are shown in Table 24.9.

It can be observed that for six out of ten combinations analysed, the t_{calc} is greater than the t_{crit} (2.23). Results following digestion methods **M4** and **MWO** are thus significantly different at the 95 % confidence level from those obtained after digestion methods **M1**, **M2** and **M3**. On the other hand, no significant difference was observed between the three methods of digestion **M1**, **M2** and **M3** and between **M4** and **MWO** indicating that the values obtained were comparable.

Zinc

The zinc levels obtained in the 7 TCM and 4 AM following the five methods of digestion are given in Table 24.10.

Zinc was detected in all the 11 traditional medicines analysed. The highest zinc level observed was in *Po Chai Pills* for the TCM and in *Diabecon* for the AM. The zinc level in the TCM *Po Chai Pills* can be explained by the presence of the ingredient *Cortex Magnolia Officinalis* which is rich in a number of trace elements including zinc [47]. On the other hand, the zinc content in *Diabecon* is mainly due to two of its constituents, namely *Yashad Bhasma* and *Shilajit*. *Yashad Bhasma* also known as zinc calx, is processed zinc which is obtained from the calcination of *Rasaka satva* which is a metallic extract of zinc carbonate/zinc oxide [48]. On the other hand, *Shilajit* is a rich source of essential and trace minerals including zinc [42]. The range for zinc in the TCM was found to be within that obtained by Shao

Table 24.10 Zinc levels (μg/g) in the 7 TCM and 4 AM following the five methods of digestion

Medicine	M1	M2	M3	M4	MWO
Diabecon	26.8 ± 0.2	24.0 ± 0.3	22.3 ± 0.2	25.5 ± 0.1	22.0 ± 0.1
Pilex	11.8 ± 0.1	8.7 ± 0.6	11.1 ± 0.4	9.7 ± 0.2	9.9 ± 0.1
Septilin	13.3 ± 0.1	10.7 ± 0.4	12.9 ± 0.1	9.2 ± 0.1	10.1 ± 0.2
Softovac	11.8 ± 0.1	8.2 ± 0.2	11.6 ± 0.1	6.3 ± 0.2	9.7 ± 0.2
Jin Qian Chao	9.6 ± 0.1	7.9 ± 0.1	9.4 ± 0.1	5.1 ± 0.2	6.5 ± 0.2
Leung Pui Kee	16.9 ± 0.1	14.2 ± 0.2	16.2 ± 0.1	13.8 ± 0.1	15.0 ± 0.3
Po Chai Pills	26.7 ± 0.1	25.0 ± 0.3	26.4 ± 0.2	18.8 ± 0.2	22.7 ± 0.1
Quiang Li Tien- Ma	13.5 ± 0.1	12.0 ± 0.2	13.1 ± 0.1	10.6 ± 0.1	11.9 ± 0.1
Saplingtan	3.7 ± 0.1	3.2 ± 0.1	3.5 ± 0.1	3.2 ± 0.1	2.8 ± 0.2
Watermelon Frost	4.5 ± 0.1	3.9 ± 0.2	4.0 ± 0.1	3.6 ± 0.1	4.0 ± 0.1
Niuhuang chiehdu Pien	8.1 ± 0.3	6.0 ± 0.2	4.0 ± 0.3	4.4 ± 0.2	5.1 ± 0.2

et al. [49] (3.2–86.2 μg/g). Similarly, the range of zinc found in the AM is within the range (4–109 μg/g) observed by Al-Omari [17].

The wet digestion **M1** using HNO_3 and H_2O_2 was found to result in the highest zinc levels followed by **M3**. Digestion method **M4** gave the least zinc level for seven out of the eleven medicines analysed. Using *t*-test, it was found that the t_{calc} was greater than the t_{crit} (2.23) at 95 % confidence for the following digestion pairs: **M1/M2, M1/M3, M1/M4, M1/MWO, M3/M4** and **M3/MWO** (Table 24.11), inferring that there is a significant difference between the results obtained after **M1** digestion method and the other digestion methods (**M2, M3, M4** and **MWO**).

Potassium and Sodium

Table 24.12 gives the potassium and sodium levels obtained in the 7 TCM and 4 AM by flame photometry following **MWO** digestion.

The ranges observed for potassium were 6,200–21,143 μg/g and 31.7–26,416 μg/g for the AM and TCM respectively. The highest potassium content was found in the TCM *Jin Qian Chao* (26,416 μg/g) followed by the AM *Diabecon* (21,143 μg/g). As *Jin Qian Chao* is composed solely of *Herba Lysimachiae*, the high potassium content of the TCM may be due to the presence of inorganic potassium chloride absorbed from the soil in the herb [44]. On the other hand, the high amount of potassium in *Diabecon* results from the ingredients *Shilajit* and *Abhrak Bhasma*. *Shilajit* is obtained from the Himalayan Mountains and is composed of a large number of essential metals in the ionic form including potassium [42]. *Bhasmas* are prepared by calcination of metals and *Abhrak Bhasma* contains trace amounts of potassium [48]. The highest level in the TCM observed by Wang et al. [11] was 11,054 μg/g which is lower than that observed in the present study (26,416 μg/g). The lowest potassium contents were observed in two TCM, namely *Saplingtan* (31.7 μg/g) and *Watermelon Frost* (70.5 μg/g). On the other hand, the range of potassium levels observed in the AM was within that obtained by Giacomino et al. [12] (44.5–21,440 μg/g).

Table 24.11 Comparison of zinc content between different pairs of digestion methods using *t*-test

Digestion pair	Mean difference	Standard deviation	t_{calc}
M1/M2	2.08	0.99	6.96
M1/M3	1.11	1.59	2.31
M1/M4	3.32	2.18	5.05
M1/MWO	2.45	1.30	6.27
M2/M3	0.97	1.68	1.92
M2/M4	1.24	2.08	1.97
M2/MWO	0.37	1.24	0.99
M3/M4	2.21	2.98	2.46
M3/MWO	1.34	1.40	3.18
M4/MWO	0.86	1.94	1.48

Table 24.12 Potassium and sodium levels (μg/g) in the 7 TCM and 4 AM after **MWO** digestion

Medicine	Potassium	Sodium
Diabecon	21,143 ± 7	5,173 ± 73
Pilex	6,408 ± 71	1,972 ± 73
Septilin	6,200 ± 3	3,500 ± 1
Softovac	6,690 ± 352	1,460 ± 2
Jin Qian Chao	26,416 ± 338	3,350 ± 4
Leung Pui Kee	9,795 ± 68	1,535 ± 73
Po Chai Pills	7,677 ± 71	1,099 ± 73
Quiang Li Tien-Ma	5,844 ± 71	1,608 ± 146
Saplingtan	31.7 ±3.5	ND
Watermelon Frost	70.5 ± 0.1	6,191 ± 73
Niuhuangchiehdu Pien	4,575 ± 71	1,099 ± 73

The sodium levels observed ranged from 1,099 μg/g in *Po Chai Pills* and *Niuhuangchiehdu Pien* to 6,191 μg/g in *Watermelon Frost*. In *Saplingtan*, the sodium level was below the detection limit of the apparatus. The highest sodium level was observed in *Watermelon Frost* (6,191 μg/g) which is used for the treatment of throat diseases, general wounds with bleeding, toothache, burns and scald, and can be accounted by the presence of the ingredient Mirabilite (also known as Glauber's salt, $Na_2SO_4{\cdot}10H_2O$) in its formulation. Due to its anti-bacterial property, sodium is also present in high amount in *Septilin* (3,500 μg/g) which is mainly used as an anti-infective therapy for the treatment of sinusitis. The high sodium level observed (5,173 μg/g) in *Diabecon* can be accounted by the presence of *Shilajit*, a major ingredient rich in various metallic salts including sodium [42].

To investigate the matrix effect in the medicines, the potassium level in the AM, *Diabecon* and the sodium level in the TCM *Niuhuangchiehdu Pien*, were determined by the method of standard addition and the results obtained compared with those obtained from standard calibration curves. The plots obtained for potassium level in *Diabecon* and sodium level in *Niuhuangchiehdu Pien* are depicted in Figs. 24.1 and 24.2 respectively.

The potassium and sodium levels obtained from both standard calibration curve and standard addition are shown in Table 24.13.

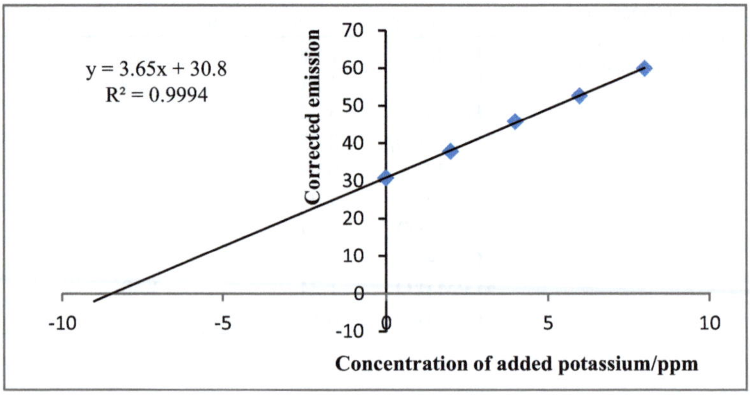

Fig. 24.1 Plot of corrected emission against concentration of added potassium for *Diabecon*

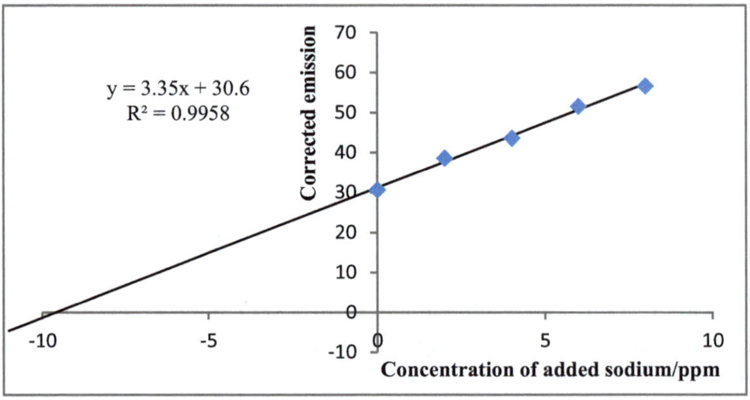

Fig. 24.2 Plot of corrected emission against concentration of added sodium for TCM *Niuhuangchiehdu Pien*

It is observed that the sodium and potassium concentration obtained from standard addition are close to those obtained from standard calibration curves with a CV of 0.22–3.9 %. It can thus be deduced that the matrix did not have much influence on the sodium or potassium content.

24.3.2 Precision of Flame Photometric Method

In this study, intra-day precision, also known as repeatability, was determined for the potassium content in the TCM *Po Chai Pills* and inter-day precision for the potassium and sodium content for all the medicines. The precision of an analytical procedure is usually expressed as the % relative standard deviation (%RSD) of a

Table 24.13 Concentrations obtained by standard calibration curve and standard addition

	Potassium level (μg/g)			Sodium level (μg/g)		
Medicine	Calib curve	Std add	CV (%)	Calib curve	Std add	CV (%)
Diabecon	21,143	21,096	0.22	–	–	–
Niuhuangchiehdu Pien	–	–	–	1,099	1,142	3.9

Table 24.14 Intra-day precision data

Time/min	Potassium level (μg/g)
0	7,610
15	7,469
30	7,610
45	7,469
60	7,469
75	7,469
90	7,469
105	7,469
120	7,469
135	7,610
Mean ± SD	7,511 ± 68
%RSD	0.91

series of measurements. Precision is concentration dependent and the acceptable precision (%RSD) for concentration ($\geq$ 100 ppm) is 15 % [50].

Intra-Day Precision

Intra-day precision expresses the precision under the same operating conditions over a short interval of time within the same day and was determined for the potassium content by flame photometry in the TCM *Po Chai Pills* by measuring the same sample ten times at 15 min' time interval. The results are shown in Table 24.14.

The repeatability of the analytical method for the TCM *Po Chai Pills* was excellent as the %RSD was found to be 0.91 %.

Inter-Day Precision

The inter-day precision of the photometric analyses and the method validity were assessed by analysing the microwave-digested samples of the 11 medicines on five different days: day 1, day 2, day 3, day 8 and day 15 for potassium and sodium. The results are shown in Table 24.15 and Fig. 24.3 respectively.

It was observed that the %RSD for all the medicines (1.3–9.0 %) for both the potassium and sodium analysis were well below the acceptable precision and thus it can be inferred that the inter-day precision for the analytical method was good.

Table 24.15 Inter-day precision analysis of potassium

	Potassium level (μg/g)						
Medicine	Day 1	Day 2	Day 3	Day 8	Day 15	Mean ± SD (μg/g)	%RSD
Diabecon	21,143	21,485	20,931	21,010	21,675	21,249 ± 285	1.3
Pilex	6,408	6,607	6,518	6,728	6,497	6,552 ± 109	1.7
Septilin	6,197	6,675	6,234	6,293	6,136	6,307 ± 191	3.0
Softovac	6,690	6,948	6,589	6,800	6,786	6,763 ± 120	1.8
Jin Qian Chao	26,416	27,628	25,191	26,448	28,541	26,845 ± 1,146	4.3
Leung Pui Kee	9,795	10,156	9,784	9,990	9,967	9,938 ± 138	1.4
Po Chai Pills	7,677	7,972	7,441	7,743	7,654	7,697 ± 171	2.2
Quiang Li Tien Ma	5,844	5,924	5,595	5,858	6,208	5,886 ± 196	3.3
Saplingtan	32	27.3	28.4	25.4	25.3	27.7 ± 2.5	9.0
Watermelon Frost	71	61	61	61	54.2	61.6 ± 5.4	8.8
Niuhuangchieh du Pien	4,575	4,423	4,175	4,274	4,835	4,456 ± 233	5.2

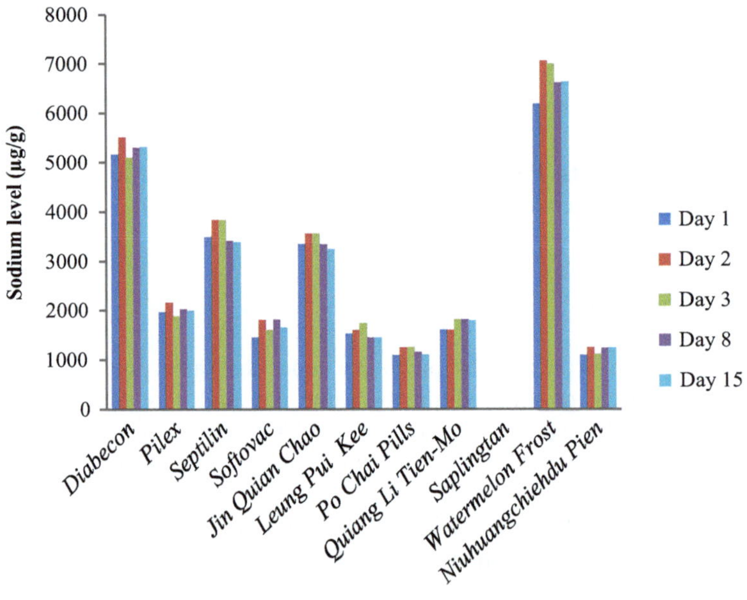

Fig. 24.3 Variation of the sodium content of the TCM and AM on an inter-day basis

24.4 Conclusions

In this study, 7 TCM and 4 AM purchased in Mauritius were analysed for their calcium, magnesium, nickel and zinc levels by flame atomic absorption spectroscopy after digestion by four wet open methods and **MWO**. The results obtained after the different methods of digestion were compared using the student's *t*-test.

In addition, the potassium and sodium levels in the 11 medicines were determined by flame photometry, whose validity was checked by analyzing its precision.

In both the TCM and AM, potassium was present in the highest amount. The highest calcium, magnesium, nickel, potassium, sodium and zinc level were observed respectively in *Niuhuangchiehdu Pien* (912 μg/g), Pilex (265 μg/g), *Leung Pui Kee* (13.1 μg/g), *Jin Qian Chao* (26,416 μg/g), *Watermelon Frost* (6,191 μg/g) and *Diabecon* (26.8 μg/g). **MWO** digestion was found to be more effective for Mg, method **M1** [conc HNO_3 + H_2O_2 (30 %)] for Ca and Zn, and method **M4** (aqua regia) for nickel.

From replicate measurements, the precision of the flame photometric method used for the analysis of potassium was found to be good as the %RSD for both intra-day (0.91 %) and inter-day (1.3–9.0 %) precision was less than the acceptable % RSD (15 %). For sodium, the inter-day precision was found to be in the range 3.0–9.0 %.

To investigate matrix interference, standard addition was carried out on the AM *Diabecon* for potassium and the TCM *Niuhuangchiehdu Pien* for sodium. A coefficient of variation of 0.22 % and 3.9 % was observed for the potassium and sodium level respectively indicating little interference from the sample matrix.

Further study is under way for the analysis of other metals, in particular toxic ones in these TCM and AM.

Acknowledgements We are very grateful to the University of Mauritius for providing the necessary facilities to carry out the project. Thanks also to the technical staff of the Chemistry Department, in particular to Mr. V Ramsahye for assistance.

References

1. World Health Organization (2000) General guidelines for methodologies on research and evaluation of traditional medicine. World Health Organization, Geneva, pp 1–73
2. Saper RB, Kales SN, Paquin J, Burns MJ, Eisenberg DM, Davis RB, Phillips RS (2004) Heavy metal content of Ayurvedic herbal medicine products. J Am Med Assoc 292:2868–2873
3. Wang X, Zhuang Z, Sun D, Hong J, Wu X, Lee F, Yang MS, Leung HW (1999) Trace metals in traditional Chinese medicine: a preliminary study using ICP-MS for metal determination and As speciation. Atom Spectrosc 20:86–91
4. Centres of Disease Control and Prevention (2004) Lead poisoning associated with use of Ayurvedic medications – five states, 2000–2003. MMWR Morb Mortal Wkly Rep 53:582–584
5. Ernst E (2002) Heavy metals in traditional Indian remedies. Eur J Clin Pharmacol 57:891–896
6. Saper RB, Phillips RS, Sehgal A, Khouri N, Davis RB, Paquin J, Thuppil V, Kales SN (2008) Lead, mercury and arsenic in US- and Indian-manufactured Ayurvedic medicines sold *via* the internet. J Am Med Assoc 300:915–923
7. Yang L-L, Gao L-R, Zhang D-Q (2003) Speciation analysis of arsenic in traditional Chinese medicines by hydride generation-atomic fluorescence spectrometry. Anal Sci 19:897–902
8. Chan K (2003) Some aspects of toxic contaminants in herbal medicines. Chemosphere 52:1361–1371
9. Koh H-L, Woo S-O (2000) Chinese proprietary medicine in Singapore: regulatory control of toxic heavy metals and undeclared drugs. Drug Saf 23:351–362

10. Wu X-H, Sun D-H, Zhuang Z-X, Wang X-R, Gong H-F, Hong J-X, Lee FSC (2002) Analysis and leaching characteristics of mercury and arsenic in Chinese medicinal materials. Anal Chim Acta 453:311–323
11. Wang C-F, Duo M-J, Chang EE, Yang JY (1996) Essential and toxic trace elements in the Chinese medicine. J Radioanal Nucl Ch Ar 211:333–347
12. Giacomino A, Abollino O, Malandrino M, Karthik M, Murugesan V (2010) Determination and assessment of the contents of essential and potentially toxic elements in Ayurvedic medicine formulations by inductively coupled plasma-optical emission spectrometry. Microchem J 99:2–6
13. Dargan PI, Gawarammana IB, Archer JRH, House IM, Shaw D, Wood DM (2008) Heavy metal poisoning from Ayurvedic traditional medicines: an emerging problem? Int J Environ Health 2:463–474
14. Espinoza EO, Mann MJ, Bleasdell B (1995) Arsenic and mercury in traditional Chinese herbal balls. New Engl J Med 333:803–804
15. Kales SN, Christophi CA, Saper RB (2005) Hematopoietic toxicity from lead-containing Ayurvedic medications. Med Sci Monit 13:CR295–CR298
16. Au AM, Ko R, Boo FO, Hsu R, Perez G, Yang Z (2000) Screening methods for drugs and heavy metals in Chinese patent medicines. Bull Environ Contam Toxicol 65:112–119
17. Al-Omari S (2011) Determination of essential and toxic trace elements in ten herbal medicines using energy-dispersive XRF analysis. X-Ray Spectrom 40:31–36
18. Lam JC, Chan K-K, Yip Y-C, Tong W-F, Sin DW (2010) Accurate determination of lead in Chinese herbs using isotope dilution inductively coupled plasma mass spectrometry (ID-ICP-MS). Food Chem 121:552–560
19. Sa F, Vong WT, Chan TM, Lam CWK (2010) Determination of heavy metals in proprietary Chinese medicines by inductively coupled plasma–mass spectrometry. Clin Chim Acta 411:896 (Abstract)
20. Yuan X, Ling KH, Keung CW (2009) The analysis of heavy metals in Chinese herbal medicine by flow injection–mercury hydride system and graphite furnace atomic absorption spectrometry. Phytochem Anal 20:293–297
21. Cooper K, Noller B, Connell D, Yu D, Sadler R, Olszowy H, Golding G, Tinggi U, Moore MR, Myers S (2007) Public health risks from heavy metals and metalloids present in traditional Chinese medicines. J Toxicol Env Health 70:1694–1699
22. Gunturu KS, Nagarajan P, McPhedran P, Goodman TR, Hodsdon ME, Strout MP (2011) Ayurvedic herbal medicine and lead poisoning. J Hematol Oncol 4:51–56
23. Lavekar GS, Ravishankar B, Gaidhani S, Shukla VJ, Ashok BK, Padhi MM (2010) Mahayograj guggulu: heavy metal estimation and safety studies. Int J Ayurveda Res 1:150–158
24. Gair R (2008) Heavy metal poisoning from Ayurvedic medicines. BC Med J 50:105
25. Dobos GJ, Tan L, Cohen MH, McIntyre M, Bauer R, Li X, Bensoussan A (2005) Are national quality standards for traditional Chinese herbal medicine sufficient? Current governmental regulations for traditional Chinese herbal medicine in certain Western countries and China as the Eastern origin country. Complement Ther Med 13:183–190
26. Mazzanti G, Battinelli L, Daniele C, Costantini S, Ciaralli L, Evandri MG (2008) Purity control of some Chinese crude herbal drugs marketed in Italy. Food Chem Toxicol 46:3043–3047
27. Koch I, Moriarty M, House K, Jie S, Cullen WR, Saper RB, Reimer KJ (2011) Bioaccessibility of lead and arsenic in traditional Indian medicines. Sci Total Environ 409:4545–4552
28. Lokhande R, Singare P, Andhale M (2010) Study on mineral content of some Ayurvedic Indian medicinal plants by instrumental neutron activation analysis and AAS techniques. Health Sci J 4:157–168
29. Lokhande R, Singare P, Andhele M, Acharya R (2010) Study of some Indian medicinal plants by application of INAA and AAS techniques. Nat Sci 2:26–32

30. Avula B, Wang Y-H, Smillie TJ, Duzgoren-Aydin NS, Khan IA (2010) Quantitative determination of multiple elements in botanicals and dietary supplements using ICP-MS. J Agric Food Chem 58:8887–8894
31. Fei T, Dehong L, Fengqun Z, Junhua L, Hua T, Xiangzhong K (2010) Determination of trace elements in Chinese medicinal plants by instrumental neutron activation analysis. J Radioanal Nucl Chem 284:507–511
32. Mahawatte P, Dissanayaka KR, Hewamanna R (2006) Elemental concentrations of some Ayurvedic drugs using energy dispersive XRF. J Radioanal Nucl Chem 270:657–660
33. Choudhury RP, Kumar A, Reddy AVR, Garg AN (2007) Thermal neutron activation analysis of essential and trace elements and organic constituents in Trikatu: an Ayurvedic formulation. J Radioanal Nucl Chem 274:411–419
34. Singare P, Lokhande R, Andhale M, Acharya R (2010) Availability of essential trace elements in Ayurvedic Indian medicinal herbs using instrumental neutron activation analysis and atomic absorption spectroscopy. World J Sci Technol Sustain Dev 7:175–190
35. Yan Q, Yang L, Wang Q, Ma X (2012) Determination of major and trace elements in six herbal drugs for relieving heat and toxicity by ICP-AES with microwave digestion. J Saudi Chem Soc 16:287–290
36. Krebs J, Michalak M (2007) Calcium: a matter of life or death. Elsevier BV, Amsterdam
37. Office of Dietary Supplements (2009) Dietary supplement fact sheet: Magnesium. http://ods.od.nih.gov/factsheets/magnesium/. Accessed 5 Mar 2012
38. Haas E, Levin B (2006) Staying healthy with nutrition. The complete guide to diet and nutritional medicine. Celestial Arts, Berkeley
39. Office of Dietary Supplements (2011) Dietary supplement fact sheet: Zinc. http://ods.od.nih.gov/factsheets/Zinc-HealthProfessional/. Accessed 7 Mar 2012
40. Elson MH (2011) Role of potassium in maintaining health. http://org/patients/potassium-health. Accessed 13 Dec 2011
41. Gittleman AL (2000) Understanding salt and sodium. http://curezone.com/foods/salt/understanding_salt_and_sodium.html. Accessed 26 Oct 2010
42. Asian Health Secrets (2010) Shilajit: Nature's Rejuvenator. http://www.asianhealthsecrets.com/letha/?p=2386. Accessed 13 Mar 2012
43. Word Press (2010) Medicinal plants, herbs, spices: Guggul. Available from: http://herbtime.wordpress.com/2010/02/19/guggul-commiphora-mukul-or-balsamodendron-mukul/. Accessed 15 Mar 2012
44. Xu L, Wang W (2002) Jin Qian Cao. In: Chinese materia medica: combinations and applications. Donica, St Albans, p 240
45. Wieckowski SK, Mezyk Z (1999) Studies of trace element content in selected medical herbs. Pol J Environ Stud 8:129–130
46. Akhtar N, Khan SA, Khan L, Hussain I, Marwat KB (2008) Profile of heavy metals in selected medicinal plants. Pak J Weed Sci Res 14:101–110
47. Guo L (2012) The producing area of Chinese medicine and famous region drug research – *Magnolia officinalis*. In: Kuang H (ed) Recent advances in theories and practice of Chinese medicine. InTech, Shanghai, pp 453–460. www.intechopen.com/download/pdf/pdfs_id/26178. Accessed 13 Jan 2012
48. Himalaya Herbal Health Care (2002) Himalaya's herbs and minerals: Yashad bhasma. http://www.himalayahealthcare.com/aboutayurveda/cahy.htm. Accessed 15 Mar 2012
49. Shao YY, Cai BS, Lin J, Xu JS (2007) Determination of trace elements in 13 traditional Chinese medicines in medicated diet for *diabetes mellitus* by ICP-MS. Zhongguo Zhong Yao Za Zhi 32:2011–2015 [Abstract]
50. European Medicines Agency (2009) Guidelines for the validation of analytical methods used in residue depletion studies. http://www.ema.europa.eu/docs/en_GB/document_library/Scientific_guideline/2010/01/WC500040499.pdf. Accessed 13 Jan 2012

Chapter 25
A Comparative Study on Preserving Milk Using Grass Species *Hyperenium Rufa* for Fumigating Milk Containers and Pasteurisation

Milton A. Wesuta and William K. Isharaza

Abstract Milk and dairy products form a major dietary component for many cattle rearing communities worldwide. The Banyankore, an ethnic group in Western Uganda, has over centuries used and established different methods of preserving milk and other dairy products like ghee and yoghurt. Fumigants derived from specific plant species such as *Hyperenium rufa*, *Rhus natalensis* and *Molle combretum* have been used when preparing vessels for storage of fresh milk and milk products. *H. rufa* is the most commonly used fumigant for temporary preservation of fresh/raw milk. To date, there is no scientific literature evaluating the effectiveness of these traditional practices. We validated the use of fumigant from *H. rufa* to preserve the quality of fresh milk using the alcohol test, 10 min resazurin test, clot-on-boiling test and pH measurements. The total number of bacteria per milliliter of milk was determined by total plate count (TPC). Exposing raw milk to fumigant from *H. rufa* in wooden milk pots curtailed microbial growth for over 24 h at room temperature before sourage occurred. Results of parallel tests on raw milk preserved by fumigation compared well with those of pasteurised milk in sterilised glass vessels. We conclude that fumigation of milk pots using *H. rufa* can preserve milk for many hours before boiling for consumption. This method would be useful in rural communities; however, it is no substitute to pasteurisation.

25.1 Introduction

Many cattle rearing communities worldwide use milk and dairy products to meet the major component of their nutritional requirements and as major sources of income.

Raw milk from a healthy cow's udder may contain some spoilage or pathogenic microorganisms at the time of milking. The chemical composition of the milk

---

M.A. Wesuta (✉) • W.K. Isharaza
Department of Biochemistry, Mbarara University of Science and Technology, Mbarara, Uganda
e-mail: apollomwes@must.ac.ug

M. Gupta Bhowon et al. (eds.), *Chemistry: The Key to our Sustainable Future*,
DOI 10.1007/978-94-007-7389-9_25,

provides an excellent medium for the proliferation of these microorganisms together with those that may accidentally or intentionally be introduced into the milk during subsequent processes. This increased microbial load leads to deterioration of the milk within hours after milking.

Not all the milk produced daily is consumed immediately. Preservation methods provide opportunities for stopping or delaying deterioration of milk. There are techniques used at either domestic or industrial production scale to preserve milk before consumption. They include modern methods such as chemical preservation, refrigeration, freezing and various forms of heat treatment. The use of activated lactoperoxidase system (LPS) treatment can enable farmers sell milk in far-off markets [1]. The dairy sector in the South West Uganda includes rudimentary pasteurisation units [2]. The Banyankore, an ethnic group in Western Uganda, has over centuries used different traditional methods for preserving raw milk and dairy products [3]. Fumigants from various plant species have been used to prepare vessels for milk storage. There is currently no documented scientific information regarding these traditional milk preservation practices among the Banyankore.

The objective of this study was, therefore, to compare the effectiveness of using *H. rufa* to fumigate wooden pots for milk preservation as done by the Banyankore and the conventional method of pasteurisation. We hypothesised that the mean values of storage by fumigation were greater than those of other storage conditions.

25.2 Methodology

25.2.1 Fumigation of "Ebyanzi" and Glass Vessels

Two wooden milk pots locally called "*Ebyanzi*" were washed with clean water, dried overnight and the interior of the vessels were carefully scrubbed to remove remains of milk particles from previous use. Four glass bottles were washed with clean water and dried overnight. Smoke produced from the burning grass was left to saturate one of the milk pots (*kyanzi)* for 5 min. One of the cleaned glass bottles was similarly treated. Another glass bottle was sterilised by autoclaving at 121 °C for 15 min. The remaining three containers were not treated before introducing the milk samples.

25.2.2 Treatment of Milk Samples

Four milk samples, 3 L each, were collected on different days from selected lactating cows and four from milk cooling centers. 500 ml portions of each milk sample were poured into six containers as represented in Table 25.1.

Pasteurisation in glass bottle (f) was done by heating the milk in the sterile bottle in a water bath at 72 °C, for 30 min.

Table 25.1 Containers and the six different forms of milk treatment/storage conditions

Container	Treatment/storage condition	Comment
(a)	Fumigated *kyanzi*	Positive test for fumigation of *kyanzi*
(b)	Unfumigated *kyanzi*	Negative control for fumigation of *kyanzi*
(c)	Fumigated glass bottle	Positive test for fumigation of glass bottle
(d)	Unfumigated glass bottle	Negative control for fumigation of glass bottle
(e)	Sterile glass bottle	Negative control for pasteurization
(f)	Pasteurisation in sterile glass bottle	Positive test for pasteurization

25.2.3 Milk Quality Tests

pH

The acidity of the milk samples given as pH was determined using a pH meter (Cyberscan pH 500, Eutech Instruments, Singapore).

Clot-on-Boiling

Milk samples of 2 ml each were heated over a bunsen burner flame for 5 s. Clotting on boiling confirmed sourage of the milk and was recorded as positive or negative for no clotting.

Alcohol

Milk is considered good when there is no precipitation with addition of alcohol. 2 ml of 68 % ethanol was added to 2 ml of milk sample to test formation or no formation of precipitate.

10-min Resazurin Test

To detect the activity of the bacteriological organisms in milk, 10-ml samples of treated milk were pipetted into sterile boiling tubes after every 6 h. 1 ml of freshly prepared solution of resazurin was added to each tube and sealed with a sterile top. The contents were mixed until resazurin was uniformly distributed in the milk. The tubes were incubated at 37 °C for 10 min. The status of each milk sample was scored and recorded according to the colour formed [4].

Total Plate Count (TPC)

The microbiological quality of milk was tested at 0, 6 and 24-h intervals using total plate count test. Serial dilutions from 1 in 10 to 1 in 10,000,000 of the milk aliquots were made in autoclaved Butterfield's Phosphate Buffer (BPB) [BDH-UK] and deep plated in duplicate on Plate Count Agar (PCA) [Oxoid] with incubation at 35 °C for 48 h [5]. The method described in ISO 6610 (1992) [6] was used to count the colonies. The average TPC were then expressed as logarithms of colony-forming units/ml (log cfu/ml).

25.2.4 Data Management and Statistical Analysis

The rates of changes of pH values were assessed by comparing the slopes of the different best straight lines of fit for the milk samples stored in different conditions. The differences in slopes of the best straight lines of fit at different storage conditions were analyzed using two-sample *t* test, assuming a one-tail normal distribution and $\alpha = 0.05$ [7].

25.3 Results and Discussion

25.3.1 pH Changes

The pH of all samples progressively dropped. The rates of changes of pH values were assessed by comparing the slopes of the different best straight lines of fit for the samples exposed to the six different conditions (results not shown).

Two of the four milk samples from selected cows on a single farm and kept in fumigated *kyanzi* had the least drop in pH value, followed by pasteurized milk, then milk kept in the fumigated glass bottle, milk kept in the unfumigated glass bottle, milk kept in the unfumigated *kyanzi* and finally milk in sterilised glass bottle. For the other two milk samples, the least drop in pH value was observed in pasteurised milk, followed by milk kept in the fumigated *kyanzi*, in fumigated glass bottle, and untreated samples. These data indicate that fumigation of the *kyanzi* offered the greatest inhibition to milk spoilage in two samples as indicated by the rate at which the milk became more acidic. Since pasteurisation offered the greatest inhibition to milk spoilage in two milk samples, but still closely followed by fumigation of the *kyanzi*, our findings suggest that fumigation of the *kyanzi* is comparable to pasteurisation in preservation of raw milk.

Pooled milk samples from collection centres showed the least drop in pH in pasteurised milk, followed by that kept in fumigated *kyanzi*, milk in the fumigated glass bottle, the unfumigated *kyanzi*, the unfumigated glass bottle and the sterilised glass bottle. Pasteurisation therefore evidently had the greatest effect in reducing milk spoilage, closely followed by fumigation of the *kyanzi*, fumigation of the glass bottle and lastly the sterile glass bottle and the untreated vessels. A bigger difference was seen between the slopes of the best straight lines of fit (not shown) for samples from a single farm for fumigation of the *kyanzi* and other milk treatments

Table 25.2 t and p test values for determining the effect of fumigation of *kyanzi* and comparing fumigation with pasteurization

Milk sample no.	t_{fk}	t_{fp}	p_{fk}	p_{fp}
1	2.823	1.390	0.013[a]	0.097
2	2.264	0.612	0.025[a]	0.276
3	1.643	0.217	0.066	0.416
4	0.748	0.161	0.234	0.437
5	0.772	0.291	0.225	0.388
6	0.441	−0.719	0.331	0.243
7	0.502	−0.505	0.315	0.312
8	0.958	−0.108	0.176	0.458

Key: *fk*-fumigation of *kyanzi* with untreated *kyanzi*, *fp*-fumigation of the *kyanzi* with pasteurisation
[a]Statistically significant values

than with the same treatments for milk from collection centers. This suggested that milk from the single farm was better preserved by fumigation of the *kyanzi* than pooled milk. The superiority is probably due to the lower level of microorganisms in milk from selected cows than in pooled milk.

The differences in slopes of the best straight lines of fit of preserving milk using *H. rufa* fumigation compared to other storage conditions were analyzed using 2-tail t-test; the t values and the corresponding p values are shown in Table 25.2.

25.3.2 Clot-on-Boiling

Clot-on-boiling tests for all samples were negative at the beginning of monitoring, indicating that the milk was stable to heat treatment. Treated samples, as in the six different conditions described, gave positive results after different lengths of storage (results not shown). The increase in acidity during storage decreases the heat stability of the milk [8]. This was observed to vary for the different conditions under which milk samples were kept.

Comparison of results for the milk obtained from selected cows and those for pooled milk reveals that fumigation of the *kyanzi* preserves the milk from selected cows longer than it does for pooled milk. However there was no marked difference in the duration of preservation by subjecting to the other conditions. These results compare well with the trend observed by monitoring the pH.

25.3.3 Alcohol Test

The alcohol test results for all samples were negative, indicating that the milk had less than 0.21 % acid at 0 h of monitoring. The duration of storage before coagulation was observed to vary with the different storage conditions and sources of the milk samples (results not shown).

Comparing results of this test for samples from the two sources showed that fumigation of milk pots was equally effective in preserving milk obtained from the two sources. Fumigation of the *kyanzi* keeps raw milk from selected cows longer than it does for pooled milk, a trend that agrees with what were obtained in the pH changes and the clot-on-boiling tests.

25.3.4 10-min Resazurin Test

The drop in the score with storage time was due to increased metabolic activity of the microorganisms in the milk samples. The changes indicate the deterioration of raw milk. This increased activity is variedly responsible for the reduction of the resazurin dye. The final colour developed on incubation of the samples for 10 min varied with the different storage conditions and sources.

Comparing the results for the samples from the two sources using the reference table for interpretation of results [4], the drop in the test score shows that the effect of fumigation of the traditional milk pots compares well with pasteurisation. The results obtained in this test compared well with those obtained from pH, clot-on-boiling and alcohol test changes (results not shown). Fumigation of the glass bottle also curtailed the spoilage of the milk when compared to its negative control, the unfumigated glass bottle.

25.3.5 Total Plate Count Test

Figures 25.1 and 25.2 show changes in TPC results expressed as log cfu/ml of milk samples from selected cows on a single farm and pooled milk respectively done at 0, 6, and 24 h of storage under ambient temperatures.

The values indicate the rapid multiplication of the microorganisms initially present in the fresh milk. The increased numbers of microorganisms produce extracellular enzymes which break down the nutrients in the milk, thereby affecting its physical-chemical properties [9]. The changes varied depending on the different storage conditions of the milk samples.

The rate of multiplication of microorganisms in the 24 h of storage was lowest in milk from the single farm kept in the fumigated *kyanzi,* followed by pasteurised milk, milk kept in the fumigated glass bottle, the milk kept in the sterile glass bottle and lastly, that kept in the untreated containers (Fig. 25.1). This gives fumigation of the *kyanzi* superiority over pasteurisation and other forms of storage. For pooled milk samples, the rate of multiplication of microorganisms in the 24 h of storage was lowest for pasteurised milk in the sterile glass bottle, closely followed by milk kept in the fumigated *kyanzi*, the milk kept in fumigated glass bottle, the sterile glass bottle and the untreated containers (Fig. 25.2). This gives pasteurisation relative superiority over fumigation of *kyanzi* and other forms of storage. These

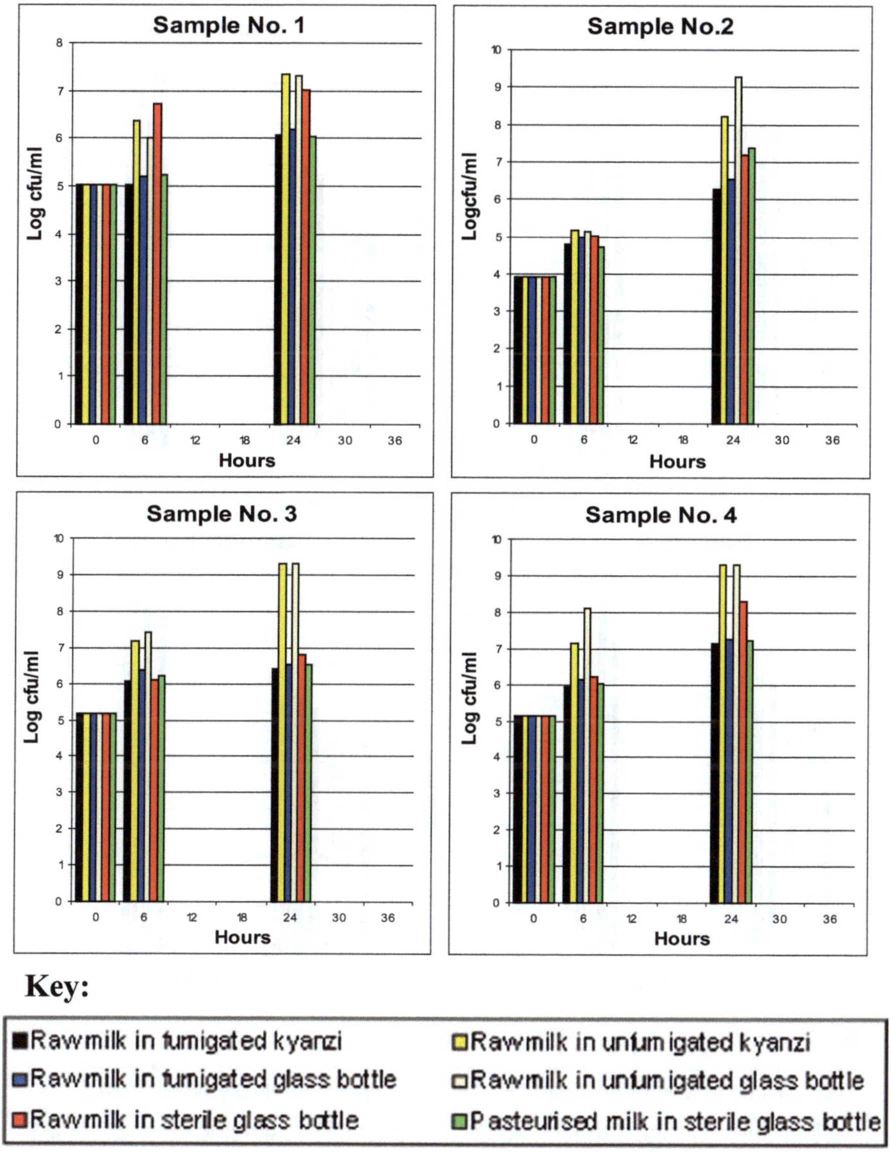

Fig. 25.1 Changes in total plate count in different fresh milk samples from selected cows during 24 h storage

results compare well with those obtained in the platform tests. Fumigation is ultimately superior to pasteurisation with respect to preservation of milk from selected animals whereas it is not the case with pooled milk.

pH values of milk samples from selected cows in fumigated *kyanzi* and those of pooled milk show a downward trend due to increase in hydrogen ions produced by

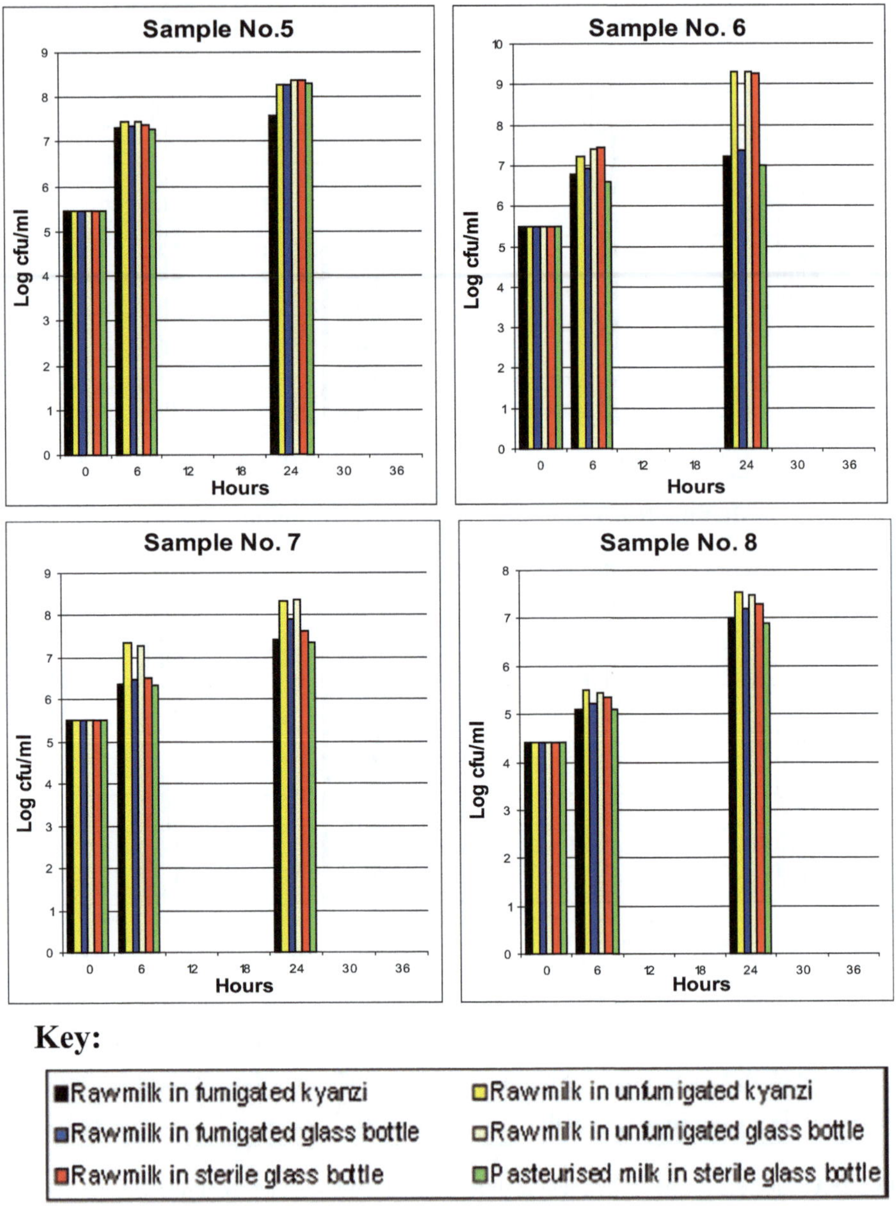

Fig. 25.2 Changes in total plate count in different pooled milk samples during 24 h storage

lactic acid, a product of fermentation of lactose by lactic acid bacteria present in the milk. This was also observed when the same milk samples were boiled and were treated with alcohol, as demonstrated in the clot-on-boiling and alcohol test results. Such milk clots or precipitates. They were negative in the first hours of storage, indicating the relative stability of the casein micelles in the milk. However, on keeping for several hours, precipitation on addition of 70 % alcohol was due to destruction of the casein micelles. The clot-on-boiling test showed the same trend. Here precipitation is due to irreversible denaturation of the protein casein molecules. Resazurin scores also lowered with time due to the increased bacterial metabolic activity that causes reduction of the purple dye through a series of colours to white. Changes in these parameters showed the same trend with those of the milk kept in fumigated *kyanzi* changing least on average followed by pasteurisation and milk kept in the fumigated glass bottle. The average TPC expressed in log cfu/ml units increased over the 24-h interval due to rapid reproduction of the bacteria by binary transverse fission, hence increasing the number of bacterial cell numbers.

25.4 Conclusions

We conclude that microbial growth that leads to milk spoilage/sourage can be curtailed for up to 30 h by storing fresh milk in a fumigated *kyanzi* using *H. rufa*. Secondly, this traditional method of milk preservation gives better results with raw milk from selected cows. However, this temporary preservation of raw milk cannot be used as a substitute for pasteurisation. This method would be convenient for the rural household where expensive modern preservation facilities that require use of electricity and other forms of energy are inaccessible and unaffordable, in that milk from the animal can be kept in relatively good quality for about 24 h, without spoilage.

Further studies are needed to identify the chemical component within grass species *H. rufa,* to establish the antimicrobial properties of the smoke from the plant, and to standardize the fumigation technique for best results.

Acknowledgements Sincere thanks are extended to Moses Mwebaze, a dairy farmer of Bwizibwera for providing "*ebyanzi*" and the 'cured' *H. rufa* for the study; Mary Kankunda, Mbarara University of Science and Technology (MUST) for her expertise in fumigation of the vessels. We are grateful to the management of Jojoma Dairy, Dairy Corporation Ltd, and milk collection centres in Mbarara town for providing the pooled milk samples.

We thank the French Embassy in Uganda for setting up the Milk Testing Laboratory at MUST. M. A. Wesuta was supported by a training grant from the French Embassy in Uganda.

References

1. Asaah NO, Fonteh F, Kamga P, Mendi S, Imele H (2007) Activation of the lactoperoxidase system as a method of preserving raw milk in areas without cooling facilities. AJFAND 7:1–15
2. Grillet N, Grimaud P, Loiseau G, Wesuta M, Faye B (2005) Sanitary quality of raw milk within the commodity sub sector in Mbarara district and Kampala city in Uganda. Revue Élev Méd Vét Pays Trop 58:245–255

3. Biggelaar den C, Mueithi W (1999) Using tree species to treat milk for palatability and preservation. Intermediate technology food chain 26 report. Available online from http://www.itdg.org/docs/agroprocessing/food_chain_26.pdf. Accessed 28 May 2005
4. Kabaterana J (2000) Manual for production of good quality milk. Country Taste (U) Ltd, Mbarara, pp 4–8
5. Samarajeewa U (1999) Manual on microbiological analysis. Uganda National Bureau of Standards (UNBS), Kampala, Uganda, pp 50–76
6. ISO 6610 IDF 153 (1992) International Standard. Milk and milk products – enumeration of coliforming units of microorganisms, colony count technique at 30 °C, 1st edn. 02–01, Geneva, Switzerland
7. Freund JE (1992) Mathematical statistics, 5th edn. Prentice-Hall International, Inc, Englewood Cliffs, pp 462–465
8. O'Connor CB (1995) Rural dairy technology, ILRI training manual 1. International Livestock Research Institute, ILRI, Addis Ababa
9. Ruas-Madeido P, Reyes-Gavilán CG, Olano A, Villmiel M (1999) Influence of refrigeration and carbon dioxide addition to raw milk on microbial levels, free monosaccharides and myo-inositol content of raw and pasteurized milk. J Dairy Res 212:44–47

Index

M. Gupta Bhowon et al. (eds.), *Chemistry: The Key to our Sustainable Future*,
DOI 10.1007/978-94-007-7389-9,

MIX
Papier aus verantwortungsvollen Quellen
Paper from responsible sources
FSC® C105338

If you have any concerns about our products,
you can contact us on
ProductSafety@springernature.com

In case Publisher is established outside the EU,
the EU authorized representative is:
Springer Nature Customer Service Center GmbH
Europaplatz 3, 69115 Heidelberg, Germany

Printed by Libri Plureos GmbH
in Hamburg, Germany